BEVERLY HILLS PUBLIC LIBRARY
3 5048 00618 8195

D0961784

The Rock from Mars

Kathy Sawyer

Random House • New York

THE ROCK FROM MARS

A Detective Story on Two Planets

Published in the United States by Random House,
an imprint of The Random House Publishing Group,
a division of Random House, Inc., New York.

Random House and colophon are registered trademarks
of Random House, Inc.

LIBRARY OF CONGRESS CATALOGING-IN-PUBLICATION DATA
Sawyer, Kathy.
The rock from Mars: a detective story on two planets / Kathy Sawyer
p. cm.
Includes index.
ISBN 1-4000-6010-9
1. Life on other planets. 2. Meteorites.
3. Mars probes. 4. Mars (Planet) I. Title.
QB54.S263 2006 576.8'39—dc22 2005046560

Printed in the United States of America on acid-free paper

www.atrandom.com

2 4 6 8 9 7 5 3 1

First Edition

Book design by Simon M. Sullivan

For John,
my rock

The author gratefully acknowledges
the support of the Alfred P. Sloan Foundation.

CONTENTS

Each atom of that stone, each mineral flake of that night filled mountain, in itself forms a world. The struggle itself toward the heights is enough to fill a man's heart. One must imagine Sisyphus happy.

—Albert Camus, *The Myth of Sisyphus*

Physicists have paid little attention to rock, mainly because we are discouraged by its apparent complexity. . . . We also may question whether it is even possible to find interesting physics in such a "dirty" and uncontrolled system.

—Po-zen Wong, "The Statistical Physics of Sedimentary Rock," *Physics Today* 41, no. 12 (1988): p. 24.

The role of the infinitely small in nature is infinitely large.

—Louis Pasteur

PROLOGUE

Some 16 million years ago, a comet or asteroid slammed into Mars with the energy of a million hydrogen bombs. The impact gouged out a crater and blasted a massive cloud of soil and rock into the planet's thin atmosphere. Most of it fell back, but some fragments were ejected at velocities as high as 11,000 miles per hour—enough to fling them out of their native gravity and off on their own orbits around the sun.

After millions of years, at least one fell into Earth's gravitational embrace. It plunged through the atmosphere trailing fire, melted a little from the heat of friction, and resolidified with a new look—a glassy black fusion crust. The rock plowed into the south polar ice cap.

Elsewhere on the rock's adopted world, mastodons, long-horned bison, and saber-toothed cats roamed. Birds of prey with twenty-five-foot wingspans drifted overhead. Beluga whales plied the seas. Horses and camels loped across plains. Vast continental glaciers covered much of what is now northern North America and Eurasia, pushing their towers of ice up to 13,000 feet high. And an inquisitive, uppity species of upright walkers crossed the Bering land bridge from Eurasia into North America.

As the rock lay icebound for thirteen millennia, the humans evolved precociously, rose beyond the day-to-day struggle for existence, and pondered concepts of justice, beauty, and their place in the cosmic scheme. They developed powerful machines that extended their reach across time and space, to the edges of all known creation. They wondered if they were alone in this daunting expanse, and at times seemed to edge tantalizingly close to the answer. All this time, a natural conveyor belt in the Antarctic was nudging the rock toward its destiny.

One polar summer day, a young woman plucked it off the ice. The rock had emerged at last into the province of human curiosity.

Twelve years later, on a muggy August afternoon in Washington, D.C., CNN, ABC, NBC, CBS, and other media outlets carried images of the rock, nested in velvet, live around the globe. Reporters and VIPs jammed an auditorium to glimpse this object and to hear the report of nine exhausted, exhilarated people who had looked into its heart and found tantalizing clues to the history of worlds.

Across town, in the Rose Garden of the White House, President Bill Clinton stepped to the microphones and said this about the alien rock: "Today, rock eight four oh oh one speaks to us across all those billions of years and millions of miles. It speaks of the possibility of life. If this discovery is confirmed, it will surely be one of the most stunning insights into our universe that science has ever uncovered. Its implications are as far-reaching and awe inspiring as can be imagined. Even as it promises answers to some of our oldest questions, it poses still others even more fundamental.

"We will continue to listen closely to what it has to say as we continue the search for answers and for knowledge that is as old as humanity itself but essential to our people's future."

What follows is an account of how a fist-sized lump fallen in wilderness came to such a pass and of the people who made it possible and of what happened next. In the end, the mysterious stranger changed their lives and much else besides.

The Rock from Mars

CHAPTER ONE

SCORE

Robbie Score had no idea what she had found, and yet it was exactly what she was looking for. She noticed the object first as a dark green blemish on the clean expanse of ancient blue that shimmered around her.

In this stretch of Antarctica, December 27, 1984, was a balmy summer day, the temperature around zero Fahrenheit, the winds abating. The sunlight shone thermonuclear white. It could play tricks on your eyes. Score was on her first trip to "the ice," cruising downslope on a snowmobile in loose formation with five others who had come here for the hunt. Inside a sarcophagus of expedition-weight clothes, she felt the painful bite of the breeze intensified by her own motion over the ground. She wore dark Polaroid glasses and three layers of gloves—glove liners, insulated gloves, and "bear paw" mittens that fit over the top of the cuff of her red polar anorak with the fur-lined hood. That and her black wind pants were standard government issue. She wore her own hat, a jaunty red, white, and black knit, which (when not covered by the hood) distinguished her from the others.

They were working in a region of soul-searing desolation known as the Far Western Icefield, whose nearest landmark was a forked ridge of rock called Allan Hills. They were a good 150 miles from the nearest outpost of anything resembling civilization. Ordinarily, the hunters would spread out in a line, about a hundred feet separating each from the next, and sweep in tandem one way and then the other across a designated grid as large as three or even five miles in one direction. Back and forth, back and forth. The downwind legs weren't bad; but heading into the wind could be brutal, the chill searing right into your face.

In places, the ice turned washboard rough, jarring the riders with ranks of long, concrete-hard dunes called sastrugi. Built of windblown ice crystals, they could stretch to hundreds of yards in length and grow to the height of a person.

To an observer looking down from a godlike vantage, the behavior might have seemed puzzling—these bright-colored motes of life sweeping out their puny patterns in terrible isolation against a continent. But they had come here with a purpose. Their begoggled eyes were scouring the bright ground for bits of dark rubble. They were hunting the fallen husks of shooting stars. Meteorites.

The exercise itself was a little like plowing a Kansas wheat field. But you had to concentrate. Done right, the job meant hour after hour of relentless, brain-numbing eyeball concentration. If you blinked, you might miss a trophy.

During these sweeps, you were alone in your own world, in your own head. You moved through a deep, oppressive silence except for the wind and the vibrating hum and scrape of the Ski-Doo as it schussed over the hard ice. No birds sang. No trees rustled. Score would think about what was going on in her life and sing the latest rock-and-roll songs to herself, using just a smidgen of her consciousness to keep a proper distance from the hunters on either side. Maintaining focus was easier when there were lots of meteorites turning up. But some days, it seemed that the people on the other side of the grid had all the luck. It was like being in the wrong line at the bank. On days of really slim pickings, she mused about what she and her roommate would fix for dinner, wondered about events back in the real world, and sang something like maybe that Hall and Oates hit about broken ice melting in the sun.

Today, things had been going so well and the weather was so fair that, toward noon, on the spur of the moment, group leader John Schutt decided to give everybody a break from the routine. He led them on an excursion to the summit of a four-hundred-foot-high escarpment lapped by giant waves of ice, like a storm surge frozen in the instant it hit the shore. They were some forty miles out from the edge of the Transantarctic Mountains. Where the hunters were now, the peaks were still buried; they had not yet pierced the miles-thick ice sheet, but the surface ice echoed their topography.

After days of feeling as if she were plying some vast and featureless sea, Score was excited about reaching this area, known as "the pinnacles." Here, fierce winds and the resistance of the obstructions beneath the glacier had sculpted the ice into strange shapes that loomed as high as twenty feet above the ice floor. In the dazzling sun that day, the effect was magical, and a welcome relief from the sameness of "the farm." Score got a little extra adrenaline kick from knowing there were treacherously deep crevasses to watch out for on this artful summit. The chasms were visible and easily avoided, but the roughness of the terrain forced the team to abandon their usual formation and pick their individual paths as best they could.

After a five-mile run along the top, Schutt turned them back toward the smooth designated search area below. Score felt the cold go right through her as she sat on the Ski-Doo, her gloved thumb taking the brunt of it out there on the chest-high control. She rumbled downhill toward the flat plain of blue ice, where she and the others formed up to resume their systematic search. Though the terrain was less rumpled now, the blue ice was still patchy, interrupted by drifts of hard snow sculpted in little bumps and ridges a foot or so high. The hunters would never have included this area in their normal search grid.

She saw the dark shape—about the size of a grapefruit—against the blue ice. It was completely uncovered. It looked strange, a vivid green color, quite different from the hundred or so other meteorites the team had collected this season. And it was on the large side for one so unusual. Still, over the years the hunt had yielded up a variety of types—black, orange, big, small, smooth as a discus, rough as pinecones. She saw nothing to suggest that, more than a decade in the future, this particular specimen would put her name in headlines and cause her to be known in certain circles as "the one."

Score got off her snowmobile and waved her arms. "Hey!" she shouted, but with the hunters spread out the way they were, sound wasn't as effective as dramatic gesturing to get their attention. As she waited in the impossible silence of the place, she stomped her feet and flailed her arms in the familiar dance that tended to accompany a rock find. Aside from any benefit to science, the event always gave the hunters a chance to rev up their blood circulation a bit.

Robbie Score was still adapting to the rigors of life here in the planet's coldest place, where the winds that screamed down the slope of the continent sometimes kept the hunters pinned in their tents for days, where the sky relaxed its guard against the sun's ultraviolet radiation, and where the air you breathed was so dry that the Sahara, by contrast, might seem like a rain forest. The dryness could split your fingers and toes if you weren't careful to keep a good supply of Bag Balm. An injudicious smile could open fissures in your chapped lips. The atmosphere was so thin it amounted to the equivalent of working at eight thousand feet or more. Here, it was easy to feel like an alien in need of a better space suit.

The fetching brunette geologist with the apple cheeks and the thousand-watt smile had actually *volunteered* for this perverse version of an Easter egg hunt. She'd *demanded* it. As a little girl, Roberta Score, nicknamed Robbie, used to fill grocery bags full of the "pretty" rocks she picked up along the shores and in the woods around the Great Lakes and the smaller lakes of Michigan where her family vacationed. Most of these finds never made it back to their Detroit home. Her dad would discard the bulk of them once he had bundled her into the car. Still, she managed to assemble a pretty good collection in the family basement.

Score graduated from high school with honors, but she lacked direction and had no mentor. She was afraid of science and started out at a local college taking liberal arts classes. At her first opportunity, she bolted from her home turf and moved to Los Angeles with a friend. She got a job as a dental assistant and, after three years, went back to school, enrolling at UCLA with the idea of becoming a dentist.

Dental students traditionally took degrees in biology. But Score found the UCLA biology department to be a forbiddingly huge and competitive jungle, and she was not particularly interested in the subject. At the urging of a friend, she enrolled in a few geology courses. She was captivated and soon adopted geology as her major. Along the way, she got to know a guy with the odd nickname of Duck, a UCLA graduate student in geology. Nine years later, he would play a key role in the destiny of Score's odd rock.

After she graduated, Score wanted a break before going on for an advanced degree. It was at this point that she dropped dentistry forever and focused her career on rock—the particular kind of rock that falls from the skies, the geologist's manna.

In 1978, fresh out of the university, she answered an ad for a job at a laboratory in Houston whose very name signified how things were looking up for the arcane enterprise of meteorite collecting: it was the new Antarctic Meteorite Laboratory at NASA's Johnson Space Center. She arrived on the scene shortly after geologists had begun to figure out the extent of the treasure from space that might be theirs, as long as they were willing to spend weeks at a stretch scuttling around on the ice in the highest, driest, coldest, windiest, and emptiest place on Earth.

The society of meteorite hunters was pretty small when Robbie Score joined up. Until recently, the cost-benefit ratio of the game had been lousy.

The notion that extraterrestrial stones could plummet randomly out of the sky had been respectable since about 1803, when the French Academy accepted it. Gradually, people realized that this cosmic rain offered unique clues to the history of the sun and planets that made our very existence possible. From the perspective of the serious meteorite devotee, however, most of this heaven-sent material went to waste. And under most circumstances, the odds of any given person retrieving a meteorite were extremely bad.

Some incoming fragments mark their arrival with brief flashes of fire against the night sky. Sometimes they arrive in "showers," as Earth passes through a lane of unusually heavy debris. Some turn into impressive fireballs that can approach the full moon in brightness, their glow fueled by incandescent gases and punctuated with a sonic boom as they plow through the atmosphere in one second, trailing a "tunnel" of vacuum behind them.

Every hour of the day, about a ton of micrometeorite dust sifts into Earth's atmosphere, spicing the crevasses and pores of the civilized world. As many as thirty thousand meteorites the size of a fist or larger make it to the surface in a year.

But because more than two-thirds of this surface is covered by water, meteorites most often splash down and sink. Even the ones that slam into solid land are usually worn down by weather so rapidly that, within months, they blend right in with the local rock population. Very few come down where there are people around to find them. And in those cases, the character of the local landscape tends to dictate how many and what kind of stones are recovered. In the rocky terrain of New England, for example, a space rock is rarely noticed unless someone actually sees it fall. In the relatively rock-free soil of the Great Plains, farmers often uncover and recognize rocks from space as they plow their fields. And then there is the possibility that, in some cases, the plummeting rocks kill their only eyewitnesses.

Because people had found meteorites all over the globe—a total of two thousand in the two hundred years prior to 1969—it had not struck anyone as surprising that explorers had stumbled across them in modest numbers at the bottom of the world, where the ice sheets provided such a clean, revealing background.

Then a few alert people figured out that there was something remarkable going on here: a secret that lay in the wild, competing forces of the ice continent could rationalize meteorite hunting by providing it an efficiency of scale.

In December 1969, a party of Japanese glaciologists found nine meteorites lying on a blue ice flat near the Yamato Mountains. The natural assumption was that the meteorites came from a common parent and had fallen together. The investigators were surprised to learn from later analyses that the stones were of distinctly different types and had spent varying periods in the ice. These facts led team members to suspect that some mysterious process was concentrating meteorites on the ice—meteorites that had arrived on Earth at different times from a variety of starting points.

In 1973, at a meeting in Switzerland where Japanese researchers described the unlikely collection of stones, a young American named William A. Cassidy was captivated by the possibility: something was causing a strange convergence of meteorites on the Antarctic ice. There must be some natural mechanism actually winnowing meteorites from the chaff of polar history and dumping them out for easy pickings.

Cassidy took up the cause in the United States, following the Japanese lead. As they recovered meteorites from the ice by the dozens at first and then by the hundreds, he used their success as a spur to his own prospective sponsors. After bouts of rejection and frustration, in 1976 Cassidy won his government's support for a modest systematic search program.

Sending teams to hunt meteorites in the Antarctic would certainly require less than a pittance of the money the United States had spent to launch the eleven manned Apollo missions between 1968 and 1972 (the equivalent of around $107.7 billion in 2005 dollars). The astronauts brought home a total of some 840 pounds (almost 382 kilograms) of moon rocks and soil. Once those flights were ended, indeed, the Antarctic would turn out to offer the most reliable and cost-effective—and, as Cassidy would put it, "mind-boggling"—source of rocks from beyond Earth. Nature delivered the specimens to the surface free of charge, without the aid of expensive rockets or standing armies of technicians and engineers. Cassidy called the Antarctic program "the poor man's space probe."

Like dashboard clocks that reveal the instant of an auto accident, meteorites are clocks stopped at different stages in the birth and evolution of worlds around the sun. Cassidy understood that these "stopped clocks" were actual bits and pieces of solar system bodies as they'd existed along this time line.

One of Cassidy's first finds in his inaugural season came from Allan Hills, not so far from where Robbie Score would kneel almost a decade later to pick up her green rock. It was inconceivable, at the time, that during the first decade and a half of operations, U.S. expeditions to Antarctic meteorite-stranding surfaces would recover more than ten thousand specimens, on top of a like number collected by Japanese and European groups. The samples would be distributed to hundreds of researchers from some two dozen nations.

Some of the rocks carried unique information that would make them celebrities of the meteorite archives, more sought after than the rest. The meteorite trove was what one rock hunter called the extraterrestrial equivalent of the placer deposits mined during the California gold rush. In both cases, it took a lot of careful and methodical sifting through vast quantities of material to pick out the gold nuggets.

And Robbie Score, this day in 1984, had come across just such a one.

For the preceding six years, as she worked in her Houston lab, Score had watched the parade of adventurers willing to brave the rigors of Antarctica for the promise of these burnt offerings from the skies. She had contented herself with studying the specimens that other people had plucked off the ice. She certainly knew her space rocks. She had worked with every meteorite collected in Antarctica, and had helped ship the supplies for each field expedition. She knew most of the hunters and had heard their stories. But people constantly asked her about proper collection procedures, what materials worked best, and so on, and there was only one way to learn that. Going to the ice would fill in a critical gap in her knowledge of meteorite science, so it was a practical move. Aside from that, she craved the adventure, and the sheer physicality of it. She had to visit the mother lode at the bottom of the world.

Having passed the extensive medical and dental checks required for expedition members (not only to avoid emergencies in an inhospitable place but because dental records could aid in the identification of badly damaged corpses), she found herself in a warehouse-like enclosure in Christchurch, New Zealand, laughing at one of the staple ironies of these treks: the team members standing there sweating like pigs in the uncomfortable heat as they got fitted out for the opposite extreme. Looking for the right sizes, they tried on the used government-issue long underwear, wind pants, wind-pant liners, flannel shirts, down vests, polar jackets, extra-heavy socks, thermal boots, glove liners, wool gloves, outer bear-claw gloves that would cover the other two pairs, balaclavas (face coverings), hats, and goggles.

At last, they boarded an LC-130 transport plane equipped with skis, for the cramped eight-hour flight to McMurdo Station, the main U.S. research base in Antarctica. McMurdo sits at the southern tip of Ross Island, about 2,400 miles south of Christchurch and 850 miles north of the South Pole. Founded in 1955, the village was constructed on barren ash and lava hills on the flanks of a steaming, active volcano called Mount Erebus, after the mythological gateway to hell. As the plane banked toward landing, Score got a heart-stopping view of the snow-covered Royal Society Range to the west.

The south polar landmass, Score knew, was twice as large as the continental United States (the "Lower 48") and covered by an ice sheet two miles thick—so massive that its weight had depressed the bedrock beneath it, forcing it deeper into Earth's crust and squeezing the planet's bottom. The continent held about 80 percent of Earth's fresh water, and 91 percent of its ice. There was virtually no atmospheric humidity. In the United States, when the humidity fell as low as 30 percent you could get a static-electricity shock from a doorknob. In Antarctica, 1 percent was considered a really damp day, and the level never got higher than 5 percent.

The dry valleys of Antarctica probably provided the closest approximation on Earth to typical conditions on the planet Mars.

On the warmest days of the Antarctic summer (December and January), when the population swelled with scientists, McMurdo's temperature typically hovered just below freezing, and some other parts of the continent *might* climb that high. The sun was up twenty-four hours a day, its angle always about that of midmorning. As for winter, the temperatures were known (rarely) to sink below minus 100 degrees Fahrenheit at the pole. (The record low for the continent is minus 126 degrees Fahrenheit.)

McMurdo—"Mac Town"—was the jumping-off point for the meteorite hunters, as well as for troops of other scientists and explorers who came here. With its cloak of prehistoric volcanic ash, during the summer snow melt it sat like a black cigarette-burn mark on the white continent. Score found it to be a motley sprawl of construction with the spartan intimacy of any frontier town. However, it had also come to embody the government bureaucracy's approach to defending itself against frontier risks: it engulfed visiting field-workers in a rigid system laced with Washington-style red tape.

Some people were happy to swap this stifling protective maze for the simpler dangers out in the field.

In 1839–40, James Clark Ross had maneuvered his ships *Erebus* and *Terror* into the McMurdo Sound here before sailing eastward along the front of the great ice shelf that now bears his name. It was from this area, in the early 1900s, that Robert F. Scott and Ernest Shackleton deployed their sledging parties.

Knowledge of the whole of the ice continent had been bought slowly and painfully—most of it coming in a burst during the twentieth century. In the earlier days of Antarctic exploration, the profits to be taken from whaling and sealing and penguin oil had lured some to Antarctica's margins. The early 1900s brought a succession of "heroic" explorers. Later, Cold War military interests looked to Antarctica as a site for bases and materials testing as well as an international competition to claim it. The ice continent became, as one chronicler put it, "a powerful magnet for human fancy."

Then something remarkable happened: *scientists* won the battle for Antarctica. In the 1950s and early 1960s, the drive to conduct research here on varied fronts, as well as a new appreciation of the fragility of the environment that made it possible, led to an international accord. The treaty prohibited military activity and the removal of specimens of any kind—except for use in scientific research. (Among the disciplines pursued here would be glaciology, meteorology, upper-atmosphere physics, astrophysics, mapping, biology, ocean sciences, biological and medical sciences, and environmental studies.) Since the early 1960s, no nation has owned Antarctica. No passport has been required to enter.

The men and women of the more or less permanent blue-collar workforce at McMurdo referred to seasonal visitors like Score as "beakers," a term of friendly derision. Some beakers managed to get along well. But they had shockingly little time for socializing; their weeks in the Antarctic would often constitute their research for the year. Their livelihood and their careers depended on how much they could get done here.

The first job facing Score and her colleagues was to gather all the field gear, spare parts, food, and fuel they would need for their six-week expedition some forty miles beyond the edges of the Transantarctic Mountains (a chain that spans the continent and therefore has two northern ends). Once in the field, they would have to be completely self-sufficient, able to make their own repairs and manage their own safety and health. To that end, geologist John Schutt—the resident field safety officer and crevasse expert—led them on a two-day mock outing to teach them minimal survival skills and how to use the research equipment. They also had to learn routine chores—how to set up a tent, light the stove, lash a sled, and avoid cold-related injury.

Schutt had fallen in love with the Antarctic—the light, the freedom from traffic jams and crowds, the landscape like no other. It was a place where, as the mantra went, "you challenge yourself every day." Schutt had been a skier since he was five, and had learned his love of mountain climbing as a boy growing up in the Pacific Northwest. He showed a special enthusiasm for the world's remote, cold, and snowy places, having worked on Alaskan ice fields and an ice island in the Arctic. He had been on mountaineering expeditions to Alaska, South America, and the Himalayas, where he had scaled the 27,800-foot peak of Makalu. Since 1976, Schutt had worked with the meteorite hunters in the Antarctic and, as their sherpa, was well on his way to becoming a legend.

His reputation had grown through stories like this: Schutt and a friend were climbing a ridge, roped together for safety, when the friend fell without warning. Seeing that the much bulkier man threatened to pull both of them into a fatal nine-hundred-foot free fall, John quickly threw himself off into space on the *opposite* side of the ridge and saved them both.

In Schutt's view, tent fires were the biggest danger for the search team members, because their stoves burned continuously to warm the interior, to melt water, and so on. People sat inches from the open flames fueled by white gas. It was prudent, he advised them, to carry a butcher's knife in case you had to cut your way out. One time, a team member violated the cardinal rule: never fill the stove *inside* the tent when another stove is already turned on. The fellow spilled some fuel and started a small conflagration. He was lucky to escape with only minor burns on his hands.

Schutt also impressed on the novices the need to treat the crevasses with respect. Where the mountain ranges or other features obstructed the progress of the ice, the resistance forced these deep, treacherous cracks in the sheet—cracks big enough to swallow a hapless snowmobile or much worse. Schutt had seen one near disaster, when a team member pulling a very heavily loaded sledge tried to cross a crevasse via a snow bridge that had been weakened by several other people proceeding ahead of him. Visibility was poor. The sledge broke through and fell into the chasm. The man just managed to pull it back out before it could suck him down.

In many of the meteorite-stranding fields, the relentless winds had hardened the snow into the consistency of concrete, Schutt explained, and these hardened bridges were usually sturdy enough to allow a safe crossing. But you had to watch out for soft spots. The options for dealing with that sinking feeling, when a crevasse opened beneath you, were covered pointedly in a field manual: "Experience and circumstance will dictate whether to brake . . . or continue driving forward in hopes of getting across. . . . In either case, a change of underwear is recommended."

Score trusted Schutt absolutely. As part of the training, she and the others followed him single file out into treacherous terrain near McMurdo, where each member of the hunt team, in turn, was lowered into a crevasse. Each was then "rescued" by the others. When Score's turn came, she descended with complete assurance.

In general, Schutt advised his charges to cultivate a healthy, controlled fear of frostbite, snow blindness, hypothermia, tent fires, flying, and injury. As Ralph Harvey, another expert meteorite hunter, remarked, "It promotes awareness and points out the true killer of Antarctica: rationalization."

By early December, the team was packed and ready to leave the relative comforts of McMurdo. It took seven helicopters to carry all of them and their equipment into the field. And they had picked a hellish day for it. As they disembarked at their destination on the heavily crevassed ice sheet, they found themselves standing in a cutting thirty-five-mile-an-hour wind. As they watched the last of their chopper taxis disappear, Score felt overwhelmed with a sudden sense of being absolutely, utterly alone.

The feeling, at least in her case, passed. The scouring wind and the biting cold, after all, were what had brought her here. She had work to do.

It was the paradox of the meteorite hunter. The wind and the cold that made the Antarctic so brutally forbidding were crucial to the success of the enterprise. This was the secret that Bill Cassidy and others had figured out, thereby handing the meteorite hounds a gift beyond price. The reason had nothing to do with the rate at which meteorites fell in this place. Earth, rotating on its axis daily as it traveled around the sun yearly, encountered a steady rain of space debris over its entire surface—a relatively uniform distribution over time, as far as anyone knew.

Cassidy and the others had deciphered the action of a majestic natural conveyor belt: over many millennia, meteorites that fell here would get buried under falling snow; that snow would get buried under yet more snow, and so on, and so on. The pressure of the accumulated weight turned the snow to ice, entombing the fallen stones. Powered by gravity, this glacial raisin cake of meteorites would creep slowly but relentlessly overland from the 13,000-foot elevation at Dome Argus, near the center of the continent. It would advance at the proverbial glacial pace of perhaps four feet (1.2 meters) each year—toward the edges of the landmass. When it reached this shore, some of it calved off into the Southern Ocean and was lost. Where this slow flow was forced through narrower passages, it accelerated. In some places, though, the flow was blocked by a mountain range—a barrier too broad for the ice to flow around it—creating stagnation points. Like surf hitting a steep shoreline, the trapped ice sheet shouldered upward at these places as its billions of tons collided in slow motion with the immovable impediment. The clean, ancient blue ice was thrust up from deep, deep under the glacial mass to the surface.

The old ice was blue for the same reason that the sky and the ocean were blue: because of the way it scattered light. Familiar ice, like the cube from your fridge, is whitened by air bubbles frozen in millions of little facets. The ancient ice was pressed clean of these "whiteners." Stepping out on an expanse of blue ice about ten thousand feet thick could seem—visually—like walking on a patch of the Caribbean. Under certain circumstances, Bill Cassidy and his comrades had learned, it was possible to melt a hole in the snow, exposing an expanse of the clear ice, and thus allow the reflected sunlight to fill a dark tent with the azure glow of a jazz club.

The blue ice here was different in important ways from blue ice in Alaska or at the margins of Earth's other great ice sheets. Where the Antarctic ice rose into the light, it was exposed to the notorious katabatic (descending) winds. These dry howlers were generated over the polar cap as the colder, denser air sank and gravity drove it downslope toward the ocean. The winds swept loose snow off the ice, exposing it. Two effects then combined to wear away the naked ice: sandblasting by windblown ice crystals and vaporization by the heat of the summer sun. As surface ice was lost in these ways, the continental hydraulics moved just enough fresh upstream ice into the stagnant zone to replace the lost ice and maintain the surface elevation. This progression gradually carried buried meteorites, stones that had once been sprinkled throughout the volume of the ice, to the surface, where they were left lying. Stranded.

In the places along the coast where the landscape channeled the winds so that they converged, the mean annual wind speed was 50 miles per hour and people had reported maximum gusts of almost 200 miles per hour. Early Antarctic explorers sometimes had to crawl through the blast before they learned the fine art of "hurricane walking" at an extreme angle.

It appeared that the meteorite conveyor belt had been working away at various sites for tens of thousands of years, hundreds of thousands of years, or, at some surfaces, much longer. Some rare Antarctic meteorites had spent as many as two million years in their slow-moving terrestrial tomb before breaking out. The ones big enough to resist being blown away remained as a lag of cosmic rubble on the ice sheet. Much more than in the driest warm deserts of Earth, the natural deep freeze and incredible dryness here inhibited the weathering away (and contamination) of the rocks.

The specialists came to understand that, given a properly meticulous and methodical search over the long term, they could assemble from these sites a representative sample of all the meteorites that fell on Earth.

Out on the ice sheet for at least a six-week stay, Score and her comrades lived in two-person "Scott tents," whose design had not changed much since they were first used early in the century. Double-walled for insulation, and nine feet on a side, the bright yellow pyramid-shaped structures were designed to withstand winds of 120 miles per hour. Boxes of frozen food (meats, seafood, vegetables, fruit, cheese, rice, breads, chocolate bars, soups, and so on) were stacked outside each tent.

Outdoor summer temperatures typically ranged between 5 above and 10 below zero degrees Fahrenheit, although a high wind would make it feel much, much colder—like living in a combination meat locker and wind tunnel. Inside, although the snow underfoot might be a subfreezing cold "sink," when the stoves were going the top of the tent could reach 90 degrees Fahrenheit—good for quick-thawing some food items before cooking them.

All water had to be harvested from the polar ice—a time-consuming and laborious process that involved hefting masses of the heavy ice around, fueling up the stove, and so forth. As one Antarctic veteran observed, you never appreciated how little energy there was in a gallon of gas until you tried to use it to convert cold, cold ice to drinkable water. It was amazing how much energy that took. The daily chore of melting just enough ice for drinking, cooking, and other basic needs probably consumed more time in the field than any other daily activity—typically an hour a day per two-person tent. One consequence was that, instead of a good old civilized thirty-gallon shower every day, people in the field tended to take five-minute sponge baths—once a week, whether they needed them or not.

This season, there had been occasional storms with winds over forty-five miles per hour, which kept the team pinned down in their tents, telling each other about their lives. In the worst years, workdays lost to high winds could add up to maybe 80 percent. On this expedition, a tolerable day or less per week had been the norm. Score passed the time by writing in her journal, reading books, playing cards, and catching up on her sleep. She never felt bored.

Score shared her tent with a woman who had recently come through surgery. When the helicopters had dropped them off in the wilderness in a howling gale that first day, the roommate had started lamenting her decision to make the trip; she'd wanted to fly back with the choppers. Aside from the woman's own plight, Schutt worried that this would throw Score off her stride just as she was getting started. But Score turned out to be resilient and relentlessly upbeat. She acclimated quickly and savored every moment. Like Schutt, she had fallen hard for the place.

Over the years, the notion of adventure on the ice had attracted applicants as diverse as Japanese flower arrangers and an occasional composer of symphonies. Some had never even been camping before. Those who made the cut, whether they tended toward the extreme-sport type or seemingly fragile geekdom, typically had a great desire to explore new places and to prove themselves at some level. When the winds rose, though, the capacity for confinement with a good book proved a more valuable attribute than the compulsion to skydive off a mountain. As the six-week tour on the ice neared its end, sometimes even the most stable among them would feel the stresses of survival and the forced intimacy with fellow hunters starting to wear them down mentally. And older hunters would find that, with faltering blood circulation to the extremities, their hands and feet would chill more easily, and the skin of their faces tended to develop the waxy-looking patches known as frostnip—the precursor to frostbite—without their realizing it. Still, generally speaking, the team members were happy to be there; they were intensely motivated; and they made it work.

Cassidy would sit in his tent in the mornings, as a first-year novice, "listening to the wind flapping the canvas and the sibilant whisper of ice crystals saltating along the surface" and cringe at the prospect of going out into it.

For Score and company, the typical workday would begin around seven-thirty A.M., with breakfast (usually a quick-and-dirty bowl of oatmeal), cleanup, and the daily radio check-in with McMurdo. The group met at nine A.M. to map out the day's strategy. If their search stayed close to camp, the team might return at lunchtime to heat up some soup. Otherwise, they would eat a cold lunch "out there" and work through until about six P.M.—sweeping back and forth across their search grid, keeping their eyes and brains focused through the long hours and the chill. When the weather turned nasty, the search would be abandoned. It would be counterproductive to go on, because the hunters would wear themselves out while also missing a lot of meteorites.

The evening meal was the biggest production. Heavy on steak and sometimes lobster, the dining was as good as it could possibly be under the circumstances. A person could burn through five thousand calories a day in the cold, and many people lost weight despite the high-fat diet. They usually ate inside their own tents, but on Christmas, they would crowd together in the largest tent for a special communal feast and celebrate with silly gifts they brought for each other. They might hang their stockings upside down, for the anti-Santa at the South Pole.

Beverage intake sometimes required special thought. Kool-Aid was popular for fending off dehydration, but there were important distinctions. The staff at McMurdo routinely held clinics on urine color. Dark brown meant serious dehydration. Therefore, when venturing into protected parts of the Antarctic wilderness—where the international treaty agreement required a person to pee into a container, not on the local ecology—the old-timers cautioned that you wanted to make *purple* Kool-Aid for drinking, so there would be no possibility of confusion.

Just about everybody agreed that the biggest pain of all, what Bill Cassidy called a character builder, was the necessity of going outside to relieve oneself—no matter what the weather was, no matter how much the windchill burned. The destination would be a proper distance from camp, usually in the lee of a Ski-Doo, where you would squat in the age-old ritual over a hole in the snow.

At "night," the trek took on an added splendor, because the sun was up all the time. And this was no ordinary light. It was (in contrast to the darkened interior where people tried to sleep) a *blinding* light like the end of creation, intensified by the glare off the ice. When someone opened the tent, there was the effect of the nuclear flash in that moment before the blast wave hits.

Robbie Score and her teammates had come into the experience with full knowledge that it would be a schizoid mission in which the wind was your enemy and it rattled your brain as it rattled your tent and, sometimes, seemed to suck every bit of heat out of you. But this same implacable nemesis was also your essential ally. It was why you had come. You relied on the wind to sweep snow off the blue ice and uncover the treasure. You were there in the hall of the winds on purpose.

December 27 was beautiful, relatively calm and mild. When Score spotted the rock, she got off her snowmobile and waved her arms. Gradually, others congregated around her, and someone put a flag in the ice to mark the location.

Schutt helped her collect the specimen. He could tell right away that this one was unusual. Some of the fusion crust—the glassy black charring from its passage through Earth's atmosphere (13,000 years earlier, as people would determine later)—was gone, so he could see into the rock's interior. It looked green to him, too. He and Score stripped off their layers of gloves. They needed the dexterity. Kneeling, they each tucked the gloves under a knee to keep them from blowing away. Score took one of the special "clean bags" provided by NASA (originally designed for collection of rocks on the moon) and used it to pick up the specimen—4.25 pounds of it, she would learn later—by wrapping the bag around it so that her hand never touched it. With Schutt assisting, and working as quickly as possible with her exposed hands feeling the air bite, she used the same care she used for any meteorite. The procedures were designed to minimize contamination by, say, nose drips, as well as from gloves, hands, or other nonindigenous sources. With the rock inside, she quickly sealed the specially sterilized nylon bag with Teflon tape. She wrote a field number on the tape. She and Schutt used a surveying instrument and an electronic distance-measuring unit to record the location (relative to survey stations located all up and down the ice field; the infinitely more convenient Global Positioning System of satellites was not yet available).

The field notes they jotted down that day described the rock as "highly-shocked, grayish-green . . . 90 percent covered with fusion crust" (with the additional comment "Yowza-Yowza"). They added the odd rock to the day's catch in one of the backpacks that a couple of teammates carried. Back at camp at the end of the day, the specimen would go into the expedition collection box.

For most rocks, the retrieval was considered a team effort and no one hunter was singled out as the discoverer. True, the search involved skills and powers of concentration, but luck also played a big part. So, together, the team would take the hits and misses, the good days and the "tent days" when the winds screamed. But Score and this rock would be forever linked. This was partly because she saw the rock as special and felt an abiding, proprietary interest in its destiny; and it was partly because she was well acquainted with others whose lives would become caught up in the fuss over its meaning: Duck Mittlefehldt, David McKay, Everett Gibson, Kathie Thomas. They would be working in the same building as she—Building 31 at Johnson Space Center in Houston—and she would run into them there and at parties, and they would chat about the twists and turns in the saga that began on this day. More than a decade later, when journalists and others around the world were clamoring to hear how the rock had been plucked off the ice, everybody would point toward this spunky woman, and toward the beguiling smile—not toward a team. She made a better story. That would be fine with Schutt. In his view, she was a most worthy ambassador for the meteorite hunt.

The team had already bagged over one hundred specimens that season, a total that would grow to three hundred or so before their expedition ended. This particular find at Allan Hills was noteworthy at the time only because of its green cast and large size out there in the clean dazzle of sun and ice.

When they were ready to head home, the team packed their collection into ice chests, in burglar-resistant containers, and put them aboard the end-of-season ship out of McMurdo for the long sail northward. To further minimize the risks of contamination, the specimens were kept frozen en route. In keeping with the international treaty, the bulk of the ship's cargo was the season's accumulation of trash and waste, all routinely carried home.

When the ship docked at Port Hueneme, California, technicians packed the rocks in dry ice for air transport to Houston, to their new home at the meteorite curation laboratory at NASA's Johnson Space Center, not far from the lab where the lunar samples from the Apollo moon missions were still carefully preserved and tended.

After the season's haul arrived, the NASA meteorite archivists followed the routine protocol by which they were able quickly to characterize and describe large numbers of new finds: they took the specimens out of the bags and put them in a nitrogen chamber. Using a stream of dry nitrogen, the staff freeze-dried the meteorites to remove any attached ice and snow, and to avoid possible contamination from the leaching effects of liquid water during transit. They photographed them and examined them carefully, using both eye and microscope. Finally, they broke off a small chip from each specimen for the initial classification by specialists at the Smithsonian Institution in Washington, D.C.

Score, returning from her first hunt, looked forward to showing off the strange specimen to colleagues. "Wait till you see this green rock!" she kept saying. But as the treasures were decanted, Score was startled to see that she had been the victim of an illusion—perhaps not the last one the rock would foster. In normal laboratory light, the rock did not look so green. It resembled nothing so much as a chunk of gray cement, although parts of it had a greenish cast. She and Schutt would talk about this optical trick off and on over at least the next couple of years. They chalked up the misimpression to the eerie Antarctic light, or possibly an effect of the sunglasses they had worn, or both. But then such distortions should have applied to other rocks as well, and didn't, so the question would remain something of a mystery.

Score was in charge of assigning the rocks their laboratory numbers, and she singled this one out despite, or perhaps because of, the trick it had played on her. She knew this was the one she wanted to look at first. So she designated it ALH84001: ALH for Allan Hills, after the major landmark near the site where she'd bagged it; 84 for the year (traditionally marked in December); and 001 because she had selected it to be the first of the season's harvest to be recorded and analyzed.

Robbie Score had become the unwitting instrument of a beginning. Many years in the future "her" rock would send scientists down the slippery slopes and into the unexplored crevasses of one of the most profound questions ever asked by civilization. The rock would emerge as the catalyst in one of the most bitter scientific debates of the century. It would lead down new avenues of understanding—and some blind alleys—in humankind's long struggle to understand what defines a living thing at the most fundamental levels, how life came to exist on Earth, and whether the terrestrial version is unique in all of time and space. The rock's mere possibilities would incite yearnings of the spirit, clashes of ego, and intrigues among the powerful. The White House would summon prominent thinkers to discuss the potential implications for religion, culture, and the federal budget. The rock would help explode a scandal that ensnared a presidential adviser and his mistress.

The rock had escaped Earth's polar ice only to be entrained in another kind of complicated glacial advance, this time moved along not by gravity but by the force of human curiosity.

But for nine more years, the rock and its secrets would rest in quiet obscurity on a shelf, a victim of mistaken identity.

CHAPTER TWO

MOON DUST

Neil Armstrong stood alone in the middle of a broad, cratered plain cloaked in dust as fine as talcum powder. The lighting was eerie, dreamlike. Blazing brilliance at ground level contrasted with hard ebony blackness overhead. It reminded him of standing in a floodlit sandlot baseball field under an Ohio night. He pulled a collapsible tool from a pocket on his thigh and, turning his back to the explosive glare of the white sun, bent awkwardly in his bulky suit to scoop up some soil and rock. When he studied the pieces close up, he saw the true shades of this shining place: charcoal brown or nearly black, like powdered graphite.

Armstrong, thirty-eight, a son of the American heartland, had just picked up the first geological samples ever taken by a human on another world. David McKay, five years younger, watched from a seat some 240,000 miles across the void that Texas summer evening in July 1969. He was the lone geologist inside the hushed inner sanctum of Mission Control, Houston, during the landing of the first men on the moon.

McKay watched with wide, owlish brown eyes, sensing in his own nerve endings the motions of those human hands at work on that alien plain, willing them to move faster, pick up more stuff. It was the ultimate field expedition, and Armstrong was, in a sense, McKay's surrogate. The geek geologist (taciturn and ironic) had helped teach the celebrated astronaut (taciturn and earnest) about the just-invented art of lunar prospecting. Now, during the surface exploration, McKay could only watch and see how those few hurried lessons had taken. The events unfolding out there were what flight director Gene Kranz (referring, of course, to much more than mere geology) had called "the final exam."

Sitting in the back row of the main floor in Mission Control, a place redolent of cigarette smoke, stale pizza, and burnt coffee, McKay could see the astronauts moving like ghostly apparitions on the big screen at the front of the room. He could hear the terse exchanges between Houston and the moon. He was at the center of Big History, and his spine tingled as much as anyone's.

But his focus was different from that of most of those around him and, for that matter, most of the global population that had tuned in to watch the grainy, jerky black-and-white images flowing from the moon. His focus was, you might say, microscopic. For him, the main show was not in the technological virtuosity of the moon landing, not in its Cold War symbolism, not in the triumph of the first human footfall on another world, and not even in the steely courage and skill exhibited by Armstrong and his crewmate Buzz Aldrin.

The stomach-clenching descent from orbit had been punctuated with alarm bells as the spacecraft headed for a landing about two miles beyond the planned aim point and an overtaxed computer threatened to abort the landing. Armstrong, with Aldrin reading out crucial data, had maneuvered the landing craft manually to avoid a treacherous crater-pocked and boulder-strewn obstacle course and make it to a smoother patch. They'd contacted the surface with about twenty seconds of fuel remaining. In the harrowing final seconds before touchdown, Armstrong had found his visibility obscured, his sense of the craft's motion confused, by the lunar dust kicked up by his blasting jets—pulsing, fan-shaped sheets of dust expanding like veils flung by a dancer and captured for posterity on a movie camera attached to the landing craft.

From the moment Armstrong exited the lander and activated a small TV camera on the side of the craft, sending the images flowing to Earth, McKay and a few others like him scattered around the world watched the dust—to see how it took the imprint of Armstrong's tread and clung to the light blue sole of his boot, how it flew off strangely in all directions like a swarm of disturbed flies (with little gravity to sort them, no atmosphere to interfere) when the astronaut tried to shovel it up with an aluminum scoop. McKay could barely stand the wait for the samples to arrive back at Earth. If he couldn't go to the moon himself, let the moon come to him. Then he would no longer be a mere spectator.

On this July evening, the meteorite trapped in the ancient blue ice of Earth's south polar region lay undisturbed some eight thousand statute miles to the south of McKay. Robbie Score, still a kid, was on a camping vacation with her family in California. They had all driven to the nearest town to see a movie (*Yellow Submarine*, with the Beatles). Robbie and her dad kept slipping out of the theater to watch the moon landing on the television set in a store window display next door.

More than fifteen years still had to pass before Score would pluck "her" rock from its resting place on the ice and start the slow unfolding of events that would lead McKay on to a frontier of a different sort, where he would be tested in ways he had never anticipated. But it was the unprecedented national push toward this summer moment in 1969, toward the imprint of that first human footfall in the alien dust, that would set the stage for McKay's unlikely turning later on.

It was Apollo that sanctified this swampy, buggy acreage of salt-grass pastureland south of Houston as the once and future temple of the stones. Apollo would draw together many of the primary players in this story. It would set the trajectory of their lives, shape their dreams, and determine that Robbie Score's rock, as it emerged from its frozen history years from now, would fetch up in their midst with all its enigma, here in this meltingly torrid precinct of south Texas, of all places.

It has been said that the advance of science is driven not so much by the quest for truth as by a scramble for jobs. Apollo in the early 1960s opened a gusher of tax dollars to fuel a crash program modeled after the secret Manhattan Project, which had developed the atomic bomb. (This one, however, would be wildly public.) The result was the largest single civilian project in history.

It proved to be a soul-stirring drama—a "triumph of the squares," a NASA official would call it—that fleetingly enthralled much of civilization and eventually seduced even the cynics and ironists. Author Norman Mailer (who had scorned the U.S. space program as protofascist) would be moved to acknowledge the moon landing as a victory for the WASP culture and a rebuke to the counterculture. "You've been drunk all summer," he wrote, ". . . and *they* have taken the moon."

David Stewart McKay was one of the squares.

The ancestral McKays had moved from Scotland to Ireland and then, in 1830, immigrated to the United States, to settle in Pennsylvania. One McKay helped build the Erie Canal; another fought in the Civil War, as a member of the Pennsylvania National Guard, and his rifle—made by Eli Whitney Co.—would become one of David McKay's family treasures. Still later, a McKay started a carriage factory with the motto "If it's McKay, it's okay." On his mother's side, David knew, a grandfather had worked for a steel mill and gotten elected county commissioner; he had good "people skills." Her uncle Derwood was a photographer who experimented in his own darkroom—the only known McKay ancestor who came close to a scientific pursuit of any kind.

David McKay was born in 1936 above the strata where the first oil well in America began pumping in the 1800s, near Titusville, Pennsylvania. His father worked for an oil company—but not as a geologist. He was an accountant.

Still, two of the three sons developed an abiding interest in how the oil had gotten there and other matters geological. Even as a boy, David was fascinated by the notion that you could go into the great out-of-doors, look around, and see coded messages about how the natural world worked. For McKay, there really was a sermon, or at least a little good gossip, in every stone.

When he was ten, the family followed the oil jobs to Tulsa. After high school, David McKay went to Rice University in Houston to get a bachelor's degree in geology. A "TRB" (Typical Rice Boy), he wore his pants high and carried a slide rule. It was geek heaven.

McKay moved on to the University of California at Berkeley for a master's but soon got fed up with school and bolted. He hired out to oil companies, doing survey work and living in a drafty trailer in the desert. He worked on offshore rigs in the Pacific near California and in Alaska's Cook Inlet. But the industry jobs, with their slow pace and funky, low-tech tone, soon frustrated him even more than school had. After about a year, he escaped back to Rice to acquire a Ph.D. in geochemistry, which he completed in 1964.

That year, McKay returned to Berkeley to do postdoctoral lab work. A bit roguish, he was lanky, with dark tousled hair, dark eyes behind black-rimmed glasses that he had worn since eighth grade. He wore a Rice class ring. It was during this period that he met a young woman who would prove both elusive and inescapable in his life. Her name was Mary Fae Coulter, and she was a sassy Phi Beta Kappa from Rice, a schoolteacher who was at Berkeley to get a master's degree in English.

They would go out to a movie now and then, but Coulter was on guard. She noted that David McKay always seemed to be on the lookout for other women. Although she thought he was cute and enjoyed his company, she was not about to get herself dangerously tangled up with this roving Romeo.

In any case, McKay soon got a call from a Rice colleague who was staffing up a geology group at the new space complex in Houston. The fellow offered McKay a job.

In the early sixties, NASA was hiring in a near frenzy, signing up almost anyone who walked in the door with a bachelor's degree—and a few Ph.D. scientists for ballast. The designation of a large cow pasture twenty-five miles south of Houston as a major hub of space operations had triggered waves of dismay and rumors of scandal at first. The choice had been determined primarily through the influence of Texas politicians, most prominently Vice President Lyndon Johnson, an ardent ally of the space program, and Albert Thomas, chairman of NASA's funding committee in the U.S. House of Representatives. An oil company (with Rice University as intermediary) had magically donated the required expanse of land, one thousand acres of unproductive oil field where cattle grazed, not far from the refineries around Galveston Bay. (The geology here had been laid down in stream deposits from the erosion of the Rocky Mountains. These sediments of sands and clays had been set down on top of decaying organic matter transformed over the centuries into oil and natural gas.)

The NASA officials who had to move there initially balked at the region's glowering climate, which was enough to make a body yearn for the rigors of outer space. "For eight months Houston was an unbelievably torrid effluvial sump," Tom Wolfe wrote. The humid air was heavy with the odors of the Houston Ship Channel, as well as of petroleum, and chemical and paper plants, and periodically hung thick with mosquitoes. And the nearest town to the space complex, Clear Lake City, was named loosely after an inlet that Wolfe said was "about as clear as the eyeballs of a poisoned bass."

But the resistance soon evaporated, once the serendipitous wisdom of the choice became clear. The move represented an escape from the "old fogies" back at NASA headquarters in Washington, people realized. It was a chance to recruit young people with creative ideas and build something truly new. Besides, the hot, humid climate had some advantages, such as minimizing the threat of blizzards and extended airport shutdowns. The clay beneath the surface of the land in these parts precluded the buildup of friction that ended in earthquakes. Instead, the region experienced only the more benign "fault creep." That, plus the city's location about halfway between the West and East Coasts, turned out to be a blessing for a workforce that (thanks to the geopolitics of space) had to travel by plane constantly in both directions—to the launch complex in Florida and to the aerospace contractors concentrated in California.

The new recruits were swept up in the chaotic excitement of an undertaking whose momentum, for the moment, seemed irreversible. The timing couldn't have been better for an impatient young geochemist.

In the summer of 1965, word went around the space center that one of the female managers, a chemist, was expecting her boyfriend to come and join NASA. That fellow turned out to be McKay.

He arrived in June 1965 at the still unfinished Manned Spacecraft Center (later rechristened Johnson Space Center), where offices at first were scattered in strip shopping centers and at the spartan, faded facilities of nearby Ellington Air Force Base. By 1967, he'd moved into the science building, the freshly built Building 31, two stories of labs and offices whose facade was a variety of limestone known as Austin chalk, from a Texas quarry. The numbers signified the order in which the buildings had been completed. In the next decade, an extension of Building 31 would become the permanent home of the Antarctic meteorite collection.

McKay became a junior member of a small, feisty offshoot of the tribe of people who made their living interpreting the language of rock. He and others in this select band had seen a rare opportunity in lunar geology, even though the field barely existed at the time. (It was being invented almost single-handedly by a contemporary of McKay's, the irrepressible geologist Eugene Shoemaker, of the U.S. Geological Survey, and a few others.) A number of scientific superstars, some of them Nobel material, felt the lure. The promised harvest of lunar samples looked to this group like the scientific equivalent of an oil gusher. It would provide a bonanza of revelations about the history of Earth's corner of the cosmos, and how we humans came to be here. And it was the chance of a lifetime to cherry-pick big-league, career-making discoveries.

McKay's first assignment as one of the new kids was dumbfoundingly exotic. He was to help teach astronauts (amazing!) how to collect rocks on the moon (more amazing!). He was one of two scientists assigned specifically to train the *Apollo 11* crew—Armstrong, Aldrin, and Mike Collins, who would remain alone in lunar orbit while the other two landed on the surface.

The job had many attractions, not least the aura of glamour and social appeal that attended anyone who worked with the astronauts. McKay was hardly immune from that. A young woman who knew McKay off and on in those years thought that his hobnobbing with the astronauts had given him a new social confidence. At Rice, he had dated nurses and townies, but not the women at the top of the campus social pecking order. Now he found that he had a certain cachet. He was a professional with prospects. He kept the requisite little black book and dated several women concurrently.

At the same time, McKay was learning to navigate the jungle of rivalries that made up the fledgling space bureaucracy. From a young age, he'd disliked confrontation and conflict. But now he had landed himself in something of a vipers' nest.

First, there were the fault lines *inside* the ranks of the scientists. Civil servant geologists working for the upstart NASA, for example, found themselves in abrasive competition with geologists from the venerable U.S. Geological Survey, which had been founded in 1879 to coordinate scientific exploration of the American West. Early in the Apollo days, the USGS folks thought they had a deal that would give them control over the returning moon samples. In their view, NASA had reneged, and they resented it. Each side had its share of aggressive people who liked to run things. Though the most serious infighting occurred before his arrival, McKay felt the tension, mostly when his team got into squabbles with the USGS over who should be teaching the geology courses. Personally, McKay got along with Shoemaker and other USGS people. He did not intend to let the turf wars divert him.

Then there was another divide—between the scientists, on one side, and the engineers, who managed agency programs and built and flew things, on the other. In the roaring, young 1960s culture of the "right stuff" that McKay had entered, his crowd was often regarded as tech support—necessary nuisances, or worse. Noting that Building 31, the science building, was the only one on-site with black columns along its facade, McKay and his coworkers used to joke that this must designate them as the black sheep of the family.

At the birth of Apollo, it was far from inevitable that any astronaut would be doing a lick of geological research on the moon, and as late as 1968, NASA managers recommended deleting most of the geology from the first landing mission.

President Kennedy had proposed the Apollo missions not in the name of science but out of an urgent desire to win a symbolic victory over the Communist enemy. The congressional mandate for the National Aeronautics and Space Administration (NASA) was to innovate, to lead the world in space spectaculars, to demonstrate the superiority of America and its capitalist, democratic system (with a lowercase *d*). The space agency was above all an engineering organization built on the bedrock of politics, with a pork barrel perpetually rolling through.

But once it dawned on them that the moon landings might actually happen, the champions of science—led by the feisty, crusading USGS geologist Gene Shoemaker—started dogging NASA and Congress for a role. They argued that if the country was going to fire a barrage of tax dollars toward the moon, those few guys who got to go might as well do something useful out there. In 1964, NASA grudgingly agreed to plan a package of lunar experiments and even to hire some actual *scientists* to fly in space (as long as they could also fly jets).

Still, the space agency retained its own special approach even when it came to scientific research. In order to force the changes required by the Apollo buildup, NASA administrator James Webb saw to it that NASA "purposefully spread the wealth," as one historian noted, "and even pioneered the noncompetitive contract in order to save time and foster specific skills" throughout industry. The agency pointedly "refused to go along with the old concept that scientific merit was the only determinant of who got a grant," sometimes favoring second-rate universities whether or not their proposals were the most deserving. "Thus, NASA pioneered reverse discrimination in order to foster expertise in more regions (and please more congressmen)."

This strain inherent in NASA's culture—a certain disregard for the usual scientific standards combined with its deference to practical politics—would generate waves of conflict and credibility problems for the agency long after its Cold War underpinnings had faded, and even at times when the scientific work in question was worthy. Decades later, David McKay would be right in the middle of an eruption of hostilities over this genetic instinct within the agency.

Some of the most flamboyant cultural clashes around McKay involved the Apollo astronauts. From their vantage point on the hazardous front lines, the space fliers viewed the rearguard scientists, medical doctors, and their ilk in an even more negative light than did the general run of engineers. Many in the astronaut camp saw the scientific people as serious distractions from the risky business at hand: "Larry Lightbulbs" and "Mad Professors," purveying "ding-a-ling stuff"—dangerous stuff, if your purpose was to conduct safe, successful operations in space and beat the enemy.

As they taught their rock lessons, McKay and the other geologists were forbidden to test the astronauts, or rate them, or do anything to trigger competition in the already hotly competitive ranks. The astronauts' energy, alertness, and reputed supercompetence might have misled some geologists into assuming the spacemen would be quite willing and able to tackle what amounted to a college-level course for geology majors. But many astronauts considered the nerdy geology lectures big yawners. Many thought the planned fifty-eight hours of lecture time was way too much. Could someone please explain why moon explorers had to learn the formula for *turquoise*? And was it really reasonable to teach them how to use an aneroid barometer? It depended on *air pressure*!

But there were exceptions. One of the astronauts who impressed McKay and others by taking an active interest in rock school was the diligent Armstrong.

And the broader tensions would ease as time went on. One reason was the arrival of a peacemaker in the person of Harrison "Jack" Schmitt, one of the few scientists selected for astronaut training during Apollo. Armed with a Ph.D. in geology from Harvard, he also had a combination of personality and political acumen that enabled him to smooth tensions on both sides. He helped scale back the irritations inflicted by fellow scientists while encouraging fellow astronauts to be more appreciative of geology's rewards.

In McKay's view, in any case, the astronauts did most of their learning not in the classroom but on field trips. That was his favorite thing about the job. The "field" was where the rocks were. Rocks and soil meant data—a record preserved in minerals. These trips brought him a kind of happiness that went beyond getting away from the urban hum and beyond standing in some majestic slice of nature aware of your own breathing, feeling how the air moved, noticing how the clouds looked, hearing what the birds sounded like in that moment. You sensed the infinite history beneath this place and every place, the past that stretched deep into the well of time and was just waiting to be deciphered.

On these trips, McKay, a handful of other geologists, and a contingent of up to a couple dozen astronauts would go off to some exotic outback to paw over the indigenous geology. They traveled to Iceland, Hawaii, Alaska's Valley of Ten Thousand Smokes, and a variety of spots in the continental United States.

In the field, neophytes tended to pick out the most unusual, the prettiest, or the most striking specimens. McKay would explain to the attentive Armstrong and other "students" that what the geologists wanted was the most *common* type, something that was characteristic of the neighborhood. Averageness was the important thing. And yet, if a Martian had landed on Earth looking for representative samples of the dominant species, by that same standard this group of *Homo sapiens* would have had to be rejected. The space center had assembled one of the most atypical collections of specimens on the planet, engaged in the most esoteric pursuits imaginable: rocketeers, moon mapmakers, plotters of interplanetary trajectories, space colony habitat designers, developers of dehydrated ice cream and other delectables for dining in weightlessness, the coolest of death-defying pilots, and this motley clutch of rock detectives.

McKay was admitted to Mission Control for this first moon landing courtesy of the space center's public affairs chief, who had met McKay on a field trip, gotten to know him, and asked for him personally, so the selection had not gone through the usual chain of command. The idea, apparently, was to make sure that, among all the technicians, flight controllers, capcoms (capsule communicators), management types, and pilots watching their screens in the terraced amphitheater and in the adjoining glassed-in VIP boxes, there was at least one person on hand who could explain to the press and the public any comments the astronauts might make about what they encountered on the lunar surface.

In his pockets, McKay carried old coins given to him by each of his grandfathers. One was a silver dollar from 1803, the other was a counterfeit half-dollar. He wanted some kind of bridge to his family's past from this historic event.

To a degree that is almost incomprehensible today, much about the moon was a mystery to everybody at that time, including the "experts." People had speculated that the moon's surface might be like any of the following: cotton candy, honeycombs, Cracker Jacks, toothpicks, or tiddlywinks. The favored analogy was "fairy castles," a loose reference to the sand structures found in home aquariums.

Despite lunar studies conducted by robotic craft, lingering uncertainties raised concerns about the Apollo missions' safety. For example, was the moon powder so insubstantial that the lander would sink in and disappear? At least one prominent scientist argued the possibility.

The "unknown" that caused the most fear was the threat of deadly moon microbes. Might the astronauts pick up bugs lurking in the lunar soil and bring them home as a plague? Alien life-forms, along with nuclear mutants, had been populating science-fiction stories for some time. (The latest blockbuster on the theme, *The Andromeda Strain*, had just been published in May 1969.) And there was little hard information to indicate that such threats were not equally abundant in reality, *out there*. But most geologists were convinced that nothing could live on the moon. There was no air, probably no water (except possibly frozen in spots perpetually shaded from the sun), and there was constant bombardment by radiation and space debris. As one had observed, if you wanted to design a sterilizing machine, you could hardly do better than the surface of the moon.

Still, there was enough concern among citizens and governments around the world that NASA officials felt they had to take precautions against the remotest possibility that a moon germ could make it to Earth. Both the astronauts and the boxes of rocks and dirt they hauled back from the moon would be quarantined in the new Lunar Receiving Laboratory for about a three-week period beginning when they splashed down in the ocean.

During his two-hour-and-thirteen-minute sojourn out on the lunar landscape, Armstrong performed like an A student. He astounded McKay and other geologists back home with his cool and competent descriptions of what he saw, starting right after he took that first "small step" into the lunar dust: "Yes, the surface is fine and powdery. I can kick it up loosely with my toe. It does adhere in fine layers like powdered charcoal to the sole and sides of my boots. I only go in a small fraction of an inch, maybe an eighth of an inch, but I can see the footprints of my boots and the treads in the fine, sandy particles." He reported that some of the hard rock samples appeared to be (yes, he actually said it) phenocrysts. For the anxious geologists watching back on Earth, it was almost orgasmic.

By the time Armstrong and Aldrin climbed back into their spacecraft, they were covered with the sootlike moon dust. One respected scientist had theorized that the stuff would explode in a fiery conflagration when oxygen hit it. When the spacewalkers took off their helmets, they got their first whiff. They discovered that the moon, the lyrical silvery muse of poetry and legend, had the acrid odor of gunpowder or wet ash.

During the first post-moonwalk press conference, carried live, nationwide, on TV late that night, the geology questions were directed to a slightly nervous but excited David McKay, seated with several others onstage in the space center's auditorium. CBS anchorman Walter Cronkite, an unabashed fan of the space program, asked McKay (by remote) to explain an astronaut's comment that some of the lunar soil they had retrieved behaved as if it were "wet." Why would the soil stick to the tools like that if it was not wet? McKay responded carefully that the soil was probably very fine grained and might have had electrostatic properties that made it clumpy, so that it *seemed* damp but was actually dry.

The ever-cautious McKay based his responses on the only information he had to go on—the pictures from the moon that he had seen that day, plus the astronauts' descriptions. He would be relieved later when his speculation was confirmed. It was the first time he had been on TV as a scientist, and he enjoyed getting calls the next day from friends and relatives who'd seen his appearance, played over and over. Even though he was a decidedly junior member of the team, he felt confident that he could handle the geology questions as well as anybody. The cameras made him uncomfortable, though not paralyzingly so. This would still be true when he made his *second* appearance on a global stage—almost a quarter century later, because of work he would do a short walk northeast of his ringside seat in Mission Control, inside a building and a culture born of Apollo.

Less than a week after the first moonwalk, the first handpicked pieces of moon reached Earth. McKay watched on TV from Building 31 when five senior geologists in sterile white lab coats and caps, inside the new Lunar Receiving Laboratory nearby, inspected that first haul. The group had kept the rocks isolated inside a vacuum chamber. A technician, operating his own arms inside rubber arms attached to the chamber in order to prevent contamination, reached in and opened the first sealed aluminum rock box. It was a moment of monumental anticipation. The geologists were a pride of Galahads approaching the Grail.

Then they saw the contents. Apparently forgetting the audience watching on live TV, one of them—a Harvard researcher—blurted: "Holy shit! It looks like a bunch of burnt potatoes!"

Armstrong had decided that he and Aldrin would shovel as much lunar soil into the box as possible before they closed it. The result was a briefly frustrating cloak of blackish dirt over everything in there.

Two nights later, when the scientists got the first chunk of lunar sample cleaned off, they recognized it as basalt, a volcanic rock. That single moment proved the value of the collection and effectively resolved a long-standing dispute about how the moon had formed. The evidence doomed the "cold mooners," led by a Nobel laureate, and handed victory to the "hot mooners" who insisted the moon had been geologically "alive."

The art of handling the first alien samples involved what NASA calls a steep learning curve. For one thing, it soon became obvious that keeping the specimens in a vacuum had been a terrible idea. Somebody was always puncturing a glove and breaking the barrier between the vacuum and the outside world. Alarms would go off. Dirt would pour into the box. And, for a time, the lab staff took to diving under tables and hiding so they would not be found and themselves sent into quarantine. The lab soon switched to storing samples in cabinets filled with nitrogen, a much more manageable medium. This system would later be adopted for Antarctic meteorite samples as well.

The rock custodians also learned early on that they had problems with potential contaminants built into the sample enclosures or mistakenly stored there. A young geochemist named Everett Gibson, who arrived at the space center on July 24, the day the *Apollo 11* crew splashed down safely in the Pacific, was among those enlisted to help clean things up. Gibson would later become McKay's partner in a public fight that would change people's understanding of contamination in rock.

After three landing missions, the civilized world was persuaded that moon stuff was free of alien killer microbes and other hazards. Humankind was not at risk from the rocks, although the "purity" of the rocks was in some jeopardy from humankind.

McKay was in the first group of investigators selected to analyze the coveted samples. Right after the first landing, he extricated himself from the astronaut-training business. For McKay, those duties had been a departure—albeit necessary and entertaining—from his true purpose. All along, he had been working in his spare time to set up a lab and conduct his own research. By the time the first lunar samples headed back to Earth, he had managed to acquire microscopes and other equipment and essentially duplicated the lab he had worked in at Berkeley.

Because McKay knew that most of the others would be working on the *rocks*, he'd decided to focus on the finer material that made up the lunar regolith. He saw the promise of a pioneering field expedition. Moon dust was the last stage in the pulverizing of lunar rocks, a tossed salad of all kinds of moon stuff, remixed by impacts. The dust had memories he wanted to unlock, molecule by molecule, a record of events that would lead him far back into the history of the cosmos. If he followed it far enough, he might enjoy a thrill akin to Armstrong's, of being the first to plant his flag on a patch of frontier knowledge.

McKay teamed up with two coworkers to write the proposal. It was accepted, but he was never sure whether the decision had been made on the basis of the proposal's scientific merit or because the bosses wanted to mobilize the talents of the young newcomers. He always suspected that it was the latter. In any case, he knew he would have to compete by the same rules as everybody else from then on.

David McKay and the other rock detectives were in a race to publish the initial results from the historic first field investigations on the moon. This meant a mad scramble to do research, write, and revise papers. It meant working through holidays and vacations. It was all so momentous. Along with the chance of a career-making breakthrough came the danger of a world-class stumble. McKay, thirty-two years old when he'd watched Armstrong's moon landing, was at the beginning of his career, while the competition included well-established heavy hitters.

Among the young up-and-comers who passed through the space center during this period was J. William Schopf, who had completed his graduate work at Harvard. He had participated in studies that had opened the way to a long-hidden fossil record of Earth's early life-forms. Not long after the first moon landing, he joined five other scientists conducting the preliminary sorting and description of the *Apollo 11* and *Apollo 12* lunar samples. During that time, he might have first crossed paths with David McKay. While Schopf would focus his career on Earth, not space, his trajectory would collide with McKay's twenty-seven years later in a very public and dramatic way.

In his lab in Building 31, McKay and his crew settled down to the tedious business of analyzing moon soils at the level of individual grains. The work could be compared to that of a crime-scene investigator: in both cases, people analyzed tiny quantities of evidence in minute detail in order to figure out how each clue had come to be there, and in the end, they would pull out the threads of hidden stories.

McKay suspected that bombardment by microscopic meteorites whipping in at over five thousand miles per hour (eight thousand kilometers per hour) had changed the lunar surface in predictable ways. He focused on certain types of particles. To name them, the team appropriated the term *agglutinates* from studies of volcanic eruptions on Earth, an interest that would take McKay repeatedly to Japan.

McKay and his staff set up a system that seemed mind-bogglingly monotonous. They would plot the data and determine the typical sizes of dust grains, and how they varied, so that one type of soil could be compared precisely with another. McKay mounted thin shavings of grains, *individually*, under the microscope and identified each particle. He kept track of them, maybe five hundred particles at a time.

In another setting, such narrow-beam focus and repetition might well be viewed as some kind of weird compulsion. To someone like McKay, his approach was a rational, logical way to mount an assault on the elusive unknown.

Although scientists like McKay were often cast as clinical, hard-eyed realists, they could also be the ultimate romantics. They tended to operate, out of necessity, with a heightened awareness of the incredible insignificance of the human life span set against the vastness of time and space. They knew not only that they would die in a blink but that the sun would die and the planets would become cinders. The cosmos would thin out and grow cold and dark (or be consumed in a crush of thermonuclear fire). They appreciated the extent to which we humans are prisoners of our senses and our language—and how much of reality still lay beyond those limits. And yet, they believed that through an accumulation of tiny steps they might somehow reach even that reality. This (along with the demands of time-limited experiments) was what kept them constantly at their benches.

Little wonder "civilians" sometimes found them off-putting, with their earnest smarts, obscure terminology, and workaholism. Geeks.

Like Sisyphus, condemned to push a stone endlessly up a mountain, many researchers devoted their lives to the exertion, not knowing if they would reach the summit. Like Sisyphus, they found fulfillment in the struggle itself—but they also wanted to get someplace. It was the sort of fanciful, audacious optimism, almost like religious faith, that in a few short centuries had carried the human species to amazing knowledge far beyond human senses—knowledge that stretched from the workings of subatomic particles to the edges of the known universe. And McKay felt this ancient momentum.

As McKay and his coworkers piled up more and more of his corpuscular information, they assembled a systematic portrait of how meteorites smashed up the soils and how natural processes gardened them over the eons. Though others worked the same vein, McKay's model, published most extensively in 1972, became a kind of index for establishing broader insights into the workings of the moon. He had been in the right place at the right time, had been able to select the lunar samples he wanted, and had had good ideas that had panned out.

Inside his tribe, the slender geologist, as calm and deliberate as a stalagmite, came to be recognized as one of the leading specialists in the field, one of the elite company of rock detectives who had managed to decode the shining face of the poets' moon.

All through this period, McKay's love life flourished. He had kept in touch with Mary Fae Coulter. The two no longer lived in the same city, but McKay's astronaut-training expeditions gave him enough mobility to pass through San Francisco to see her a couple of times.

Coulter eventually took a teaching job in Kobe, Japan, and invited everybody she knew to come visit her. David McKay—who, after all, had his own moon-related interest in the violent eruptions of Japan's volcanoes—took her up on it. He stayed for two weeks. They saw more of each other in that exotic fortnight than in all the previous years of their acquaintance. She was impressed with how much they had in common: Both came from authoritarian Presbyterian families. Both were brainy and enjoyed the same quick, ironic humor. She felt that they both saw the stratigraphy of hidden meanings beneath the surface of a conversation.

They rode trains all over Japan and had a wonderful time. They'd be rumbling along, watching the scenery go by, and she'd say, "David, what if I showed you a rock blue and green with little gold specks, what would it be?" He would murmur sweet nothings, like: "Porphyritic pyroxene granodiorite." Or perhaps: "Diatomiferous kimberlitic lherzolite." She loved it.

On their Japanese train tours, McKay would get passersby to take photographs of the two of them at various shrines. But she noticed that, sometimes, he would arrange to get a shot of himself alone—for his other female friends back home, she assumed. She also noticed that he was buying gifts for about nine women in Houston. He had brought along his little address book, and she discovered that it contained the name of one of *her* friends, whom

McKay had met at a dinner party at *her* house in San Francisco years earlier. She couldn't believe it.

One day, McKay asked Coulter directly why she didn't just move back to Houston. She gave him an equally direct look and said, "I don't like standing in lines." He said, "Oh, but you'd be *ichiban,*" using the Japanese word for "number one." She said, "Well, if I'm *ichiban* out of nine or ten, I'd probably only get about two dates with you a week." He noted that they'd been together twenty-four hours a day during his two-week visit, which put him about a year's worth of dates ahead of the game. And so it went.

Then one day, before he left Japan, McKay asked Coulter to marry him. She hesitated. It wasn't just his wandering heart that stopped her. She had yet to convince herself that this Japanese idyll wasn't something akin to a shipboard romance—born more of the setting than of true, deep feeling. She put McKay off.

But she was pleased when his first letter arrived after his return to Houston and he hadn't changed his mind about her.

In 1971, when Coulter moved back to the United States, she and David McKay were married. After a honeymoon in Mexico, the couple settled down in a town house near the space center. Actually, they were a threesome. Living with the newlyweds was a cat he had been keeping for one of his other woman friends—a coal-black animal that McKay liked to call Snowflake.

Apollo was one of those rare thunderclaps of history. There was never any doubt about its importance, no question that it would be the stuff of textbooks and museums and time capsules. But many people were mistaken about its robustness. At the time of that first landing, it was easy to think that this was just the beginning. Groups of geologists busied themselves drawing up grand wish lists and debating which landing spots would be the most rewarding for the next lunar digs. As they received the rain of lunar treasure, McKay and the others felt infused with a sense of wonder at the miracle times they lived in.

But the more politically alert among them knew that the brash American offensive in space was already running out of fuel. When Neil Armstrong set that first tractor-tread footprint into the lunar talc, he effectively stamped out Apollo's reason for being. The space race was won. Mission accomplished. Americans, beset with body counts in Vietnam, assassinations in high places, a decaying civil rights movement, riots and demonstrations in the streets, a general disillusionment with government, politics, and the "establishment," and assorted other social upheavals, possessed quite sensible and practical reasons for shifting focus.

There would be five more moon landings after the first one, sustained by the dwindling funding in the Apollo pipeline. In a late victory for the geologists, the last three Apollo missions would be devoted mainly to field investigations, each more complex than the one preceding it. On the last three, the astronauts would get equipment upgrades that included a battery-powered moon cart that resembled a dune buggy. Riding in this contraption, they could cover much more ground. They traveled hours from base camp, venturing into the moon's central highlands. But it was not until *Apollo 17* that the moon would see its first (and, so far, only) professional scientist. Geologist Harrison Schmitt won his flight papers just in time.

On Tuesday, December 12, 1972, Schmitt found himself and crewmate Gene Cernan afoot in the Valley of Taurus-Littrow, on the southeastern shore of the Sea of Serenity, which forms the left eye of the "man in the moon." It was, as anticipated, a geologist's dream. Their haul would include one of the oldest fragments found during Apollo—possibly as much as 4.5 billion years old.

Toward the end of an exhausting day of digging and collecting, the pair discovered a bright orange, red, and yellow substance. It would turn out to be made up of tiny beads of glass—evidence of a fiery volcanic fountain that had jetted out of the young moon into the lunar sky some 3.5 billion years earlier. For the poetically minded, such a flare might be taken as a burst of celestial jubilation heralding a notable development over on the cooling planet that waxed and waned in the moon's sky: life had sprung up there. It was primitive life, to be sure, but the cosmos, perhaps for the first time, had taken a step toward becoming conscious of itself.

The last Apollo astronauts would come home on December 19, 1972, just three and a half years after David McKay watched Armstrong scoop up that first bag of moon dust. The six landing parties on the lunar surface would collectively haul back a total of 842 pounds of moon stuff. It would be enough to keep a few people, including McKay, gainfully employed for decades.

The formerly supercharged Houston space complex aged through the 1980s, developing the outward ambience of a quiet, rural college campus, where the outrageous swamp climate and the surreal routines of human space flight were gentled with a landscaping of duck ponds and shade trees.

Out along the approach road reclined the symbol of NASA's faded glory: a giant Saturn moon rocket dismantled into pieces for tourists to inspect, like lengths of fossilized bone from a mythic biotech dragon. Rimming the campus were strip malls, waterfront boating attractions, and tidy residential neighborhoods with in-ground swimming pools galore. A nearby McDonald's sported a supersized, fiberglass astronaut thirteen feet tall, whose outstretched left arm beckoned with an order of fries. A new kid-oriented, space-lite tourist attraction was taking shape just west of the government complex. Along NASA Road 1, which linked the space center to the interstate, the ratio of vacant lots and shuttered buildings to funky space-souvenir shops, light industry, motels, and restaurants rose and fell with the economic tides.

By 1993, David and Mary Fae McKay had raised three daughters, two of them already independent young adults. David McKay's dark mane had thinned and turned white, making his dark brown eyes appear more prominent behind his glasses. Approaching sixty, he was coping with a bad heart, a trait inherited from his father. Barely visible in his ear was a hearing aid.

McKay had now spent so much time at his scanning electron microscope that he was acutely allergic to the Polaroid coating used on the films. He'd gotten that way from using a squeegee countless times over the years to smooth out the photos (before digital technology came along and eliminated that step). If he so much as touched a photograph, blisters would pop up on his fingers.

He had become a master of the scanning electron microscope, having developed a deep understanding of how the software talked to the hardware and an almost uncanny feel for the physical nature of electron bombardment. This understanding was valuable if one was to extract maximum information from the instrument. Some who worked with McKay thought there must be few, if any, better at the technique—or, rather, the art—and certainly no one better in NASA. Once, in the early 1990s, McKay and a coworker were frustrated with the images, not able to get the quality they wanted. The colleague, much impressed, would remember years later how McKay decided to fetch the engineering drawings that had come with the manuals, traced the problem back to its cause, and tweaked the microscope accordingly.

McKay's natural bent was to go beyond the measurement and analysis of his microscopic samples and try to reproduce those results by conducting an experiment, if possible. He wanted to show how the items under the microscope had come to be made in nature. Among other things, this led him to design experiments that helped show how future explorers might use the moon's natural resources to manufacture oxygen on the lunar surface.

Inside Building 31, as the cracks in the linoleum spread and the direness of money struggles rose, McKay's little corner of the world took on the musty ambience of a backwater. But within this domain, he had risen to senior status and accumulated professional honors. His office wall was covered with award plaques and certificates. He had accepted the vicissitudes of office politics good-naturedly, always the good soldier, passionate about his work but unobtrusive.

By now, McKay had written a couple of hundred papers on lunar topsoil alone, including the major breakthroughs published in the early 1970s. In a sense, you could say that McKay had spent most of his adulthood on the lifeless surface of the moon.

His reputation for taking excruciating care in his work was as persistent as bedrock—but maddening, at times, to friends. Wendell Mendell, a space physicist and lunar specialist down the hall, occasionally found himself gritting his teeth at McKay's infernal caution. But he also admired McKay's focused and principled approach to his work. One day in the late 1980s, the two of them were walking to lunch while discussing some big policy controversy the physicist had gotten himself involved in. McKay rebuked him.

"You've got to stop this. You've got to get back to writing proposals and doing research or else you'll never be competitive. You'll never get back into the competition against other scientists because you'll lose your edge, you'll lose your expertise, you'll lose your knowledge." McKay's overriding message was "Stop diddling with all this hand-wringing stuff and get back to good, hard, honest work doing science."

Mary Fae, with a Ph.D. in English, worked as a technical writer and editor at the space center, and was known to be an aggressive (some said "difficult") advocate of her point of view. Since 1980, McKay's brother Gordon, also blessed with the geology gene, had worked in the same building. Gordon was nine years younger than David, and some of their coworkers perceived at least the ordinary level of sibling rivalry there.

David and Mary Fae had moved the family and their cats into a three-story modern glass-and-cedar sanctuary hidden away in a cul-de-sac. They'd bought three acres of woodland, dense with oak and southern pine trees, which backed up to Clear Creek, a tributary of Galveston Bay. They deliberately left the lot wild. It had been affordable on their government salaries because it was part of a hundred-year floodplain. Their architect, a friend, told them it was the most modest home he had ever worked on. He designed it to withstand the one flood they thought they might get during an anticipated thirty-year tenure in the house. They relegated the ground floor to serve as a kind of basement, with a shop and laundry room.

In July 1979, two months after they moved in, they played host to a five-hundred-year storm. A record rain sent eight feet of water surging through the property. That same summer, they had a second flood that crested at four feet. Having learned their lesson, they removed the carpeting from the wood stairs and converted them to tile-covered steel. They made a few other adjustments and settled in, with the expectation that every so often tons of water would course through their quotidian underpinnings.

Once, they found a family of armadillos living under the front porch. They caught glimpses of opossums and other creatures in their woods. And one time, a large pileated woodpecker attacked the cedar house, pecking through a wall and wreaking considerable damage.

Their second floor, at a safe altitude, held the living and dining rooms, with a soaring cathedral ceiling over a low fireplace, and a gourmet kitchen where they liked to entertain. They covered the floors with muted Oriental rugs. Walls of glass on two levels made the thickets of trees part of the decor. A corner of one other house, barely visible at some distance through the foliage, was the only sign of neighbors. A Houston magazine did an article on the place.

The house abounded with evidence of the McKays' ongoing romance with Japan. On the walls of the staircases hung brilliant Japanese wedding kimonos; a shelf next to the kitchen held trompe l'oeil plates of pasta and other artificial foods manufactured in Japan. On the flood-prone ground level, they had installed a Japanese-style hot tub. And central to the main level was a tatami room (named after the straw matting that covered the floor), a clever conception the McKays had learned to appreciate. On a raised platform that could be closed off by sliding wooden doors (*fusuma*), the room held several low tables and stacks of floor cushions and backrests. According to the need, it could be a family TV room where the kids could sprawl or it could provide relaxed dining for six to eight guests. Many an evening, the house in the woods bustled with the McKays' guests, chatting and sipping margaritas or beer.

Johnson Space Center, like any closed society, had its share of jungle rivalries and cliques. Civil servants, for instance, were in a loftier caste than contract employees, and the former did not invite the latter to their Christmas parties, and vice versa. Newcomers considered the Apollo-era crowd quite the old-boy clique.

At the same time, the atmosphere was one of intellectual curiosity and collegiality. On Fridays, a group of planetary scientists from Building 31 would meet at a small Vietnamese restaurant up the road from the space center. As they shared entrées, they would also trade far-out ideas and indulge in freewheeling "what-iffing," about scientific possibilities, the potential for surprises in nature, and the like. In this setting, they could let their hair down. As one occasional participant put it, "they would pluck ideas up like Clean Wipes"—tissues widely used in laboratories—"and throw them away."

The McKays became good friends with a sunny young woman named Robbie Score—not only through their work in Building 31 but also on the party circuit. By the early 1990s, Score had gained considerable influence in the archives up on the second floor, where the growing Antarctic meteorite collection—the bounty of the "poor man's space program"—was housed. Well versed in the rocks, she was also famously generous in assisting those who wanted to study them—her "dudes and dudettes," as she sometimes called them.

The desperate dyspepsia in the space program had grown so deep that NASA had almost nowhere to go but up. In 1986, the *Challenger* tragedy, besides killing seven astronauts, exposed flaws and fissures that branched throughout the agency. The staggering fallout of that event—emotional, political, technical, and otherwise—finally obliterated the flickering halo of infallibility that had lingered from the Apollo triumphs. The thrust into space had long since metamorphosed from a national security imperative to a discretionary budget item, competing for money head-to-head with war veterans and cancer research. And NASA was racking up a record of failed projects, delays, and enormous cost overruns.

The space shuttle program was plagued with technical glitches and delays. Congress seemed on course to cancel the presumed centerpiece of the agency's future—the costly project to build a space station to serve as a laboratory complex in low Earth orbit. NASA engineers had launched the long-awaited, "revolutionary" Hubble space telescope into orbit with a devastating flaw in its lens, making the project an object of derision. A megamission arriving at Mars was lost to another human error. And in 1989, when the first President Bush proposed that the United States revive the exploration of the moon and Mars, NASA and its private contractors responded with a plan so expensive, self-interested, and unimaginative that the initiative sank like a rock.

A fed-up White House had recently fired the NASA chief and brought in a new one, who vowed to reform the agency, shake it out of its defensive crouch, and inject it with new energy—no matter how many enemies he might earn.

There had been times, especially during the 1980s, when McKay and his coworkers wondered whether the whole agency would be shut down. He kept up his contacts with universities, on the theory that he might have to go job hunting.

But McKay preferred working for the government. The pay was reasonable. The benefits were good. And his senior status gave him job security, as long as the agency itself survived. He had money for the research he enjoyed, and he didn't particularly want to teach. Besides, he had friends at universities who'd convinced him that the politics of academe were even more horrific than those in the NASA sandboxes. McKay once remarked, "I always go along to get along, then I do what I want to do."

By the early 1990s, McKay had concluded that, in more than one sense, the moon was deader than Elvis. McKay's work had earned him a reputation as one of the world's masters on his chosen subject, and he had published some two hundred papers, but he had become increasingly aware that his was a narrow audience—maybe a couple of dozen people who read his work carefully.

It was important to him that his achievements be known and respected within the tribe. He was doing the "good, hard, honest work" of science. But he was getting tired of doing research so far out of mainstream geology. In the perpetual battles for funding, more alluring celestial bodies and rival NASA groups had wrested away much of the money and attention.

In short, McKay felt he had been on the same dusty path too long. As he would remark one day, "I've studied agglutinates until I'm sick of them."

McKay was expanding into new territory: He had taken up meteorites and, well, yes, more dust. But it was a different kind of dust—*cosmic* dust, the dust that rains through space. He had been casting about for something unprecedented, something fresh to restoke his cooling fires.

Very soon, a couple of the guys down the hall would come to him with just the thing.

CHAPTER THREE

ODD DUCKS

"SHERGOTTY, NAKHLA, AND Chassigny. Shergotty, Nakhla, and Chassigny." The names rang in the brain like some incantation from the netherworld. For David "Duck" Mittlefehldt, the words provided the essential key that would loft Robbie Score's rock out of obscurity and expose its special gifts.

The year was 1993, and another muggy Houston summer was yielding to fall. Building 31 was so quiet you could almost hear the roaches ticking along behind the walls. Just about everybody wore crepe-soled shoes or sneakers, so there was no echo of footfall; only the white noise of humming machinery. People here often worked behind doors with combination locks and multiple warning signs: DANGER: HIGH VOLTAGE and DANGER: POISON GAS. Exposed wire bundles were routed high around the walls, and in some places dust bunnies collected in the corners behind space-age machinery on well-worn linoleum floors.

In one of these quiet labs, Mittlefehldt, a wiry guy with a beard, a receding hairline, and a resemblance to the actor Michael Keaton, sat and stared at his problem. Mittlefehldt was in the middle of one of those pivotal moments when a human mind, revved on a combination of its own accumulated knowledge, frustration, learned skills, instinct, competitive drive, anxiety, and imagination, finally uncoiled for a leap of insight that would send ripples through the world.

Or maybe, as he would say with a shrug, if he hadn't thought of it, somebody else would have, and besides, it had taken him long enough.

Either way, Mittlefehldt's train of logic would lead to the unmasking of the mother of all planetary rocks. The leap would change the lives of David McKay and several other people working, unawares, in this building and in other places thousands of miles away. It would trigger no end of public fuss. And, as so often happens in these matters, this wasn't at all what Mittlefehldt had set out to do—which could help explain why it had taken him so long.

He had acquired the nickname Duck in college, and it had stuck. Scattered around his small office in Building 31, Mittlefehldt kept a collection of duck magnets, plastic ducks, duck pictures, and duck postcards, including a depiction of fat, feathery duck bottoms sticking out of water. There was a stained-glass wall hanging with a couple of ducks taking wing. Friends noted with amusement that he would refer in a published geology paper to "duck-shaped" features.

Flanked by his feathered support group, Mittlefehldt pondered the infernal microscope images of tiny grains from a meteorite. They had vexed him off and on for some years now, and here he was again. They were odd; they didn't fit in. They had threatened to mess up his major project. He was a stuck duck.

A geochemist, Mittlefehldt had gravitated to meteorite studies on a whim back in 1973, as a graduate student at UCLA. It was at about that time that he became friends with a younger geology student named Robbie Score. Now both of them were in Houston and she was on the staff of the Antarctic meteorite lab that supplied his rock samples. They both worked here in the cloistered preserves of Building 31, staring at minuscule, pedestrian crumbs of what seemed the humblest matter. But their shared preoccupation was with vast sweeps of time and space.

Mittlefehldt believed firmly that humans should explore space, not only to enrich human understanding of nature but as a matter of sheer survival, in case the species should need a second world to live on. (Remember the dinosaurs!) He had been at work for several years on what he hoped would be a valuable contribution to the emerging picture of what was "out there" around Earth, and how it had gotten there. It was a major project to decipher the hidden record carried inside certain asteroids, odd-shaped, beat-up rocks that swarmed along an orbital track concentrated between Mars and Jupiter.

In 1988, in pursuit of that goal, Mittlefehldt had applied to the meteorite curators down the hall for a sample of a resoundingly ordinary family of meteorites that had most likely been knocked off the asteroid 4 Vesta in a series of collisions. Vesta's diameter of more than three hundred miles (almost five hundred kilometers) made it one of the largest known asteroids and a rich source of the space debris that rained steadily down on Earth.

The rocks that were widely presumed to be spawn of Vesta had been subjected to such intense heat early in their existence that they had started to melt. Mittlefehldt hoped to learn what had produced this prodigious heat in the embryonic solar system. Happily, the Antarctic had coughed up a number of new specimens in this family, and Mittlefehldt intended to study as many as he could get his hands on. His goal was to produce the first systematic study of its kind.

One of the supposedly plain-vanilla meteorite samples that Robbie Score's lab sent Mittlefehldt was actually from the odd rock that Score had picked up in 1984. The world knows now that it did *not* come from Vesta. But to find that out, Mittlefehldt was building on a whole history of little advances by other people who were intensely fascinated by rocks.

On Earth, rock is the cool skin that insulates life on the surface from the inferno deep inside the planet; it is the foundation upon which civilizations were built, the universal benchmark of stability. Rock distinguishes the inner planets. Except for the little oddball Pluto, the outer planets are big gas balls with no defined surface.

In the minerals that make up rock—natural crystalline assemblies of chemical elements—geologists can read the chronicle of Earth's history. And in the years leading up to Mittlefehldt's epiphany that summer day, scientists had begun to see in the rocks that fell from space, as one put it, useful "keys that can unlock the vaults of cosmic memory."

If Mittlefehldt's timing had been a little off, his efforts might not have triggered such a remarkable train of events. But his unmasking of the rock happened to fit nicely with changes in the portrait of nature that had only recently begun to surface. So, thanks to his aggravated—but unhurried—curiosity, the rock's days of obscurity were numbered.

Once launched, however, the investigation into the rock's hidden past would be as unguided as the rock's own journey had been. The quest would turn into a kind of sporadic human relay race with no starting gun, no master, no one who knew where the finish line was or what it would look like. You might say it took on a life of its own.

But first, Mittlefehldt had to figure out the salient fact: somebody had misidentified this sample.

·

After Robbie Score's hunting team shipped the rock home from Antarctica, it arrived in Houston (along with the other frozen samples in the 1984–85 season's haul) in a big shipping container. Each specimen was sealed in its own sanitized bag and wound like a mummy with tape. The specimen then joined the growing ranks of space rocks on the second floor of the Building 31 complex.

The meteorite suite featured a Class 10,000 clean room designed to minimize contamination. It had a special air-filtration system, and the rules required that anyone entering take off all metal jewelry, don a surgical-type cap, gown, and shoe covers, and pass through an air shower.

The complex had a culinary air about it, with its bright lights, pristine tables, cabinets like industrial refrigerators, and ovens. Some of the samples in their plastic wrap rested in metal containers that resembled lasagna pans. But then there were the arms. Disembodied black-rubber arms, with hands, waved and swayed organically in the moving air currents as if beckoning—or cautioning. The lab technicians would put their own arms inside, reversing the gloves to the inside of the chambers, in order to work on the rocks without breaching the sterile environment.

Here, the curators weighed the rock from Allan Hills and recorded a deliberately superficial description. A technician equipped with a small silver hammer and a rock saw, operating inside the rubber arms, attacked the rock from the angular, "hackly" end (as opposed to the smooth, blocky end) and split off one-half gram for dispatch to the Smithsonian Institution in Washington, where it was to be analyzed and classified in more detail. The mother rock stayed in Houston, in the archives, usually put away in a nitrogen cabinet and sealed in a nylon bag.

In late spring of 1985, the chip from the Allan Hills rock—resting in a tiny box like an engagement ring—arrived at the Smithsonian's National Museum of Natural History. Upstairs from the exhibitions and down some long pastel-drab corridors that resembled those of Building 31 in Houston was an office and laboratory complex off-limits to tourists.

There, the task of classifying the Allan Hills rock fell to a young meteorite curator named Glenn MacPherson, a relative newcomer to the job whose boss was away on a sabbatical. With Robbie Score's rock, which was not designated as a high priority, he did the same thing he would do with some six hundred such samples flowing through his lab that season. In due course, a techician sliced, ground, and polished the sample into sections so thin they were transparent (thinner than a piece of paper). These sections could be used over and over by different researchers, like a library book.

After the usual prep work, MacPherson examined the shard through an electron microprobe that showed its chemical composition. After about half an hour, he decided the rock was a piece of an ordinary asteroid. His finding appeared in the August 1985 *Antarctic Meteorite Newsletter*, a description of the season's haul circulated to those who might be interested in requesting samples for research.

The newsletter first presented Roberta Score's eyeball assessment: "Eighty percent of this rectangular-shaped [meteorite] is covered with dull black fusion crust. . . . Areas not covered by fusion crust have a greenish-gray color and a blocky texture. Cleavage planes are obvious on some large crystal faces and the stone has a shocked appearance." She had noted small areas of oxidation and abundant small black grains scattered throughout, adding, "Small fractures are numerous." She had judged it to be an achondrite, a rare kind of stony meteorite.

Following that paragraph was MacPherson's microscopic analysis, classifying the rock as a "diogenite" (i.e., a common igneous rock from an asteroid, most likely 4 Vesta). He got the basics right: "The meteorite consists of orthopyroxene . . . that forms a polygonal-granular mosaic. . . . Veins of intensely granulated pyroxene cross cut the section. . . . Other phases include minor chromite and irregular patches of a featureless and isotropic maskelynite. The section . . . does contain patches of brown, Fe-rich [iron-rich] carbonate." He attributed this odd rusty character—odd for this type of meteorite—to weathering that had occurred on Earth.

The rock had fooled him, true. But this was in part because MacPherson felt honor-bound under the ground rules laid down by meteorite investigators *not* to learn too much about the meteorite. It was standard practice for the person classifying a new sample to do the minimum necessary. This prevented him or anyone else in the curator's lab from skimming off the cream of information on incoming specimens, thereby robbing all the hungry investigators out there of the prizes that had lured them to study such rocks in the first place—the unprecedented insights that led to published papers and enhanced reputations.

Years later, despite his conviction that he had done his job properly, MacPherson could not help feeling a touch of chagrin when this unremarked event, this routine encounter with one little wafer among the many, became a footnote in a very public and ferocious feud.

• • •

The meteorite curators in Houston had based their selection of Duck Mittlefehldt's sample on this initial misidentification. At first, Mittlefehldt accepted the assessment written on the label. To be sure, he noted that the "asteroid" sample included some weird signs and portents, considering its humble parentage. (There were a lot of carbonates, for instance, which were previously unknown in this ordinary family of meteorite.) But Mittlefehldt, like the few others who had studied pieces of it at this point, at first assumed that these were the result of weathering during the rock's long sojourn in Antarctic ice. Otherwise, his initial bulk analysis of the chip's chemical composition showed no significant conflict with a relatively ordinary origin on Vesta.

Finding himself with traces of the Allan Hills rock left over after these studies, Mittlefehldt glued the remnant grains—only a few times larger than the period at the end of this sentence (about a square millimeter)—to a glass slide and polished them flat. In the spring of 1990, he put the grains under an electron microprobe, which fired a narrow stream of electrons at them. (Their atoms would give off X-rays with an energy signature unique to whatever element was in the target item, and with an intensity that indicated the amount of that element.)

"This can't be right," he thought. The results indicated properties and interactions that seemed impossible on the parent asteroid.

Growing up in Jamestown, New York, Mittlefehldt had discovered as early as third grade that he had a natural affinity for science, even though there was no familial goad in that direction. His mother worked for an insurance company, and his dad was employed by the local bank. But he somehow always knew more about scientific subjects—anything at all to do with science—than the other kids.

But now his confidence in his own instincts wavered. He was still assuming that the curator back in Washington had been correct, and therefore his analysis must be in error even though he had carried it out with his usual rigor.

Mittlefehldt kept an open mind. He knew this was often the way you learned new things: you focused in on these little oddities, things that didn't add up—the geological odd ducks—and you chased them down. They were usually trying to tell you something important. But it took a focused effort to climb out of the old sucking sump of conventional thinking.

Mittlefehldt dropped the matter and moved on to other things. That was the way he liked to work: keep several projects going at once, put the riddles aside, and let them simmer and churn in his subconscious for a while. They would get sorted out in their own good time.

In a way, that was how he had gotten his nickname—by putting something off. As an undergraduate at the State University of New York at Fredonia, he'd waited until the last possible moment to do a laboratory assignment due the next morning. A substitute professor had assigned a huge batch of rock samples he'd (wrongly) assumed the students could analyze easily in four hours. Mittlefehldt and several others started after supper that night. Four hours passed, and they were still there. Things started to get crazy toward the shank of the night, and Mittlefehldt admitted to his fellow sufferers that he was "cracking up." Someone turned that into "quacking up." And then they started calling him Duck. ("You really had to be there," he would say later, shaking his head.) But he finished the work and was one of the first to leave, at around three A.M. No problem. Lab class wasn't until eight A.M. He had not only completed the assignment, he had acquired a lifelong monicker.

It followed him to UCLA, where he earned his Ph.D. in geochemistry, to the University of Arizona, and through a sojourn in Israel's Negev, on the faculty of Ben-Gurion University. He was still Duck when he arrived in Houston, in September 1985.

In July 1993, he was forced to end his laissez-faire approach to the mystery of the rock's lineage because of what happened when he finally wrote his paper on the origins of the common meteorites, the paper he had been working on for five years. One of the anonymous experts assigned to evaluate Mittlefehldt's paper before publication, as part of the routine process, criticized the work, chastising Mittlefehldt for his handling, or nonhandling, of the questions about the weird signature in that particular sample.

Mittlefehldt realized he had to crack the case. He went back to the electron microprobe with his little glass-mounted bits and set the magnification as high as he could. He checked and rechecked, and rechecked again, all through that summer. He compared this sample with other meteorite samples he had analyzed on the same day, in the same way, as this rebellious one; the others all had the "correct" signature and this one (still) did not. He finally convinced himself that his analysis had *not* been in error.

The signature in the specimen was truly different. Now he had to confront the implications.

Mittlefehldt thought over the whole sequence of experiments—and tried to liberate his mind from those ditches of conventional thinking, to sort through his wider mental inventory.

About three years earlier, Mittlefehldt had read up on Martian meteorites just to make sure he was current. He knew NASA was gearing up to send a new generation of robotic geologists to Mars with the ultimate goal of bringing back pieces of Martian rock and soil, and he wanted to be in a position to do some of the analysis on the Mars samples. He had kept up with the subject. You never knew when something important, something you could use, was going to turn up.

Now it struck Mittlefehldt that there was one family of rocks where this frustrating, weird one would fit in nicely, geochemically speaking. The SNCs. Of course. SNCs (pronounced "snicks") were a tiny family of stones that shared characteristics with one another but contrasted dramatically with the dominant meteorite population. The SNCs were Martian.

The acronym SNC refers to the lyrical names of the sites where three of these stones had been collected over the span of almost a century: the French village of Chassigny (1815), where people heard loud sonic booms and saw a stone weighing about nine pounds (four kilograms) fall from the sky; the town of Shergotty, in India (1865), where people heard similar booms and saw an eleven-pound (five kilogram) object fall; and the village of El-Nakhla, in Egypt (1911), where witnesses reported the impacts of multiple fragments that fell in a shower from the explosion of a single meteorite higher up. People reported that one of the objects killed a dog. In that case, scientists recovered some forty stones with a collective weight of about twenty-two pounds (ten kilograms).

Shergotty, Nakhla, and Chassigny . . .

Scientists had not arrived easily at the recognition that the SNCs came from Mars. Many found it preposterous to imagine that a chunk of that planet might reach Earth's surface on its own—a gift of nature—in a relatively unaltered state. Surely any impact violent enough to knock a rock off a planet would alter the specimen beyond recognition, or vaporize it completely. In the early 1980s, the experts had to stop scoffing when geochemist Donald Bogard and his coworkers finally found convincing evidence to the contrary.

Bogard, working at the Houston complex, wondered why this group of meteorites stood out from the crowd. In a landmark study of one such specimen, he and a colleague heated black glass (formed inside the rock by a violent impact or shock) and analyzed the trapped gases that bubbled up. Clever detective work led them to discover that the chemical composition matched perfectly that of the Martian atmosphere—which was unlike that of any other known body. And how had Bogard known what the Martian atmosphere was like? The robotic U.S. Viking spacecraft that landed on Mars in 1976 had sent a direct analysis of it back to Earth.

Through "guilt by association," as one scientist put it, the other oddball meteorites with similar compositions to this one (and to Mars) were deemed to have made the trip from Mars as well.

This amazing pilgrimage of stones was part of a portrait then emerging of the solar system as a vast pinball gallery, a tarantella of spheres and rough chunks sweeping around the sun on random tracks that sometimes brought them to the same point in space and time. Humanity had finally begun to grasp the profound importance of cataclysmic impacts in shaping the history of the solar system, of Earth, and of all life, including human evolution. The killer rock that wiped out the dinosaurs—an extinction scenario that just recently had won scientific acceptance after years of controversy—was the most celebrated instance.

Human awareness of the glittering firmament as a threatening presence would take another great leap in 1994, with the unprecedented "live" spectacle of shattered comet fragments ripping into Jupiter's gaseous surface, kicking up towering plumes and leaving dark bruises in its swirling pastels. The event would be recorded by most of the world's telescopes, heralded by the media, and followed intensely on the emerging Internet. Afterward, visions of lethal celestial "incoming" would proliferate in books, movies, magazine articles, television documentaries, and the public imagination.

Fortunately for civilization, most of the cosmic rubble that peppered Earth was small. In this context, respectable scientists had long speculated about whether the planets and other cosmic bodies might have "swapped spit" and seeded one another with living organisms. Recent and ongoing spaceflight experiments indicated that at least some hardy microbes could survive a journey across the vacuum of space—if they were somehow shielded from deadly, DNA-wrecking radiation.

As Mittlefehldt fretted through the waning summer of 1993, of all the thousands of meteorites studied, only nine—the SNCs—had been identified as Martian. Sitting there in the hush of Building 31, he strongly suspected that he was looking at the tenth known piece of Mars.

Mittlefehldt was still wary. His insight resolved only part of the mystery. This rock was weird; its composition made it some kind of strange outlier—even among the SNCs. He wanted to do a bit more sorting out. Meanwhile, he worked on his main project—and a funny thing happened.

He had accumulated a good many samples of asteroid-spawned meteorites. As he studied them, he saw one with a familiar label. It was another presumed piece of old friend Vesta, picked up in Antarctica. To his astonishment, this sample showed some weird signatures, including one that matched up with what he saw in the Allan Hills rock. The readings now were also quite different from his readings the last time he had studied a sample with this same label.

Then he saw the whole picture. He was in the lab, looking at columns of atomic ratios, and seeing far too much sulfur for any run-of-the-mill meteorite. But the reading was typical . . . in *Martian* meteorites. (He checked the calibration of the microprobe, just to be sure the reading was no mistake. It was fine.) He backed up and took another look at the sample, imagining it as a whole individual, personality and all. Now he noticed that the texture was "wrong" for an ordinary Vesta-type meteorite. In fact, the texture was just like that of the meteorite from Allan Hills, which he had seen intact in pictures. It hit him: this *was* an Allan Hills chip, had to be. He had a *second* mislabeled sample.

That fluky reading was the clincher he had been looking for. "Everything clicked," he would say later. He all but screamed "Eureka!"

He knew he not only had a Martian meteorite but a special one at that. He would recall later with a grin, "It was the most satisfying experience of my life . . . knowing that other people had studied [the rock] and hadn't tumbled to it."

Mittlefehldt, like many others, had been boning up on Martian meteorites for practical reasons. After a hiatus that began in the mid-1970s, U.S. robotic missions to Mars were about to resume. Whatever else it might or might not have, Mars had funding.

The last rocky planet out from the sun before the asteroid belt and the succession of giant gas worlds, Mars was a miserable, dry, dust-blown, cold, and barren desert world. And yet, for some, it sang like the Sirens. It was the most Earth-like body known to exist anywhere. That plus its nearness made it seem most likely to end the cosmic aloneness of Earthlings. (The surface of Venus was hot enough to melt lead and shrouded in a poisonous atmosphere with one hundred times the atmospheric pressure on Earth; Mercury, closest to the sun, was as airless as Earth's moon, its surface cooked by solar radiation with ten times the intensity of that on the moon; at least one satellite of Jupiter might conceivably harbor life, but the Jovian system generated belts of powerful, deadly radiation; and the intriguing moons of Saturn were a daunting billion miles from Earth, with just 1 percent of the sunlight.)

Mars was our sister world, the familiar red beacon in the night sky. Although it appeared at best only one-hundredth the size of the full moon, it was the only planet whose actual surface could be seen with the naked human eye. (The surface of Venus was obscured by that atmospheric shroud; that of Mercury, by its proximity to the sun. The other planets were too distant to see unaided, and most had no defined surfaces.)

Mars and Earth lived in the Goldilocks zone, where the sun's heat was neither too strong nor too weak to sustain life (as we know it). The Martian day was only slightly longer than Earth's (although its year was almost twice as long). The Martian night could be worse than the Antarctic, at minus 125 degrees Fahrenheit.

The low temperatures and pressures on Mars meant that any liquid water reaching its surface would rapidly boil off, freeze, or poof into vapor. But the daytime sunlight—about half as strong as that on Earth—would sometimes heat rocks and soils above freezing. And the north and south polar frost caps, showing water ice exposed at the surface, might someday be converted to provide an abundance of the most basic of human needs. Scientists theorized that water must also be locked up in permafrost not far beneath the Martian surface, within reach of drills, and that the soil might contain water bound up in mineral grains.

Its similarities to Earth and its neighborly nearness made Mars the most rational next destination for human space explorers and, Duck Mittlefehldt thought, "one of the more plausible future habitats for humanity." Beyond its practical appeal, Mars was the object of a yearning—for the promise of kinship with other living beings in the vast and empty blackness, for a mirror to be held up against the infinite wilderness to show us who we are. Or are not. And what our future might hold.

The human romance with Mars had flickered and shifted over the centuries. Early telescopes showed linear tracings on the reddish surface that prompted speculation about artificial construction by unknown engineers; seasonal changes suggested the blossoming and retreat of vegetation. Even then, water was the overarching enigma, the thread wound throughout the Martian mystery. In the speculations of Percival Lowell, canals carried water from the poles to nourish an advanced civilization concentrated in the equatorial deserts, where the climate was no harsher than, say, that of "the South of England." Life-giving waters flowed through the Mars novels of Edgar Rice Burroughs, and H. G. Wells conjured a Martian race of "intellects vast and cool and unsympathetic." Poet of the possible Ray Bradbury, writing in the 1940s, described doomed yellow-eyed Martians keeping house beside an empty fossil sea on that "dead, dreaming world."

Duck Mittlefehldt was about to drop a catalyst into the drama of the search for life on Mars, which had been a driving force of the U.S. space program since almost the beginning. A landmark report by independent scientists in 1962 chose "the search for extraterrestrial life as the prime goal of space biology." Not since Darwin, and before him Copernicus, they wrote, has science had "the opportunity for so great an impact on man's understanding of man."

With the dawn of the space age had come the disillusionment of icy fact—the day in 1965 that Mariner 4 swept over a frosty slice of ancient cratered terrain to provide Earth's first fleeting close-up of its neighbor, a snapshot of a world bleak and dead. Mariner 9, the first Earth ship to slip into orbit around Mars, brightened the outlook a bit with evidence of a more interesting and changeable world. Water had once flowed there, the polar caps had ice in them, they expanded and retreated as the seasons changed, and there was water in clouds that drifted in the Martian skies.

Encouraged, scientists pressed more aggressively on the big question: Was there any hint of life? To the young scientist and soon-to-be-celebrity Carl Sagan, the Mars of today strongly evoked an Earth he had seen in Colorado and Arizona and Nevada. And he noted that, for much of history, "those regions of Earth not covered by water looked rather like Mars today—with an atmosphere rich in carbon dioxide, with ultraviolet light shining fiercely down on the surface through an atmosphere devoid of ozone." Though large plants and animals had come along in the last 10 percent of Earth's history, "for three billion years, there were microorganisms everywhere on Earth. To look for life on Mars, we must look for microbes."

With the two Viking missions of the 1970s came "a giant and abrupt escalation in Mars exploration." Among other things, the Viking landers conducted the first probes in search of organics—signs of possible biology—in the Martian surface soil. Almost everybody concluded that the answer, at least for that time and in those two spots three thousand miles (five thousand kilometers) apart, was no. They'd found no organic molecules and unexpectedly sterile soil. If Earth had a mountain three times as high as Mount Everest, its peak rising 100,000 feet into the stratosphere, exposed to ozone and ultraviolet radiation, conditions there would be Mars-like.

Still, the two craft had landed in flat desert terrain, where they were more likely to survive the touchdown but arguably less likely to find organic material. And the sensitivity of the instruments was far less than what could be applied in current state-of-the-art Earth labs.

The news was agonizing for some, in part because the evidence was so tantalizing. A few scientists would insist that Viking said something closer to *maybe*. Indisputably, the experience provided an early lesson in the difficulties of distinguishing signs of alien life-forms from unpredictable chemical reactions on an alien world.

Still, accumulating evidence indicated that, billions of years earlier, the young Mars had been even more Earth-like, with flowing waters and a warmer climate. In the 1990s, NASA would adopt a single organizing principle for its Mars explorations: follow the water. This would be the overarching theme of the struggle to understand the planet.

And now, as the United States readied a new wave of Mars-bound robots, an important new clue presented itself courtesy of Duck Mittlefehldt.

In October 1993, Mittlefehldt revealed the discovery of a new Martian meteorite.

The rock Robbie Score had plucked from the ice almost nine years earlier was hailed as extraordinary even in the exclusive coterie of SNCs. The news spread rapidly by word of mouth in Building 31. Score, working in the meteorite archives, was thrilled when a coworker mentioned it—but not all that surprised. She had always known that the rock was weird.

Down on the first floor, David McKay's ears perked up when he detected the hallway buzz about the new find. While still working on his lunar soils, he had also branched out into studies of meteorites and cosmic dust. In fact, he was for the first time getting money from a division of NASA devoted to studying the possibility of extraterrestrial life. McKay had won the grant for his proposal to analyze cosmic dust for carbon (a building block of life). He and a coworker, Kathie Thomas, came up with results that showed a much higher abundance of the stuff than anyone had expected.

McKay volunteered to join a little consortium Duck Mittlefehldt had formed to share his samples of the intriguing new Mars rock.

As more and more people got their hands on bits of that rock in the months that followed, they learned how special it really was.

This was the first rock ever studied that appeared to have formed *beneath* the Martian surface, probably on the floor of a magma chamber. Investigators therefore anticipated a rich vein of new information about the geological processes that had helped shape the red planet. And the rock's unusually high concentration of carbon compounds, possibly from molten subsurface volcanic flows, provided what Mittlefehldt considered "probably the first convincing case for [a tangible storehouse] of primordial carbon inside Mars." Not that the thought of Martian biology was on his mind at this point, but still—carbon was the congenial element that provided the essential framework for all known life. Requests for pieces of the rock poured in, dozens of scientists around the world began to scrutinize it, the march of revelation was under way, and the "batons" were flying like mad.

The next stunner was the rock's age.

Geologists in Germany used a tried-and-tested geological clock—the processes of radioactive decay—to show that the Martian material in question had crystallized 4.5 billion years ago.

The rock had hardened out of a molten volcanic flow as it cooled down on an infant Mars still forming in the wan light of the newborn sun. Almost as old as the solar system, the rock was more than three times the age of the next-oldest known Mars meteorite. And the oldest native Earth rocks ever examined dated back no more than 3.8 billion years.

The rock from Mars, in short, was the oldest known from any planet. It was rivaled in age only by the meteorites that came from asteroids. Like a message in a bottle, the pilgrim stone carried an unprecedented record of eons of Martian geological history.

Here was the Ur-rock of planetary memory.

Excited investigators soon showed that the rock had started its road trip some 16 million years earlier as an unguided ballistic missile—blasted off Mars by an incoming asteroid or comet. They did this by measuring the effects of the high-energy cosmic rays that had bombarded it as it traveled. They estimated how long the rock had been imprisoned in the Antarctic ice sheet (13,000 years) by measuring the radioactive decay of products of that long pounding in space.

Some researchers studied impact craters on Mars with the goal of figuring out where the rock had been situated at the time of the jolt that liberated it. The theory was that, given the rock's age, its home neighborhood was in the heavily cratered highlands of the southern hemisphere, the most ancient terrain on Mars. An impact scar in the Sinus Sabaeus region had the right age and characteristics.

The record that most fascinated Duck Mittlefehldt and others had to do with a major impact that had traumatized the rock, possibly within a billion years after it had first cooled and hardened. At that time, the rock had been severely shocked and had cracked, presumably when an asteroid or comet plowed into the surface nearby. The resulting fissures and crushed places allowed the ephemeral chemistries of the changing Martian environment to infiltrate and leave their signs.

The rock seemed to offer clues, in other words, from a time not much less than 4 billion years ago when the worst barrages of violent collisions that had wracked the early, rubble-filled solar system had abated and the climate on Mars was relatively warm, with liquid water flowing or standing. The conditions on Mars in that season might not have been so terribly different from those on the toddling Earth of the same period, where primitive life might already have formed.

Accordingly, in late 1993, those who liked to delve into the histories of worlds delighted that here, in this single lump, they had found a key that could help them decipher Mars's most deeply buried secrets. And in time it would become arguably the most intently studied of all known meteorites. What they didn't suspect was how much the rock would reveal about Earthlings. While they were looking into the rock, it would be looking back into them.

Mittlefehldt and others focused on the rock's unusual and puzzling abundance of carbon compounds, especially its carbonates. They made chemical maps showing that the compounds were quite complicated and varied in composition and, as far as Mittlefehldt could tell, distributed all through the rock. But many looked distorted, ruined. The ones with a certain rounded shape, some big enough to be visible to the naked eye, seemed to have been deposited in high concentrations along fractures in the rock. He visited the rock in its permanent quarters in the meteorite lab (to look at "raw" pieces not yet prepared for analysis) and noted that whenever he saw a face that was one side of a fracture, it was loaded with those rounded carbonate globs.

The rock detectives knew that carbonates, such as limestone (chalk), formed in water and were most commonly a by-product of sea-dwelling *organisms* (though they could also be formed in purely chemical, nonbiological processes). Carbonates were found in beds of fossils laid down in the slow accumulation of shells from defunct sea life. Oceans? Animals? What were these carbonates doing inside a very old *volcanic* rock that, moreover, came from Mars?

Mittlefehldt knew just who to ask.

Like several others in Building 31, Mittlefehldt had developed an admiration for a young man named Chris Romanek. Even though he was barely out of school, Romanek, thirty-five, had gained a reputation as an impressively sophisticated specialist in carbonates.

Late one evening in November 1993, Duck Mittlefehldt appeared in Romanek's doorway. "Hey, Chris, you want to see some really neat pictures of a meteorite I'm working on?"

"Sure," Romanek answered, looking up from his work. He walked across the hall with Mittlefehldt to the other lab. There, on the little CRT screen on Mittlefehldt's electron microprobe, Romanek stared at the images of tiny circular globs—a firmament of orangey rosettes arrayed along the rock's fissured surfaces, flecks big enough to be visible to the naked eye but only about the diameter of a coarse human hair.

"That's fascinating." Romanek said. "What is that?"

"Carbonates."

"Whoa, carbonates?!"

Mittlefehldt explained that this was a sample from a Martian meteorite. Romanek asked which one, unaware that any of them had such a rich lode of his favorite stuff. Mittlefehldt told him it was a new one that he, the Duck, had just unveiled. (The intriguing carbonate globules would be variously referred to by researchers as "rosettes," "orangettes," "orange spheroids," "rounded zoned blebs," "disk-shaped concretions," and "spheroidal aggregates," among other things.)

"Well, Duck, you've got to get a piece of that for me." Romanek felt confident he would be able to pry into this mystery and find out how the carbonates had gotten in there.

Research on the strange rock might have simmered along indefinitely were it not for young Romanek. But he had grabbed the baton, and he would carry it in a direction that no one expected.

Slender, athletic (a runner), newly equipped with a Ph.D. from Texas A&M, Romanek was a "low-temperature geochemist." (The term wasn't a hip reference to his personality or his metabolism; it delineated the particular focus of his studies.) When he'd been in high school in the foothills of South Carolina, an alert teacher had recognized the young man's affinity for science and helped him set his course. Eventually, Romanek had focused on what happens in the oceans of water that, so far, were not known to exist anywhere in the universe except Earth.

In graduate school, Romanek concentrated on clams that live in the middle of the Pacific. He studied the carbonate shells in an effort to understand how these particular bivalves grew their shells and what kind of information was locked inside them. He planned to use that work to study bivalve shells of fossils from extinct organisms, to see if he could understand the biology of some creature that no longer existed.

At a certain point, Romanek realized that he really didn't know much about the "tool" he was using, courtesy of the clam. So he went to Texas A&M to work on his Ph.D., and, instead of having a clam grow a shell for him, he grew his own crop of the common *calcium* carbonates that form on Earth—the cements that hold certain rocks and sandstones together, the components of seashells. He watched carbonates precipitate directly out of a lab solution so that he could understand what was happening to their atoms (the isotopic ratios) as the substance moved from the liquid to the solid phase. That helped him understand which types of information you could extract from a carbonate in the natural environment, whether it had been formed by living things or in a nonbiological natural process such as in a hot spring deposit.

What captivated his mind in all this was the realization that when you looked at the natural world around you, objects had stories to tell about how they had formed and about the influences that had altered them. The shells formed something like tree rings. An organism was born. It grew a shell through its life, and it died. Where others saw a plain old shell, Romanek saw a tape recorder, and it was recording everything that shell experienced in its life.

He found it remarkable that when you collected a living organism, a living bivalve just out of the ocean, you could take the shell and analyze it and figure out, well, this organism is five years old and has lived in this area for a certain fraction of that time. You could also take a shell from a deposit that was 300 million years old and learn something about its daily life 300 million years ago. The idea blew his mind.

When he arrived at the NASA space center in Houston, in 1991, on a postdoctoral fellowship, Romanek had never met a meteorite. His main assignment was to devise a new way to sample rocks and minerals in order to study them with pinpoint precision.

His goal was to shoot a very tight laser beam—perhaps a few thousandths of the diameter of a fine hair—onto a rock surface and measure the properties of its atoms (its isotopic ratios) in order to learn at what temperatures they formed, and in what kinds of processes. An isotope of an element—say, carbon—represents a variation in its atomic recipe that matters only under certain circumstances. The heavier isotope of a Toyota, for example, would be the same model of car but with a permanent extra load—the equivalent of extra neutrons—in the trunk. The car, like the isotope, would have all the same functional properties but with more mass.

At first Romanek blasted away at earth rocks, since they were plentiful and he didn't want to waste good meteorites as he was still working out the glitches. His original intent had been to use the new technique to look at very old Earth rocks, to study their carbonate minerals for signs of biological activity.

But by the time Duck Mittlefehldt showed him the carbonates in the Mars rock, Romanek had not only gotten the new technique working pretty well; he had developed an appreciation—even enthusiasm—for space rocks.

Romanek's mentor at NASA was the geochemist Everett Gibson. One of the old Apollo hands, he was now a senior scientist in the isotope lab. When Gibson and Romanek had first talked a couple of years earlier, Romanek had said, "You know, I'd like to come down. I know that you got this laser, I know you've got this wonderful isotope ratio instrument, and I'd like to come down and try to interface these things." Gibson had said, "Yeah, Chris, well, maybe you can work on some Martian meteorites."

"Yeah, right" had been Romanek's unspoken reaction. "How do you know there *are* any meteorites from Mars? How could you ever figure that out, anyway?" It was only after he'd arrived at the space center and started talking to folks like Don Bogard, who had actually done it, that he changed his mind.

In fact, he was hooked. When Duck Mittlefehldt first summoned him to see the intriguing images of the Mars rock that November evening, Romanek was busy shooting his laser at a piece of a well-known meteorite that had fallen in Murchison, Australia, in 1969. The Murchison meteorite was a piece of the primeval rubble that coalesced at the birth of the solar system, just over 4.5 billion years ago, and had remained relatively unaltered since then. Other scientists had revealed that it was packed with amino acids—more than seventy different kinds, most of them not from Earth—and other molecules important for the development of living things. But the evidence also indicated that these basic building blocks of life had *not* been left in the meteorite by any living thing. The material lacked the specific signature that goes hand in glove with all known life—a distinctive chemical "handshake," a form of molecular right- or left-handedness.

Now Romanek interrupted his assault on the celebrated Murchison meteorite in favor of Duck Mittlefehldt's alluring newcomer. With his laser blaster, he would try to figure out what had been going on in the Martian environment when those carbonates formed, to unspool the history recorded in the rock's little "tapes" just as he had once done with those clam shells.

Romanek, in this case, would be studying the ratio of light carbon to heavy carbon found in the rock, and the same for its oxygen. The proportions amounted to messages recorded by nature. These isotopes were not the unstable type that decayed radioactively—the kind used to determine the age of the rock. These were stable sorts, and they told a different kind of story. By studying how nature had divvied up these variations in the atomic properties of a chemical element, a specialist could tell something about the temperatures present at the divvying, and possibly about how fast it had happened and in what sort of event, and whether the process had involved biology or not. (As is the case with most records, these signs could be ambiguous, open to differing interpretations.)

The isotope business, as it happened, was booming. Scientists were using isotopes to determine the frequency of certain volcanic eruptions and landslides; they were studying isotope ratios in molted bird feathers and butterfly wings to determine the points of origin of these migratory creatures; they were using isotopic measurements to trace the spread of contaminants in natural food systems; they were studying isotopes in grizzly bear food sources and bear hair to monitor changes in the grizzly diet; and the like.

If carbonates were typically formed from solutions, Romanek thought, then in the case of the Mars rock maybe waters flowing on that planet, saturated in carbon dioxide, had deposited the minerals. Enticed by Mittlefehldt and with Gibson's blessing, he followed his impulse—to go into the rock with the nicely focused, modest aim of "just understanding the chemistry of the carbonates."

In the course of the isotope work, Chris Romanek remembered having heard about another technique, one used recently to study carbonates found in hot springs on Earth. Simply put, you used acid to etch away surface material so that what was buried farther down—the internal structure—could be studied. He realized that the carbonates in this rock were plentiful enough, and large enough, that he could use the same technique.

He used water to make a dilute acid, etched the first sample, and popped it into the scanning electron microscope. What he saw there stopped him cold. He stared at the magnified spectacle for maybe half an hour, not believing his eyes. Could it be?

He was seeing "these bacterial forms all over the place," he would recall later. "I was wondering, you know, am I going to get sick from touching this?" A thought about the fictional Andromeda strain might have flashed through his mind. He headed toward Gibson's office.

He felt a little bit frightened, he had to confess, because these things—they looked exactly like living organisms.

CHAPTER FOUR

E.T.'S HANDSHAKE

Contamination.

Everett Gibson, the senior isotope specialist working across the hall from Duck Mittlefehldt's lab, stared at Romanek's buggy little shapes. Gibson spent a moment scratching his head before he realized what had happened. He had been seeing it for a quarter century.

Gibson, fifty-three, was an affable man with a sprawling West Texas accent, a mustache, and a sculpted wave of hair peaking atop his face. His ready grin was sometimes undercut by a tension around the eyes.

Gibson had joined NASA's geochemistry team in Houston in 1966, about a year behind David McKay. In contrast to McKay, Gibson was bold, feisty, eager to "push the envelope," and he loved to talk about his work. Gibson was known as something of a free spirit, prone to enthusiasms that would sometimes color his judgment.

When Gibson had arrived at NASA, he'd planned to study water in lunar rocks. It turned out there wasn't any. So instead, he'd taken an interest in more general studies of the moon's geochemistry. For sport, Gibson served as flight engineer in a B-17 Flying Fortress that he and others had rebuilt.

Like many others in this summer of 1994, Gibson was intrigued by the hot new Mars meteorite ALH84001, Robbie Score's rock from Allan Hills. In the months since its unveiling, researchers had quickly shown it to be the *oldest* rock known from any planet, three times the age of any other sample from Mars and the first Martian fragment to come from *beneath* the planet's surface. It was the only one laced with a high concentration of carbon compounds and therefore possibly an unprecedented indication of primordial carbon stored on Mars.

These attributes made the rock an object of intense desire, a rock star. In the months since Mittlefehldt had revealed its true identity, labs in Germany, Switzerland, India, Denmark, Austria, Japan, England, and all across the United States had requested samples of it. In fact, of the 150 samples of the rock distributed in the nine years since Score had retrieved it from Antarctica, curators had distributed well over half within about the last six months—not even counting the allocations to Mittlefehldt himself.

David McKay, too, had begun to take a hard look at the rock. In the years since Apollo, he had "done" moon dust and moved on to tackle cosmic dust—the itinerant particles that drift through "empty" space. He had been analyzing that material for carbon—searching not for signs and portents of *actual* biology but for carbon itself, because of that element's importance as a building block for living things, as a precursor or enabler of life.

And that had led him to meteorites, including some from Mars. Now he had access to this complicated, fractured-up newcomer. His initial intention was to study this one in search of . . . more dust!

Under Gibson's guidance, however, Romanek's creative scrutiny of the carbonates was turning their particular investigation onto a radically different path from all the others. Soon McKay would make the same turn.

Gibson and his young assistant stared at the infestation in Romanek's microscope pictures, and even before Gibson could react, Romanek realized his mistake. He had used laboratory water to dilute the acid he'd used to etch the rock sample. He realized that he had not filtered the water, and it was swimming with bacteria—*Earth* bacteria.

There may have been some sighing, some rolling of the eyes. In any case, Romanek wheeled around and went back to his bench for a do-over. This time, he filtered the water (and when he examined the filters, he could see that, as expected, they had trapped a mess of bacteria). When he looked at the microscopic landscape again under the scanning electron microscope, he no longer saw teeming bacteria. But . . . the shapes that remained still seemed eerily familiar.

He went back to Gibson with the cleaned-up images. When Gibson saw them, he thought to himself, "This is just bizarre." If this were any terrestrial rock, you'd look at these features and say, oh, yeah, of course, that's the influence of biology.

Romanek, the new kid, and Gibson, the seasoned veteran, were flirting with a wild notion. *They might, just might, be staring at evidence of once-living Martians.* It was enough to trigger butterflies. Romanek's momentary diversion by the errant lab bugs might have been a symptom of this edgy mind-set.

In the preceding months, Romanek and Gibson had been getting a second line of encouragement for this outlandish idea. When Romanek was not etching the rock with acid, he was blasting it with his laser. He would train the beam on a fractured surface of rock sample where there was a concentration of the carbonate globules. His target area was about half the diameter of a human hair (20 to 40 microns). The laser would hit the spot and melt the mineral, releasing its carbon dioxide gas. The lab team would capture the gas and analyze its isotopes against accepted international standards.

Gibson considered the results nothing less than shocking.

The carbon signature, it had turned out, was so unearthly that it confirmed for Romanek and Gibson that the carbonates had indeed formed on Mars and not on Earth during the rock's long sojourn in the Antarctic ice. It also seemed to rule out an association with biology.

As for the oxygen part of the story, Gibson and Romanek found that the carbonates had formed at temperatures in the biological Goldilocks zone—not too hot, not too cold, and, conceivably, just right for life to exist.

These results were so unexpected that Gibson sought independent confirmation. He knew that a group of geochemists and meteorite specialists in Britain had a sample of the same rock. One was a longtime colleague named Colin Pillinger who specialized in isotopic analysis. Gibson made the transatlantic call to his friend:

"Colin, have you analyzed eight-four-double-oh-one for its carbonates?"

"No we have not, it's in our queue."

"I suggest you move it up in the queue."

"Why?"

"I'd rather not tell you our numbers, but I'd like for you to, if you would, move it forward."

Ten days later, Gibson got the call back from England. Neither party wanted to divulge first. Gibson thought the Alphonse-and-Gaston routine was kind of funny. "What did you get?" "Well, what did *you* get?" Eventually they told each other their numbers.

The numbers were identical.

When those carbonate globs, about five times the diameter of a human hair, had been deposited in the rock, fizzy, carbonated liquid water was flowing on Mars at temperatures moderate enough for life to exist. That was the scenario the British numbers seemed to confirm.

The journal *Nature* would publish a paper by the Gibson and Pillinger groups in its December 1994 issue. It drew only minor attention, even though it was the first significant analysis ever done on carbonates in Martian material. The news reports focused mainly on the similarities between water from Mars and the fizz in soda drinks. And the insider reaction to the joint paper was: Wow, that is unusual carbon—probably a signature of the Martian atmosphere.

Everett Gibson had grown up in West Texas. When his family moved to the town of Hamlin, north of Abilene, he had camped and hiked in the mesquite river bottoms and worked the oil fields, too, for four summers. He saw what was beneath the surface. He saw cores come up from oil wells. He saw people get hurt, and he saw what a rough life it could be. He worked the oil fields in order to pay his way through Texas Tech, where he earned bachelor's and master's degrees in chemistry, but he didn't want to make it a way of life.

Having developed an interest in the space program, and in meteorites and other extraterrestrial materials, Gibson went to Arizona State University for a doctorate and then joined NASA.

Gibson was proud to be one of the few lunar sample investigators in the most competitive science program in the history of the world. The NASA scientists had to compete for the same research dollars with counterparts at MIT, Harvard, Princeton, Caltech, and other major universities. The proposals were all thrown into the same pot, and reviewed by outside experts—peers. It was cutthroat, and he survived.

During Apollo, Gibson and the other rock detectives were working in such virgin territory they were obliged to develop new tools. A traditional Earth geologist could drive out into the field, bring back a truckload of rocks, and break them up with a hammer for inspection. Nobody knew how much the Apollo astronauts would bring back, or how much material any researcher would get his hands on. But it was clear that the allotments would be measured in mere milligrams. A fraction of an ounce of sample would be considered elephantine. The space-age investigators had to learn how to play Sherlock Holmes with the most minuscule of crime scenes.

Accordingly, Apollo spun off a fountain of money to address the need for a radical new approach. It was a source of great satisfaction to Gibson that he contributed to the creation of new techniques and tools (specific element detectors, gas chromatographic techniques, mass spectrometers, and lasers) that were high-tech equivalents of a precision hammer and chisel. These techniques were applied, among others, to the new Antarctic meteorite collection that began to accumulate upstairs in the mid-1970s. "If I have a strength," Gibson would say, "it is the development of unique analytical capabilities."

The Apollo program also taught Gibson a hard early lesson about the difficulties of coping with contamination in laboratory research, and the special effort it took to isolate a specimen entirely from the teeming, crawling, infested, blooming workaday Earth. The kind of contamination in question here was not simple befoulment; it was a kind of lie. This contamination could fool you, distort your reading of the evidence.

In the laboratory, a research specimen might pick up contaminants from water pipes and commercial fluids (as Romanek's *Andromeda Strain* moment would remind him). And, depending on the techniques, contamination could come from cigarette ashes, human hair, dandruff, nose drips, dust and spores in the laboratory air, lint fibers from cleaning rags, fragments scraped off tools, or dead insect parts. Biological or otherwise, contamination was a constant concern; precautions against it, and efforts to identify it when it occurred, were part of the research routine.

One of Gibson's first civil servant jobs was to help clean up the cabinets where the first Apollo samples were stored in the new Lunar Receiving Laboratory across the parking lot from Building 31. Not many people knew it, but in those early days, there were serious problems in NASA's handling of the precious specimens. Gibson would recall that experience often in the months and years to come, when contamination would become an issue in his life again, when in fact he'd find himself under attack because of this rock and he'd want to impress on a listener his hard-bought understanding of the pitfalls.

No matter what sample you were analyzing, and no matter how precise and accurate your analysis, he knew you always had to consider how the sample might be trying to fool you.

In the mid-1980s, Gibson decided to learn more about this business of stable isotopes and went for a year of study with Pillinger at Britain's Open University. Pillinger's strength was in isotopic analysis, which Gibson aimed to couple with laser techniques in meteorite studies. When he got back to Houston, he started putting together a stable-isotope lab.

Johnson Space Center, however, was still geared to human spaceflight as its primary function. Meteorites and isotopes were a ways down the list. Accordingly, Gibson did not limit his focus too much. For a time, he partnered with a nutritional expert to figure out how much energy an astronaut uses in the weightlessness of space so that the space shuttle flight planners could set the proper exercise regimens and food requirements for orbiting crews. The project was a success and, most important, enabled Gibson to buy the equipment he needed to do what he really wanted to: study meteorites.

A natural storyteller, Gibson was known to twirl the dial on a combination safe in his lab complex, reach in, and pull out a piece of fallen sky for a visitor to hold—for instance, a heavy chunk of black rock that fell in February 1969 in Mexico. That meteorite had dumped about five thousand pounds of material on a village, he would explain.

"The local priest called all the people together in the church," he'd say, "and asked them to pray for forgiveness. Then outsiders started showing up and acquiring the material, giving them rewards. So the priest called the people back together in the church"—here he would smile—"and they prayed for another one to fall."

It was only when Gibson took up the study of *Martian* meteorites that he finally struck water—the commodity that had eluded him on the moon. His team heated samples from six of the SNCs—the Martian meteorites—in a series of steps, in a small vacuum system, to extract trace amounts of water. They then hand-carried the water samples to the University of Chicago, where researchers analyzed the oxygen isotopes.

Gibson and the others announced their discovery in a 1992 paper, which created a ripple among planetary scientists. He kept in his office a photograph of an actual drop of the "Martian water" just one-sixty-fourth of an inch in diameter. He showed it off proudly: an image showing a glass tube with a drop of water inside, and an arrow pointing to it.

Spacecraft investigations suggested that Mars might once have had a water-rich atmosphere and flowing waters on its surface, but that water had mysteriously disappeared. On the chill and arid planet Mars had become, the consensus remained that any flowing water that reached the surface could not stay liquid.

Gibson's isotopic studies of the meteorite water droplets indicated that there were two separate and distinct populations, or reservoirs, of oxygen on Mars. By contrast, on Earth the oxygen is all essentially the same, whether in the atmosphere, oceans, or rocks. It has been homogenized by the mixing and churning of plate tectonics (the stately grinding of pieces of planetary crust, into and over one another, a process that unifies and explains much of Earth's geological evolution). On Mars, the water apparently had a different parent source from the oxygen found in the rock. That source could have been the Martian atmosphere, an ancient Martian ocean, or even a comet that struck the planet. There was, then, apparently no mixing system, as on Earth, to blend the two types of oxygen. This was the hypothesis that Gibson's paper pointed to—no plate tectonics on Mars.

That soupçon of Martian water seemed like a pretty big deal at the time, but Gibson was about to wade into a veritable Martian tsunami.

Now, as Gibson and Romanek stared at his cleaned-up images of Martian carbonates, Romanek had in mind a 1992 geologists' meeting he had attended in Cincinnati. Robert Folk, a sediment specialist from the University of Texas, had caused quite a stir by claiming discovery of a new kind of dwarf bacteria—"nanobacteria"—and proposing that such organisms were the principal agents in the formation of Earth's carbonates and other liquid-water deposits. In fact, Folk argued, these critters were everywhere, influencing a variety of processes, but had previously gone unnoticed because of insufficient microscope power.

It was from Folk's talk that Romanek had drawn his idea of acid-etching the Martian carbonates. When Folk had etched away the superficial layers from hot springs deposits, he'd uncovered minuscule ball- and rod-shaped structures that he characterized as fossilized remains. (The term *nanobacteria* came from a unit of measurement—the nanometer, a billionth of a meter. The period at the end of this sentence is at least a hundred thousand nanometers.) While the smallest known Earth bacteria measured 200 nanometers across, Folk's entities commonly ranged between 50 and 100 nanometers (with some as small as 20). This meant his nanobacteria were only about one-thousandth the *volume* of an ordinary bacterium, a size most scientists insisted was too small to contain even the most basic machinery of life.

Folk's claim, controversial in its own right, would become intertwined with a much more abrasive contentiousness soon to engulf the rock from Mars.

When Romanek stared into the fractured landscape of the Mars rock, acid-etched the same way Folk had described (now with the lab bacteria filtered out), the shapes he saw put him irresistibly in mind of Folk's putative nanofossils. They *looked* similar. This was the first inkling of this kind of structure in the rock, whatever the structure represented. Romanek felt certain that something strange and interesting was going on here.

This was a turning point in the saga of the rock. It was young Romanek's vision of something that might have undulated or wriggled through some ancient Martian aquifer that turned the rock definitively on its route toward a special dimension of fame and controversy. His instinct gave birth to the secret collaboration that, less than three years later, would turn this charred and fractured lump into a catalyst of human ambition, bureaucratic maneuvering, professional rivalries, personal jealousies, and audacious hopes for something historic and transcendent. Some people would come to believe that, finally, through this interplanetary message in a bottle, we might learn our place in the grand scheme. This rock might help us learn who we were.

When things got ugly later on, and some people accused NASA of dreaming up this shaggy rock story just to increase the agency budget, Romanek would bite his tongue, suppressing the urge to jump up and down and shout, "Wait, no, there was no political agenda here—it was my idea!" (Romanek had been hired under a grant from the National Research Council to use NASA office space and instruments for work of mutual interest, and he considered himself independent of NASA. By the time the rock made its public splash, he would no longer be working at NASA.)

Romanek and Gibson had a mutual-admiration club going. Romanek would grind away for weeks analyzing something, then spend five minutes briefing Gibson. The older man would amaze Romanek by turning around and delivering, with complete accuracy, the same level of detail in a presentation to some third party. Romanek thought Gibson must have a photographic memory. He was grateful for Gibson's complete faith in his data. And Gibson was so experienced that he could plug in the information and see the context and the implications both clearly and quickly.

Gibson, for his part, thought Romanek was a real "find," an intellectually advanced young guy with an unusually broad background, and in many ways ahead of other people at the same stage of their careers. Gibson badly wanted to hire him as a full-fledged NASA civil servant, but because of hiring freezes and red tape about job "slots" and such, he was unable to work it out.

"If any one person gets credit for conceiving the idea" of a possible biological fingerprint in the rock, Gibson would say later, "it's Chris."

But was Romanek's thinking, the very hint of the idea that he might be seeing evidence of once-living Martians, some kind of crazy leap of imagination? Was he taking a stroll with the lunatic fringe? Shouldn't his more seasoned mentor have laughed him out of the lab?

In other years, the answers to those questions might well have been yes.

There were times when, if word of a notion such as Romanek's had gotten out, it could have earned NASA a rebuke from Congress, at the very least. The marquee aliens—the proverbial intelligent civilizations out among the stars—were faring especially poorly. Congress banished them politically at about the time Romanek was delving into the Allan Hills rock. Led by a single irate congressman, the people's representatives booted out of the federal budget the program known as the Search for Extraterrestrial Intelligence (SETI), less than a year after it had begun scanning the skies for radio signals from remote civilizations. A deadly giggle factor had set in. For the deficit-ridden taxpayer, it was just too silly. Outside of Hollywood films and other sci-fi outlets, E.T. seemed worse than defunct; he was politically incorrect.

Legitimate research into the prospects for even primitive extraterrestrial life had faced ridicule over the decades as a "pseudoscience" whose subject did not exist. And although the faithful few remained undeterred, the two Viking spacecraft had reinforced the scoffers' views by failing to detect any persuasive signs of even the faintest trace of organic molecules in the Martian topsoil. The evidence, instead, proclaimed that the modern-day surface of Mars was thoroughly barren of life as anyone defined it. Lacking a layer of protective ozone in the atmosphere, the surface was exposed to high levels of ultraviolet radiation from the sun, which tore down the chemicals essential for life.

But Chris Romanek sensed the ground shifting. Mostly below the pop-culture radar, a quiet revolution had transformed human understanding of life's fundamental nature. Those who were paying attention knew that, in a sense, E.T. had already been encountered right here on Earth, in the form of creatures as alien as anyone could have imagined, in conditions as otherworldly.

In 1977, three human explorers had descended into a shockingly unearthly world a mile and a half below the surface of the Pacific Ocean. "Debra? Isn't the deep ocean supposed to be like a desert?" came the call from the small research submarine to its waiting mother ship up on the sea surface. "There's all these animals down here."

There was no live, global TV coverage of this event and little public awareness at the time. But some people would later compare it to the landing of *Apollo 11* on the moon: daring explorers using life-support systems landed in a treacherous, unexplored realm and returned with the first bits and pieces of a previously unsuspected reality. And, unlike the Apollo astronauts, they *did* find aliens.

The divers aboard the submersible *Alvin* that day, not so long after the Viking crafts had landed on Mars, discovered the first warm-water spring ever known to exist on the chill, sunless floor of the ocean. Cramped inside their life-support sphere, its inner wall dripping with condensing moisture, its spotlights casting the first light into the virgin darkness, the explorers peered out tiny portholes and glimpsed a stunning oasis of bizarre life-forms thriving in a setting that, until then, everyone had presumed was inhospitable to living things.

Over time, the lessons from that deep-sea encounter, and others to follow, would reveal previously unsuspected truths about the nature of life—about where and by what means living creatures on Earth (and possibly elsewhere) might sustain themselves. People learned from these explorations, for example, that some living creatures could thrive without light from the sun. The microorganisms at the base of the food chain here survived in darkness on sulfur and methane—poisonous to oxygen breathers. Some creatures lived at previously unthinkable temperatures (up to 235 degrees Fahrenheit) and crushing pressures.

Further up the food chain were crabs with the wizened faces of apple dolls, huge white clams with blood-red flesh, mussels, and anemones—and, most striking of all, vast, undulant fields of eight-foot tube worms, their white stalk bodies topped with bright red plumes. The Daliesque worms had no mouth or digestive tract. Instead, they drew nourishment from bacteria that lived in their tissues.

In the last quarter of the twentieth century, some five hundred new species would be catalogued at vent communities throughout the world's oceans. And in other settings around the planet, people armed with new technologies started turning up signs of biology in other extreme environments—two miles deep in oil wells, in desiccating salt marshes, in the polar ice, and in rock within Earth's crust. Scientists had begun to study the possibility that a hidden biosphere below the planet's surface and seafloor could equal or surpass the total mass of all surface life. (More than a decade later, a Stanford University group would report that some hundred trillion bacteria live in each human gut—10 microbes for every human cell, 395 different strains—and are so vital to the host's well-being that they could be considered as one more body organ.)

Some of these organisms, astoundingly, would turn out to represent a previously unsuspected domain on the tree of life: an ancient group distinct from bacteria, plants, and animals, and linked to the first known organisms on Earth. An extreme environment, then, could have nurtured life's genesis.

And there were other developments that informed the actions of Romanek and Gibson. One was the emerging evidence that life had sprung up on Earth with remarkable swiftness after the initial violent bombardment by space rubble that filled the young solar system. This primeval barrage should have boiled off Earth's oceans, wreathing the planet for millennia at a time in vast black thunderheads shot with lightning, until they released their water, the rains raised new oceans, and the process was repeated again and again. And yet it seemed that within a relatively short time—possibly as little as 400 million years after Earth had formed out of that same rubble—life had become diverse and widespread.

Just months before Romanek first peered into the Mars rock in 1993, a paleobiologist named J. William Schopf, of UCLA, described a startling fossil find in Australia. He reported signs of eight previously unknown species of microbes, some of them amazingly complex, from as early as 3.465 billion years ago—when the planet was less than a billion years old.

At the same time, sky scientists had combined clever techniques with new technologies and (after a frustrating string of false positives) finally obtained convincing evidence of planets in orbit around stars beyond our sun. Among its many ramifications, this confirmation seemed to boost the statistical odds that there was an abundance of worlds where life might have found a toehold.

People who considered all these developments realized they were witnessing a paradigm shift—a major change in a shared system of belief and in the choices people like Romanek would make about which paths of exploration, and which career paths, might be most fruitful. Many people had been schooled on textbooks that said the emergence of life out of lifeless nature must have been so complicated and difficult that it had required a special set of conditions, which the Eden of the young Earth had secured as a rare blessing. Now here was a spate of evidence that, given the most minimal necessities (liquid water, any old energy source, whether sunlight or something else, and certain organic ingredients), the impulse for nature to come alive was irrepressible. Living things—at least *primitive* ones—could arise with astonishing speed and ease, and in almost any miserable, toxic, hellishly hostile place. Were a planet's surface to be sterilized, life could retreat underground.

Life seemed to be a terrestrial imperative, and possibly even a cosmic imperative. Maybe it was only *intelligent* life that was so difficult and rare.

All the ferment on the nature of life served to ratchet up interest in the odd alien rock—and to lend respectability to young Romanek's novel line of inquiry. If microbes could thrive in toxic, extreme, and, yes, "alien" conditions on Earth, it seemed less extreme to speculate about the possibility of microbes on a neighboring planet with the right specs—if not on the surface, then underground, in wet, warm regions of the interior. So the short answer was, no, this was not a lunatic excursion.

Romanek and Gibson kept going. Romanek took the new batch of images, along with the published pictures of Bob Folk's putative dwarf bacteria—the nanofossils—to Gibson and asked, "Can you see the difference?" Gibson's eyes lit up. The similarities were hair-raising. He knew he had a real mystery to work on here.

In addition to these images from Romanek's etchings, which seemed so strikingly suggestive of biological influence, the men also had the results from Romanek's laser-blaster work, which indicated that the carbonates had formed (a) on Mars and (b) in hospitable conditions. Right then and there, Gibson told Romanek they were going to go down the hall and talk to David McKay. It was the second week of September 1994.

Romanek knew McKay as a gray eminence who had a reputation for being very smart and very careful. McKay was also a recognized virtuoso at the scanning electron microscope, with a blister-inducing allergy to the film-coating chemical to show for it. And it was this skill on the microscope that initially brought the two men to McKay's door.

McKay, of course, had already been inspecting the same Mars rock in the course of his unending soil studies. Duck Mittlefehldt, back before he'd turned the meteorite into a rock star, had requested a fresh allocation of samples from it and received a handsome quantity. One of the major problems in geology is that a practitioner needs to get a sample that is representative of the whole rock in question. For rock as coarse-grained as this one, ideally you would want to grind up a whopping 55 pounds (25 kilograms) for study. Since the whole mother chunk weighed only 4.4 pounds (2 kilograms), that was not going to happen. The goal, in meteorite work, is to find a proper balance between the smallest sample that (you hope) will give you good answers and the largest amount that will not unduly waste material. What Mittlefehldt had was, under the circumstances, a virtual truckload.

Mittlefehldt and meteorite curator Marilyn Lindstrom had invited anybody who wanted to work on the rock to join a consortium and use Mittlefehldt's samples. Because his general motivation was to learn, and in this case specifically to learn about Mars, Mittlefehldt preferred to disperse his samples among people he thought could do "something neat" with them.

This sharing would speed up the investigation. The ritual of request and response was ordinarily handled by a committee of meteorite gatekeepers from the Smithsonian, NASA, the National Science Foundation, and various universities. It could take months—even for those whose labs were right there in Building 31 where the meteorites were housed—to get a new allocation.

Curators typically shipped the samples in one of two forms: rough "chips"—essentially crumbs, or broken-off pieces—which could be shot with lasers, crushed, or otherwise destroyed, and "thin sections," slices of rock mounted for microscopic study, which could be checked in and out of the repository like library books and used over and over.

In most cases, Mittlefehldt would grind up and homogenize a sample and give away to others "splits" of that material. An added benefit of this approach was that the researchers could compare and use one another's results from the same sample.

This afternoon in the late summer of 1994, when Gibson and Romanek strolled down to McKay's bright corner office, they sidled warily up to their topic. They were not about to admit the true nature of their speculation about the rock: that dancing coyly in the backs of their minds was the incredible possibility that they were peering at the first known signature of extraterrestrial life. They could just imagine McKay's reaction.

Instead, they closed the door and spoke, gingerly. "Dave, look at this." The visitors showed McKay Romanek's pictures. Gibson mentioned Romanek's limited expertise on the scanning electron microscope, which Romanek readily acknowledged. McKay had been in the dusty trenches with the thing for decades and knew firsthand all the techniques they needed in order to get the best possible pictures. He responded with equally guarded interest. "I've been thinking about looking at these carbonates, too. I think we really have something here."

McKay had merely set out to chase his dust trail. But he found himself, like the others, distracted by these beguiling orange carbonates. They were everywhere in the fractured rock, some squashed flat like pancakes, some spherical. Almost like little Dalí moons. You could see them with your bare eyes.

The three men formed a pact: "Okay, we'll work on it, but we won't tell anybody about it." Well, hardly anybody. They had created the nucleus of their secret society.

They wanted the search for answers to be on their own turf, so to speak. The trio had started thinking about the rock as a whole, a system, rather than as isolated crumbs under a microscope. They wanted to make sure they knew which part of the terrain they were studying. They had all been working with Duck Mittlefehldt's samples. They needed their own piece of the rock.

Romanek had developed a rapport with Robbie Score, and she became a key contact for his dealings with her meteorite lab. He found her wonderfully friendly, and happy to answer his questions about how the curatorial facility worked, or to supply lists of rocks that might be useful, or to inform him how he and his collaborators could get more samples of that rock she had plucked off the ice in 1984.

One day, Romanek and McKay walked upstairs and entered the kitchen-like meteorite archives, where the mother rock rested inside its nitrogen cabinet. They went through the protocols—took the air shower, donned the surgeon's cap and mask, and so forth. They intended to select exactly the chips they wanted from specific locations in the rock. There it sat: the mother rock with smaller pieces that had broken off, or been knocked off, and canisters containing grains of it, like sand.

Romanek had visited the mother rock several times since Duck Mittlefehldt had first introduced him to it. He'd seen it initially as two big chunks that seemed kind of glued together with all this crumbly material in the middle, like two small bricks cemented with mortar. He and his coworkers decided they wanted to study that place where the "mortar" was, because that was the zone with considerable fracturing and crumbling and milling of the parent rock material. (Curators would refer to this light-colored band, less than half an inch wide, or about a centimeter, as the "crushed zone.") These traumas had most likely happened when something slammed into Mars near the rock's native site, or in a marsquake. The two "bricks" of rock had actually rubbed up against each other and broken up. That was exactly the sort of place—full of cracks and crevices—where fluids would have moved through. And these fluids would have deposited the carbonates as they flowed through this rubble.

When the staff tried to saw the slab from the middle of the rock for Duck Mittlefehldt's consortium, it broke apart at "a porous band of weakness." He got the resulting pieces.

The curators used a numbering system for each piece that came off the mother rock. (The first sample was "01," the second "02," and so on.) They could, in theory, reconstruct the original rock, a more fortunate version of Humpty Dumpty, if they could recover all the pieces.

Few if any of the keepers of the rocks understood what Romanek and McKay were seeking on this foray. As the staff brought out sample batches, the visitors would put them underneath the microscope and say, "Look, we want pieces that look like this, with the little orange disks on them."

McKay and Romanek sorted through roughly sixty-five bits and pieces. Their arms inside those black rubber gloves, they used the curators' tweezers to pick out specific *grains.*

Some people would later accuse the McKay group of getting special treatment that enabled them to look at the rock ahead of everybody else. But by the end of 1994, the curators had distributed samples from the rock to more than forty groups around the world. And it was not uncommon for researchers to venture into the meteorite lab to pick out specimens.

In due course, the official allocators approved Gibson's request, and the new allotment arrived downstairs in its Teflon plastic wrapping. It was only a few grains. Two grams, a fraction of an ounce. The work the McKay group would do on the rock from late 1994 through the time they announced their findings to the world would be based on this iota.

Soon, this impromptu archipelago of Martian landscape was consuming more and more of their lives—days, nights, and weekends. Riding on beams of electrons and amplified light, they inspected the microscopic turf that would later turn into a geochemical Hamburger Hill—the first Martian battlefield.

CHAPTER FIVE

THE CONVERT

KATHIE THOMAS SUSPECTED that the three men had gone a little over the edge. She couldn't believe what she was hearing, what she was seeing. She had been vaguely aware of some stealthy commotion over Duck Mittlefehldt's Mars rock, but nothing had prepared her for the surreal proposition these guys had just put in front of her.

She was at work in her microscope lab this particular afternoon in mid-September 1994 when Everett Gibson appeared at her door. He invited her to walk down the hall to David McKay's office for a chat with the two of them and Romanek. They sat around a big table in McKay's corner office, daylight streaming through the windows on two sides. Romanek's images of the rock were arrayed in front of Thomas.

At the time of this meeting (as Thomas would soon learn), barely a week had passed since the three men decided to collaborate in secret. McKay had asked Gibson and Romanek if he could bring Kathie Thomas in. They'd responded, in essence, sure, we don't have a clue where this project is taking us, anyway. We could use the help.

The three men outlined what they had found so far in the rock, showed her the images, and then got to the point. McKay asked her to drop what she was doing and join their clandestine project. She heard David pose the question: "Could this be life in these carbonates? Could this be microbial life?"

These guys wanted her to help them look inside a rock for microscopic fossils of ancient Martians?!

McKay, she kept thinking, was usually sensible—known for his caution. She had worked for him for years; she respected him and thought she knew him. She knew Romanek as a coworker, and Gibson, too, though not as well. She wouldn't have pegged any one of them as a nut.

The men were well aware that they had virtually no evidence. They knew their idea sounded far-fetched. They understood that Thomas thought they sounded "wacko." But they were following an instinct, taking a creative leap based on a vivid sense of the natural world and its workings. Einstein, among others, had shown this to be a legitimate approach to advancing human knowledge—to dream up theoretical frameworks that extended well out beyond the hard data. But ultimately, they knew, the leap of imagination, in order to be validated, would have to come down in a foundation of evidence that would hold up under rigorous challenge and could be tested, and reproduced, in experiments conducted by others. That was how it worked.

The McKay team's lack of such a foundation was a big reason, if not the main reason, they wanted to work in secret. They weren't even telling Duck Mittlefehldt, who had brought the rock to Romanek's attention.

Kathie Thomas was reluctant. She was in the midst of a project that had just begun to pay off for her. And within the last few months, she had barely emerged with her skin intact from a painful fracas over some startling results she had come up with. For some time now, people had been telling her *she* was crazy. She knew what it was like to be at the center of a skirmish, and she was not eager to jump into another. She told the three guys around the table that she would think about their proposition. She wanted to talk it over with her husband, Sean Keprta, who also worked at the space center. He was in charge of health and safety. They had met on the job.

She went home that night and told him about her day. She said she had serious misgivings about getting involved with this nutty-sounding hypothesis. Her husband pondered her dilemma. He said, "You know, if they're not right about this, they need someone to keep them grounded. So you could go in with the attitude of proving them wrong. Then none of this foolishness will ever come out."

Kathie Thomas would later explain that she had gone into this inquiry as "a *doubting* Thomas."

An attractive, coltish blonde who favored jeans and sneakers, Thomas kept a doll propped on a shelf in her office: Paleontologist Barbie. As a girl growing up in the Midwest, she had been an ambitious student. She didn't care for English, preferring to compete with the boys on "their" turf, science and math. She delighted in "setting the curve" for the class and credited a few crucial teachers for egging her on. She had a chemistry teacher who, even if she answered every question correctly on an exam, would give her a 99. Why not 100? she would ask. The response, which she remembers to this day, was

"Until you know everything there is to know about the subject, you get a 99." From her grown-up perspective, she thought that this had been a valuable (though frustrating) prod to keep her always learning, moving, improving.

She arrived at the space center in 1984, the year Robbie Score plucked the rock from the blue ice of Allan Hills. Like many of her colleagues, Thomas was not a NASA civil servant but a contract employee, working for an aerospace company that would become Lockheed-Martin. That meant living with one-year contracts and the possibility of getting sacked with little notice.

Thomas had not been trained as a geologist. Her specialty was in carbon chemistry, and that fit in perfectly with the work her new boss, David McKay, was doing at the time. More important, she had become skilled in the uses of the electron beam to penetrate hidden realms of nature.

While McKay's main game had been lunar topsoil, he was branching into cosmic dust and put Thomas to work on that. He posed a question: How much carbon can be found in an interplanetary dust particle?

Thomas was fascinated with the study of these vanishingly small grains of dust that drift among the planets. She had spent a decade with them. The government regularly sent aircraft, usually spy planes, into the stratosphere with special devices that captured infalling particles before they burned up. Of the many tons of interplanetary dust that falls on Earth annually, most was thought to be the residue of comets and asteroids. As such, these particles would be remnants of the primordial solar system, kept orbiting in cold storage since the time before the first planets formed out of this same rubble.

Thomas developed a technique that used the electron beam, making it possible (she argued) to analyze, quantitatively, these incredibly tiny particles. But it was a tricky job, and she and McKay had a hard time selling the idea. People questioned whether anybody could measure such small entities. But Thomas pushed ahead, once again looking to "set the curve."

The results were such a surprise that they only encouraged Thomas's opposition. Until then, most people had supposed that cosmic dust particles were, by weight, maybe a couple percent carbon. But in some of the dust Thomas studied, the particles contained a whopping 30, perhaps even 40 percent carbon. This was more than twice the proportion found in the type of asteroids that, up to then, had been considered the most carbon-rich solid objects in the solar system. Space, it seemed, was *raining* a basic underpinning of life.

The unexpected findings upset people. "No way!" was the reaction she got. "You're crazy!" She found her published work under assault.

Thomas began to feel disadvantaged by what she saw as her innate midwestern aversion to conflict, a thoughtful, slow-talking, "nice" approach to human interaction she had learned growing up. She had chosen a career in which she needed to be a black belt in the game of intellectual assertiveness. Once again, she was pushing herself. Somehow, she had to grow a thicker skin.

Her analysis held up, and over time the crowd began to tip toward wide acceptance of the new numbers. (Years later, in the spring of 2002, Thomas would attend a conference on interplanetary dust and be thrilled to hear colleagues quoting as conventional wisdom her old carbon numbers.)

Now—here was the rub—Thomas was just gaining this new foothold in her field, when McKay came along with his dubious Martian diversion. She was in no mood to go through another prizefight, so she decided that if she was going to associate herself with this madness, she would for darn sure make certain she and her collaborators were right.

She would run into another complication along the way. Because of the secrecy the project required, and her resulting failure to report how she was really spending her time, her employer (contractor Lockheed Martin) would end up giving her the lowest possible raise (about 0.5 percent) based on a yearly review, and this reflected poorly on her record. All would be forgiven when the truth emerged, but only after a year of this low regard. She came to feel the project justified such risks.

In the days and weeks after that initial meeting, Thomas started getting to know the rock. She took a look at it under a microscope. She took another. She read some papers on related topics that Romanek gave her. She began to see what had triggered the whispering, the excitement. The rock had her.

Thomas found the landscape of the rock entrancing—aesthetically, if nothing more. The magnified surfaces had a grayish tint to them, silvery almost, with hints of green. And they sparkled in the light. But the carbonate moons—they were spectacular! Looking at one of them under magnification was like staring into a headlight on a dark road. Here you were, touring this silvery-greenish-grayish textured landscape, and suddenly you encountered this bright, vibrant orange mass circled in black and white, as if drawn in ink by some Lilliputian cartoonist. The orangey moons were fairly circular, and had finely differentiated concentric rims: black, white in the middle, then black again, all the way around. It was extraordinary.

And, as McKay would later point out to her, there was another significant thing about them. The moons, or "headlights," were not sitting on *top* of the silvery-gray rock. They were embedded *in* the rock. So some of the silvery-gray material had been worn away, or pitted, and then the carbonate globs had been deposited in the hole that was left.

She started out using the scanning electron microscope, just to get familiar with the target rock. With this instrument—David McKay's primary instrument—she could make out surface patterns and textures, and she could see some little pebble-like objects scattered everywhere, especially in the black-and-white rims. Both the rims and the orange moons they encircled were complicated, not a solid lump of a single element but changing from one kind of mineral stew to another in concentric rings, almost like tree rings. The scenario of evolving natural conditions and events on Mars that would explain all this could not be simple either.

Thomas knew that Gibson and Romanek had concluded that the carbonates were deposited under relatively low temperature conditions, most likely by currents of fizzy, carbonated water that had flowed or percolated down through the rock as it lay beneath the Martian surface. Now Thomas was taking the investigation to a new scale and depth.

To carry out the crucial task of selecting and extracting the "best" carbonate globule, she would first spend several days looking down the eyepiece of an optical microscope, aided by generous rubs of Ben-Gay for neck tension. She would begin with several rock chips and search around until she found one on a fairly flat surface with enough underlying material so the tweezers could hold it. Extracting a chosen carbonate typically went like this: She placed the naked chip of rock underneath the microscope in a clean tray. She held it with one set of tweezers and then, using a second set of blunt-end tweezers, tapped around the outside of the carbonate moon, typically 300 microns across (about four times the diameter of a fine human hair and much too huge for her purposes). She tapped one edge, then another, working her way around the black-and-white Oreo rim. After she had tapped long enough—usually about two hours—she managed to pop the whole thing out of its setting. Then she crushed it and embedded the tiny pieces—no bigger than interplanetary dust particles, say 10 to 20 microns—into epoxy.

For this task, she avoided drinking coffee, and she usually scheduled key phases for a Sunday morning, when no one else was around and she wouldn't have to worry about vibrations in Building 31.

Next she prepared the sample for deeper penetration by the big gun: the transmission electron microscope. It took about a week to get a sample ready.

She mounted the hard-won crumbs in a drop of epoxy atop a plastic "bullet." After a bit of shaving and shaping, she inserted the bullet into a round holster that secured it and proceeded to cut off extremely thin sections of the sample—tissue-thin to the point of transparency—using a keen blade made of diamond.

She slid each thin section off into a "boat" of water and picked it up on a tiny copper grid that, to the unschooled eye, resembled a contact lens for a bumblebee. That little grid fit into a holder on the microscope, in the bull's-eye of the electron beam.

The samples had to be transparent, thin enough for Thomas's electron beam to pass through them. When explaining her approach to the uninitiated, she compared the rock samples to a loaf of bread. With the simpler scanning electron microscope, you would see the outer crust, its shape and surface textures. You could, for example, distinguish between Wonder bread and a baguette. But the view you got from the big transmission electron microscope was as if you removed a slice from the middle of the loaf: you could see the interior, assess the number of raisins or pecans; you could gauge the density of the mixture, expose the lurking weevil or the engagement ring that had fallen into the dough vat. That's what this microscope gave you.

Thomas was essentially taking a three-dimensional object and transmitting its image onto a two-dimensional surface. That meant she had to know the orientation—the tilt—of her sample in order to interpret the flattened or foreshortened images.

When she was ready, she sat at the microscope, a device mounted on a countertop with a monitor attached to a tower, like a cathedral spire, that rose several feet above her head. She took the sliver of rock she had shaved off—some five hundred times smaller than the diameter of a human hair—and fired her electron beam down through the apparatus toward the target. The beam, acting like an X-ray of the structure, revealed internal features thousands of times smaller than the diameter of a fine human hair, right down to the regular patterns formed by the atoms in the crystals.

She focused on those little pebbles, or grains, she had noticed earlier, scattered all around but concentrated mainly in the black-and-white rims whose cross section was reminiscent of an Oreo cookie.

They were, she discovered, *magnetic* crystals. And they were quite different from the types she had seen in other space rubble.

Until this point, Thomas had been focusing on technique. She was aiming simply to describe the samples. She had put the notion of Martian biology out of her mind. That now changed.

Reading up on the subject, she learned that there were several possible explanations for the magnetic crystals. One was that they had been manufactured by bacteria. Microbiologists had known since 1975 that certain remarkable swimming bacteria on Earth grew magnetic crystals in their insides. These crystals (whose magnetic properties came from iron atoms) served the bugs as internal navigation devices—compasses. And what Thomas saw in the Mars rock looked a lot like those little compasses.

Keenly aware that they were roaming far beyond their own expertise, McKay and the other collaborators studied the writings of experts in microbiology and the prospects for life on other planets, particularly the burgeoning studies of extremophiles. And they reached for more direct help.

McKay approached Hojatollah Vali, who taught at McGill University in Montreal but happened to be spending a year or two at the Houston space center on a fellowship. The Iranian-born Vali's specialty was the structure of cells and the minerals formed by living organisms. At McKay's invitation, Vali examined the samples on the penetrating transmission electron microscope and conceded that the hypothesis was not completely insane: the magnetic minerals *might* have been formed by bacteria.

One night, feeling guilty because her husband was unhappy with her all-consuming work schedule, Thomas was laboring over the rock samples with Vali when they found what appeared to be a second type of magnetic crystal, also similar to those formed inside Earth bugs. They called Dave McKay in to have a look.

The discovery gave Thomas the strong sense that they had nailed the case. She would say later, with a rueful smile, "I can remember walking out to the parking lot that night—it was late, it was pitch-black, my husband was mad at me—and wondering where the bands and the parades were."

"That's where she took off," Romanek would say of Thomas. "She ran with that."

The magnetic crystals were so small a billion of them would fit on the head of a pin, but they would become arguably the most abiding and intriguing players in the drama of the rock.

One of the things that impressed itself on the collaborators, thanks to Vali, was that the coexistence of this particular menu of iron compounds (iron oxides and iron sulfides) with the carbonates seemed highly unlikely—unless living organisms had intervened to bring them together.

The array surprised Vali. Those magnetic compounds, he told the others, could certainly be found together in nonbiological systems, but then they should have been formed at extremely *high* temperatures. If the carbonates in the rock had been deposited under relatively *mild* temperatures, as Gibson and Romanek had concluded, this particular combination of minerals was difficult to explain. Also, these compounds would not ordinarily form out of the same type of solution that produced carbonates. It all seemed a little like stumbling across a ski slope in Florida in August, not knowing that some eccentric amusement park entrepreneur had gone to considerable lengths to manipulate the environment for profit.

Scientists historically have favored the elegance of simplicity. They have tried to avoid contorted reasoning, twisted and bent to fit a preconceived notion. The McKay group figured that, although there were certainly alternative possibilities, the odd confluence of the lines of evidence they were finding in the Mars rock could be explained most simply by the presence of living things. Others would interpret the same set of facts differently.

McKay and his team knew that biology flourishes along the boundaries where contrasting environments meet. Living things find ways to harness these "edge" contrasts to their benefit. The human stomach, for example, is home to billions of bacteria that would die if exposed to the outside air breathed by the stomach's owner. Terrestrial life had shown itself to be uniquely capable of manipulating its environment, and the same could be true of some primitive, long-extinct species of Martian.

At the same time, however, the McKay group was taking care to study the alternative story lines that would explain what they saw without resorting to Martian biology. Well aware of the risks of embarrassment and ridicule they could face when or if they went public, McKay and Gibson were painstaking about this sort of double-checking.

What they did next would provide fodder later on for one of the most bitter and personal feuds that would arise out of the investigation. The McKay team turned for help to a man widely recognized as a supreme authority on ancient microfossils: Bill Schopf, of UCLA. Gibson called and invited him to come down and take a look at their evidence; they arranged for their department to pay his way.

Schopf had rewritten the textbooks on the origins of life on Earth by claiming discovery of the oldest fossils known—the remains of microbes 3.465 billion years old. His memory of his visit to Houston would differ from that of the McKay group, beginning with who issued the invitation. (Schopf would later state that the call had come from "NASA administrators.")

Romanek was given the responsibility of putting together a presentation for the prominent visitor. He got busy arranging for the graphics and data that would enable Schopf to understand quickly all that the collaborators had seen in the rock. They wanted to leave plenty of time for Schopf to put some thin sections under a microscope and see for himself, and then for them to sit and discuss the whole matter in detail. Romanek prepared foam-backed poster boards and glued all the team's microscope images onto them, showing the textures and fabrics they had found in the meteorite, with the sizes of the features in the carbonate indicated.

When Schopf arrived at the space center on January 24, 1995, he found the researchers "flushed with excitement." They met in David McKay's office. The mood was cordial, by all accounts. The group laid out the posters; then Schopf looked at thin sections and chips from the rock.

The group had sworn Schopf to secrecy, and he would be true to his word. But when they took him to the meteorite archives, Robbie Score was there. She had studied under Schopf at UCLA and "put two and two together," as Gibson would recall. The group asked her to keep her surmise to herself.

Romanek would later remark that, as Schopf inspected a thin section under the microscope, "he pulled his eyes back and looked kind of shocked. But one second after that, like a true scientist, he was thinking, 'How can I explain this by anything except what you're telling me you think it is.' " In the end, Schopf pronounced the fossil-like features "very interesting" but the evidence far from convincing.

Schopf told them, "You've got some real interesting stuff there, but you're not going to go anywhere with this until you can show me organic matter in the structures." Schopf's view was that they had to find organic matter to suggest any kind of fossil at all. And they must show varying shapes and sizes, presumably representing juvenile and adult phases of growth.

Schopf would write later that the Houston group had suggested that the carbonate globules, with their black-and-white rims, resembled the disk-shaped shells of a particular type of protozoan (one-celled organism) and might be a fossilized Martian version. (The McKay group would deny that they had said or thought that. Romanek could not recall ever having heard the term in that context. McKay and Gibson said that, in an effort to take full advantage of Schopf's expertise and to overlook no possibility, they had asked if certain features in the rock could be protozoan-*like*—just to rule it out.)

Schopf told them they were wrong, that the size range did not fit biology, and that the objects lacked telltale features—"pores, tubules, wall layers, spines, chambers and internal structures—that earmark tiny protozoan shells."

Gibson and his wife drove Schopf back to the airport at the end of his two-day stay. As they dropped him off, they were struck by a comment he made: "Today I saw enough interesting structure and other things indicating possible biology in those thin sections that, if it was early in my career, I would choose to spend more time on this rock."

That same day, Gibson asked his wife, a biologist, to sit down and write an account of what Schopf had said. Gibson and McKay also wrote a memo summarizing their discussions with Schopf—part of an ongoing record of dates and developments in the project that Gibson, as the group's unofficial recording secretary and historian, was keeping.

Schopf's most disheartening point, Romanek thought, was that the McKay group would never convince anybody that these mineral structures were of biological origin unless they could find complex organic material associated with them—something like a remnant of bug slime, products of cell decay, or other recognizable biological signature. "I'm thinking, well, I could argue with Bill that there was organic matter there and it was oxidized," Romanek would say later. "But that was a very big letdown . . . because we 'knew' there was no organic matter on Mars." Many people "knew" this, based on those Viking experiments that had helped exile E.T. and paved the way for the rise of the dreaded giggle factor.

Even if their instincts were right, the McKay team realized that their punch line lacked a proper buildup. For a claim to be at least provocative, if not persuasive, it had to include a chronological narrative that provided context into which individual clues fit and made sense. There had to be a plausible *story*.

Romanek feared that the project was at a dead end.

But Kathie Thomas, for one, was not about to give up so easily. In the coming days, she would summon up a troop of California cavalry to ride in and save the day—laser cowboys from the wild west of Silicon Valley.

CHAPTER SIX

MICKEY, MINNIE, AND GOOFY

In late 1994 and early 1995, as David McKay and his collaborators puzzled over the Mars rock, Richard Zare, fifty-six, reigned as master of the eponymous Zarelab. The complex, with twenty-five researchers, was housed in a corner of the palmy, well-trimmed campus of Stanford University, in the heart of Silicon Valley.

A world-renowned laser chemist—a "towering figure in the field of chemical physics," in the words of a Harvard Nobel laureate—Zare was known for devising ways to watch how molecules "dance," break apart, and recombine in chemical reactions. He had developed various practical tools and acquired almost fifty patents. He seemed on track to receive his own Nobel.

Zare had never been a Trekkie, a space cowboy, or a stargazer. Growing up in Cleveland, he'd thought that just finding the Big Dipper was enough of an accomplishment. So he was naturally amused when a close brush with a chunk of Mars turned out to be the thing that brought him his first Warhol moment. As he would write later, that tiny piece of Mars "fell into my life, and changed it forever."

Zare looked Mephistophelian, with his heavy, dark eyebrows and salt-and-pepper mustache and goatee. And he was driven—an overachiever, haunted by a painful childhood and a memory from when he was four years old of his grandmother telling him he was "going to be judged." He thought it somehow had to do with being Jewish, and representing his people well.

But a wicked grin softened the edges of his intensity. As a professor of chemistry, Zare carried to class his boyish sense of mischief. He had enlightened generations with his glowing-kosher-pickle routine—a demonstration that sodium could be used to generate light. *Physics Today* had devoted a cover story to his learned thoughts on beer foam.

Zare was also what he laughingly called a "big shot." Gifted at navigating the treacherous political waters of Washington, he was the head of the National Science Board, which set policy for the National Science Foundation.

(The foundation is a primary dispenser of funding for American scientific research, including the meteorite hunt and other Antarctic programs.)

A few years earlier, a fateful late-night encounter had changed Zare's attitude toward the cosmos and put him on course for his liaison with the McKay team and the secret project. He had worked so late that, as he walked home from his lab, he saw not a single car moving on the usually busy campus streets. Overhead, he watched the eucalyptus leaves tremble in a breath of wind and smelled a trace of their dusky scent. The air was so clear that the stars shone as if they'd been scrubbed and polished for some special occasion. He heard footsteps, and another solitary walker crossed to his side of the road. They fell in step and started chatting, telling each other about their work. The other man, it turned out, was a meteorite specialist named Peter Buseck, visiting from Arizona, who began describing to Zare the chemical messages contained in the rocks that fall from space.

Zare told Buseck he was intrigued. In his lab was an instrument that he was certain could help decipher those messages. He felt a powerful inclination to tackle the problem, even though he knew nothing about meteorites. (It had always been his tendency to ignore the arbitrary boundaries and labels that separated one scientific specialty from the next. They blocked creativity.) As usually happened when his mind was seduced by such a dare, the sensation on this occasion was almost physical, like a buzz in his head.

Early in his career, Zare had invented a laser device that could coax out the identity of a chemical substance even when only a few molecules of it were present. After his epiphany under the stars, he and his graduate students and postdocs started using the technique to poke around in meteorites.

One of his grad students, an industrious Englishman named Simon Clemett, grew particularly enthusiastic about the work. The experiments he developed worked so well that soon far-flung researchers were sending the Zarelab samples of meteorites, interplanetary dust, and other "cosmic schmutz."

Among those who had been shipping space droppings to Zarelab was Kathie Thomas, initially as part of her work on McKay's cosmic dust project. The Houston lab sent Zare samples not only of dust but also, for purposes of comparison, of meteorites from asteroids and in some cases from Mars. The people at Zarelab learned that only ten or so of the thousands of meteorites collected to date—the SNCs—bore the unique characteristics of Mars.

Thomas enjoyed her liaison with Zarelab. She and Clemett became friends. In the gloomy wake of Schopf's visit in January 1995, Thomas told her collaborators in Building 31 that she intended to call up her buddy Simon in California. "Let's just see if there are organics associated with this meteorite."

Romanek and Thomas had already sent Zarelab samples of the rock a couple of months earlier by Federal Express. The package carried sealed containers with two tiny chips of the rock (one about .08 inch, or 2 millimeters across, the other about .04 inch, or 1 millimeter).

But in keeping with their vow of secrecy, and in order to avoid any chance of prejudicing the results, Thomas and Romanek had not told Zarelab that the material was Martian. They referred to the chips instead by the code names Mickey and Minnie.

Accompanying the samples was a letter addressed to Clemett. It included crude, hand-drawn maps of Mickie and Minnie with instructions. "Minnie has a smooth face without any blemishes except for one prominent orange structure on the surface which is almost round. . . . If you can target the orange bleb . . . separately, that would be great." It went on, "When you have finished with Minnie move on to Mickey. He is much more complex but also more interesting." The letter concluded: "Good hunting!"

Clemett found the whole thing a little weird. In fact, the lab's interest in the chips, at first, "hovered between zero and negative some number." Mickey and Minnie sat on a shelf for months.

Now Kathie Thomas's pleas grew pointed and urgent. "We want you to look at these real bad. Real bad, okay?"

Clemett finally went ahead and did the analysis. Like Kathie Thomas before him, he considered these samples, tiny as they were, to be giant boulders compared to the incredibly minute specks he was accustomed to studying.

The Zarelab technique was revolutionary and dramatic—a combination shooting gallery and Pachinko machine. Clemett called it "chemistry without the chemicals." Chemists traditionally had to grind their samples to powder, or assault them with chemicals. Zarelab's technique harnessed laser beams to replace liquid solvents. This enabled Clemett and his coworkers to examine certain features in their native setting, detect much smaller traces, and eliminate much of the danger of contamination.

In simple terms, the device—a laser mass spectrometer, the most sensitive instrument of its type in the world—shot a laser beam at a precisely chosen target the size of a pencil dot on the rock sample, which sat in a vacuum chamber containing an electric field. The first beam heated the target at a rate of about 100 million degrees per second, but for only ten-millionths of a second—so rapidly that the molecules poofed, undamaged, into a cloud of vapor.

When Clemett pushed a button, the second beam—an ultraviolet laser that could punch a hole in human skin—stabbed across the cluttered lab unshielded (as he sat very still, careful not to expose an arm or other body part) and through the rising gas plume. This beam was sixty times brighter than the first but fired for only one-billionth of a second, giving selected types of molecules a positive charge by ripping off an electron. In a split second, the positively charged molecules accelerated into a negatively charged plate five feet away, the smaller ones outsprinting the fatter ones. By measuring the time of flight, the team could identify the molecules by their mass.

The laser technique could reveal details at much smaller scales than any system yet brought to bear on the rock—features so small they could be as hard for the human mind to grasp as the diameter of the entire universe. The team was analyzing down to billionths of a part per billion. (The McKay group in Houston referred to the Zarelab equipment as "the best gun in the west.")

The Houston team asked Zarelab to zap, and map, the mystery rock, moving from the outer skin (with the blackened fusion crust) down into the interior, including an area where one of these orangey globs was embedded.

Zarelab sent the results to Houston.

Kathie Thomas phoned Zare back, sounding very excited. "Richard, please drop everything else you're doing," she said. "We just FedExed you another sample from the same rock, this one code-named Goofy. I know it's asking a lot, but we need your analysis ASAP."

Zare balked. "You can call your rocks anything you want, but we're not analyzing any more of them until you tell me what the fuss is all about. What's going on?"

Thomas hesitated. If the new results hold up, she told him, "You have found the first known organics from Mars."

Then she said, "The fragments are from ALH-eight-four-zero-zero-one, a meteorite found in Antarctica a decade ago. We are confident the meteorite originated on Mars." Of the other known meteorites from Mars—by now there were twelve—this one was by far the oldest and contained a record from a time when the red planet was relatively wet and warm.

Another pause. She took a breath, and said, "We think the meteorite may contain evidence of life."

Zare froze momentarily, wondering if he might not be dealing with a nut-case here. But he was not completely shocked. Based on the nature of the evidence they were finding, he and Clemett had begun to suspect that this interstate dialogue might be pointing toward a biological signature—one from inside what they presumed to be some kind of space rock. Still, when Zare heard Kathie yoke together the words "Mars" and "life," he felt a chill run all the way down to his ankles. He listened as Thomas went on to tell him that her group had already identified some features in the rock that they suspected might be fossilized traces of ancient bacteria.

Now the Zarelab results had supplied a whole new line of evidence that greatly bolstered the case. No wonder the folks in Houston were excited, Zare thought. So was he! He ran up the stairs to the lab to tell Simon the news, but as he ran, he sensed little red flags sprouting up around the edges of his enthusiasm. This instinct kept him from getting completely swept away.

Zare knew that most people regard scientists as resolute skeptics. That was only half right, in his view. To be really good, you had to have a split psyche, had to believe in your work at every step while at the same time never believing in it at all. A complete skeptic would be terrified of taking a chance on anything short of a sure thing while, on the other hand, belief unfettered by doubt would leave you in dire peril of fooling *yourself*—a cardinal failing. Zare believed it wouldn't do for him to arm himself with just a small shot of skepticism, or a smidgen of belief. You needed both going full blast all the time.

As a boy, Zare would read under the covers with a flashlight—about chemistry. By age nine, he was catching buses to the library, alone, to read and play chess. In junior high, he challenged an inexperienced teacher who claimed to know "facts" that Zare understood were actually still matters for debate. The principal, desperate for relief from this upstart, arranged a scholarship to a nearby private school, where Zare blossomed.

By the age of twenty-four, the precocious student had a Ph.D. from Harvard. While there, he gravitated toward bold thinkers, joining a laboratory that was known at the time as "the lunatic fringe." (Zare's mentor there later won the Nobel Prize.) Zare worked at MIT, the University of Colorado, and Columbia before the West Coast lured him in 1977.

Now (in keeping with his rule of split-psyche zeal), having found possible evidence of the first sign of life beyond Earth, Zare put his team to work trying to prove that they had found nothing of the kind.

They had to check for the familiar villain, contamination. Mickey, Minnie, and Goofy had been transported around Building 31 from the upstairs curatorial lab, and had sat in little vials before they went to Zarelab. Because organic molecules are everywhere on Earth, it was possible the samples had gotten contaminated along the way—despite all precautions.

Back in Houston, the McKay group was humming with adrenaline, a blend of tension and excitement. When they had sent the samples off to the Zarelab, they'd expected that if there was any significant organic stuff to be found, it would be most abundant close to the rock's surface, because the organics would have penetrated inward from Earth's atmosphere or oozed in from the Antarctic ice melt as terrestrial contamination. They felt pretty sure that any such readings would decrease going in toward the heart of the rock.

They were stunned by the Zarelab results. Not only were organic compounds (molecules with a crucial carbon-hydrogen bond) present in the rock, but their distribution was the reverse of what the team expected. Simon Clemett and other Zarelab workers found the organics to be sparse on the rock's surface, but as they penetrated its interior, they found an increase in the abundance of organic molecules whenever they came across a carbonate moon. As they pushed past the carbonate, the abundance would go down again.

Zarelab produced a map of the rock's geography, showing that the organic molecules were concentrated in "hot spots" around the carbonate moons, in the regions where the magnetic crystals and other suggestive features were also thickest. This was consistent with the notion that the organics were byproducts of a fossilization process, whispers of long-dead Martians.

It was this spatial relationship between the organics and the carbonate moons that gave the Houston team goose bumps.

What were these organic molecules? The team members knew they were greasy hydrocarbons of a type found on Earth wherever life has existed. Called PAHs (polycyclic aromatic hydrocarbons), they could be formed by a variety of biological processes, like the ones that form petroleum and coals. They were part of the yummy black residue on a burnt steak. Most typically, they were the product of cellular decay.

The investigators also knew that the mere presence of the organics was not proof of life. This same substance could be created without a trace of biology, if things got hot enough. All you needed was a little "flame chemistry." The stuff could be found in the smoke from a cigarette, in car exhaust, in interstellar gases, and in other places known to be quite devoid of anything alive, such as cosmic dust and bits of an ancient asteroid.

In March 1995, Kathie Thomas and other members of the group attended a planetary science conference in Houston. They caught up with Michael Meyer, NASA's chief exobiology scientist. (Exobiology referred to studies of the prospects for life beyond Earth, and up to that point had been focused heavily on the study of the origins of life.) "Wait, wait, come here, we have some really interesting stuff!" they enthused. Meyer could tell they thought they had something pretty hot, but he cautioned them, saying that although the work was indeed interesting, "I'm skeptical, and you should be very careful. Because if you make an announcement about this, the press is going to run wild with it, and it doesn't matter how many caveats you put in."

In their presentation to the gathering, Thomas and the others described in general terms their discovery of possible organics from Mars. They acknowledged that they had more work to do to rule out the possibility that the organics had been added after the rock's arrival on Earth. The crowd agreed. The reaction was one of subdued interest. *The Houston Chronicle* and other publications reported aspects of the Zarelab findings, noting that these organics were sometimes linked to the presence of living things.

The two laboratories went to work developing a stringent set of tests, controls, and clean-room procedures. Zarelab then repeated the experiments.

Simon Clemett and his coworkers spent months in a slough of tedium, trying to see whether or not they could rule out every alternative explanation for the presence of the organic compounds other than that they had been put there on Mars. Zare knew they couldn't aim for complete certainty, since there was no such thing. But he wanted the maximum obtainable.

Zarelab acquired samples from other meteorites that had survived the same odyssey through the Antarctic ice as the rock from Allan Hills, the same laboratory processing, the same storage. The lab team did not find the same pattern in these other samples.

In the end, Zare and company agreed with the McKay group in Houston that they had a viable case to put forward.

• • •

For the McKay regulars, the Zarelab results marked a critical turning point. The high numbers on the organic molecules enabled the team to fill in and firm up their story line. The organics in the rock appeared to have come from Mars. They seemed to be of the same type and pattern as those produced by primitive microbes on Earth—and different from those measured in non-Martian meteorites.

Zarelab had found the first significant evidence of complex organic material from Mars—a major discovery by itself. It cast the 1970s Viking experiments in a new light and gave the first clear indication that the planet had the sort of chemistry out of which life could have arisen—whether or not it ever had.

At the same time, the McKay group was learning more about those spine-chilling little shapes and textures that so intrigued Romanek. (And they double-checked their findings by examining a second SNC—the Shergotty meteorite from Mars. Although they found no carbonate deposits, they did find in it forms similar to those nanofossil-like shapes they saw in the Allan Hills rock.)

In early 1995, just as Zarelab was injecting fresh oomph into the project, David McKay took over the heavy lifting on the scanning electron microscope. Chris Romanek had taken hundreds of pictures of the Martian samples, but in March that year, his fellowship at NASA ran out and Gibson was unable to hire him as a civil servant. Romanek left Houston to take a job with an ecology lab in South Carolina run by the University of Georgia, but he kept up consultations with Building 31 by phone.

Coworkers started to notice McKay working late at night in the e-beam facility—the first-floor complex that housed the various electron-beam instruments. To the concern of his wife and daughters, he would stay many nights until midnight or beyond, peering at the rock samples, too excited to go home.

He followed a familiar routine, moving efficiently among instruments and cabinets as he sat enthroned in a rolling lab chair. Wearing white latex surgeon's gloves, he started by giving a white porcelain specimen dish a blast from a can of compressed gas called Dust-Off. He fished around in a plastic bag for a stainless steel container with Teflon lids, which might hold roughly a gram of meteorite. He opened it and tipped out into the specimen dish a single chunk, along with a skittering of loose dust grains, and inspected the chip to see if he needed to break it more. Then he slid the sample under a small binocular microscope with a fiber-optic light beam. The crumb of meteorite looked like a small mountain sitting in the white porcelain.

Sometimes, if the sample came from close to the meteorite surface, he could see a patch of the black fusion crust left by the rock's violent passage through Earth's atmosphere. From time to time, he made notes in his lab notebook, where he kept track of all his work.

McKay picked up a vacuum tweezer and sucked up a grain or chip of meteorite dust—a tiny speck to the naked eye, the most negligible-looking bit of nature. Next, he prepared a stub, made of carbon, aluminum, or brass, on which to mount the crumb. He used pliers to pry the top off an epoxy container, then squeezed out dabs of resin and catalyst onto a big rectangle of folded aluminum foil, using a toothpick to blend them. With his eye to a binocular microscope, he could watch as he dotted the mixture onto the stub. Then he took the vacuum tweezer (after blasting it clean with Dust-Off), and used it to pick up a speck of sample.

The next step was to coat the sample surface, so that it would show up properly under the scanning electron microscope. Because the instrument used electrons, not light, to "see" things, the process required its user to coat each sample with a conductive material—McKay often used gold and palladium—to prevent an unwanted buildup of electrical charge that would fuzz up the image. He always tried to avoid a clumsy application of coating, which could create lumps—shapes resembling bacteria, for example—that might be mistakenly interpreted as native to the sample.

Finally, the sample was ready for the scanning electron microscope. McKay passed the bits of meteorite, now mounted and coated, through an airlock inside the microscope complex—an assemblage of cylinders coming together at angles. He could lean down and look inside, through a porthole, to see the sample sitting there, at the base of the vertical column that contained the electron gun. He manually adjusted the tilt and focus, turned the voltage up to 15,000, and then rolled his chair a few feet to position himself in front of the computer screen where the magnified view of the sample would appear. Staring at the screen, and the menus of possible adjustments, he used his left hand to tweak knobs on a control board or to scoot a mouse around on the pad, fixing the alignment or correcting a blur.

Sometimes, the image on the monitor would show the aperture itself—a round black tunnel entrance. Then, with some tweaking of the focus, McKay would "dive" through to hard magnification, down the rabbit hole to the alien target terrain.

He used an X-ray device to study the composition of the material in the beam. When the device was on the "compo" setting, a spectrum-analysis graph appeared on the screen: bright red vertical lines—like flaming prairie grass growing along the bottom of the image—set against a brilliant blue field. The heights of the grass blades showed the abundances of various elements.

But the unpracticed eye could be misled. McKay knew that the spectrum was only partly due to the composition of the material under study. It was also, in part, the result of the geometry of the instrument setup, and the signal would get absorbed to a greater or lesser degree depending on how the sample was tilted.

On-screen, the scan appeared as a faint line moving down the image, like a veil slowly falling away. McKay liked to take a slower picture than most. He used the mouse to change the scan rate. When he was ready to capture an image, he punched a button labeled "Freeze," the equivalent of the shutter button on a normal camera.

It was possible to go into a kind of reverie, working in the dim glow of the monitor, enveloped by the white noise of humming machinery. Hour after hour could pass, each moment full of anticipation for the next scene. Sometimes, McKay played music—Ireland's ethereal Enya or other Celtic sounds, or a piano accompanying the sigh of spring rain.

McKay started "seeing things." He glimpsed shapes that were just beyond the vision of his microscope, which was some fifteen years old and essentially obsolete. He felt frustrated. He knew that the engineers over in Building 13 had acquired an improved version. In late 1995, he talked to the branch chief there and worked out a time-share.

He had to compete with the engineers in the other building—all spit and polish compared to his cramped and threadbare complex—for microscope time. The engineers tried to be accommodating, but they were reluctant to let McKay and his team use the facility after hours or on weekends, citing security and safety concerns.

This convinced McKay to press for a similarly advanced instrument for Building 31, although it would cost more than half a million dollars—about triple the cost of his current scanning microscope. He would succeed, but not for two more years.

The instrument McKay coveted was a rare state-of-the-art electron microscope with a field emission gun, which fired a tighter beam. It was a brand-new technology, and the space center was one of the first places that bought one. The engineers used it to analyze cracks and other flaws in space shuttle flight hardware. Because the field gun generated an enormous beam of electrons (and the more electrons you fired, the smaller you could make your target), the instrument allowed you to see extremely fine details. It was a slower process than McKay's microscope. Both operated at vacuum, but because you had to get more of the air out, this one required considerably more time pumping down the chamber.

Still, depending on how well the advanced instrument was tuned, McKay (the master tuner) could see things three or four times smaller than with the old microscope he had been using. Very nice, he thought.

McKay was among the first to use this technology for geological study, let alone biological investigation. Because this Martian meteorite was one of the first rocks ever looked at with this instrument, there was no record with which to compare his observations—no database of terrestrial rocks, or lunar rocks, or other meteorites. No way to know whether these features were common or rare. Some people would later contend that because McKay was treading such new ground, descending into a scale of smallness that was unprecedented in such a context, it was all too easy for him to misconstrue the meaning of what he saw there.

When he looked at the rock samples, McKay saw complicated forms, a lot of very fine textures. He could see variations of both the chemistry and the textures over a few hundred nanometers. And he was constantly thinking, as he always did when looking at a body of microscopic evidence, "What is this stuff? How did it form?"

One night in January 1996, Gibson joined McKay as he worked late with the powerful field gun in Building 13. McKay was zooming in on the black part of an Oreo rim when he saw something that jumped out at him. He moved aside so that Gibson could take a look. "Wow!" Gibson murmured excitedly. "What in the world . . ." They saw a shape that took their breath away. It looked segmented, like a Tootsie Roll or a worm.

The evidence had been mounting steadily. The team had begun to put all the pieces together by now—the greasy PAHs, the complex chemistry of the carbonate moons, the magnetic minerals, the suggestive mineral shapes. Now this. They were peering into a mineral landscape so incredibly small that the scale alone made it alien territory for the human eye. They were able to see details as small as 30 nanometers. The "segmented structure" (as they would call it) might have been a million times smaller than an ordinary bacterial cell. And yet—it looked more like the fossil of a long-dead Martian than anything they had yet seen in the rock. They looked at each other. This could be . . . something.

Over the succeeding months, McKay and Gibson would spot other shapes that they thought resembled Earth microbes—tubular or egg-shaped structures, or sometimes what looked like a school of microbes frozen in tandem undulation. They ranged in size from one-hundredth to one-thousandth the diameter of a human hair.

But this one would haunt them. After their spontaneous reaction wound down, McKay and Gibson set out to methodically document the size, location, and other aspects of this . . . segmented structure. Gibson would store the sample in a safe in his office.

That night, when they first saw the shape, Gibson in particular felt awestruck. When he got home his wife, Morgan, asked, "Is something wrong? Something at work?" He put her off, wanting to save it for later, after he had digested the implications. He had trouble sleeping that night. He would tell a vast audience about it the following year, saying it was "undoubtedly the most exciting thing I've done in my twenty-seven years as a scientist. I have to admit it does beat Apollo and the excitement there, and that was tough to do."

Some days later, he laid the photograph of the "worm" down on the kitchen table casually, without saying anything, and his wife, with a background in biology, remarked offhandedly, "Hmmm, that looks like bacteria."

The team members began to ask, "Why don't we write it up?" "Isn't it time?" In early 1996, they decided to publish their results.

The little band knew that they had not nailed their hypothesis definitively; nor were they sure that they ever could. They had no way of knowing even when they would take another major step closer to that goal. Although there were gaps and weak points, the team felt strongly that they finally had the underpinnings of a plausible, even compelling narrative: Once upon a time, tiny Martian organisms swam in fizzy Martian waters, their life (and death) processes spawning the combination of bug shit and slime and decaying matter and other stuff entrained in the carbonate globules as the water receded or evaporated, depositing these blobs in subterranean fissures in the rock.

The orangey carbonate blobs must have formed on Mars, because they were cracked in ways that could not have occurred when the rock hit Earth. The carbon in them bore the distinct isotopic signature of the Martian atmosphere, and the oxygen isotopes suggested the carbonates had been deposited in warm, wet conditions. The Oreo rims bore the mineral signatures of both oxidization (including rusting) and its opposite (known as reduction). The formations would require extreme conditions if produced nonbiologically, but were common on Earth as a consequence of biological activity. When the hypothetical organisms died, their decomposition could have produced the greasy organic residue—the PAHs. And in some cases (perhaps), just as fossils form in limestone on Earth, their remains could have been sculpted in those provocative mineral shapes, visible only under a powerful electron beam and eerily similar (the team thought) to the fossilized remains of Earth bacteria; even their incredibly small size was almost the same as those controversial, hypothetical, but provocative terrestrial nanofossils.

But there was no smoking gun.

The McKay team knew there were other plausible story lines for the rock that did not involve biology. They found the biological explanation by far the simplest and most elegant and therefore the best. It was Occam's razor.

At least some of the people whose names would be on the paper—there were nine—had been feeling the emotional shear of almost unbearable bottled-up excitement alternating with a kind of gut-fluttering dread. This would not be the first time somebody had announced the discovery of possible signs of life in a meteorite. And on each earlier occasion, the evidence had turned out to be some kind of Earth contamination or, worse, an illusion born of unchecked desire. The stomping grounds of science were littered with the embarrassed and ruined career carcasses of those who had dared to make claims about extraterrestrial life, or about ancient Earth fossils. The collaborators knew they might well reap a harvest of humiliation from the very people whose good opinion they most coveted.

Or their work might change the way people regard their place in the universe.

At least by publishing, they could end the secrecy and move on to the next essential step in the dance: a squad of their most skilled counterparts, many of them rivals for the same prize, would get a crack at proving McKay et al. wrong.

The secrecy itself was becoming unpleasant—and impossible.

McKay and the others had several reasons for hiding their results. They knew that dozens of other groups were studying pieces of the rock. They didn't want to start talking openly about what they were seeing, thereby attracting others to follow the same line. They didn't intend to get "scooped," as McKay put it.

Then there was the turf problem. Within the NASA family, Ames Research Center in California, not far from Stanford and Zarelab, was the traditional stronghold for research on extraterrestrial life. The tack the McKay group was pursuing in Houston was a cross-border intrusion. The McKay group wanted to marshal as much evidence as current technology allowed before they showed their hand.

They also knew that if they let the putative aliens out of the bag prematurely, they would blow their chances of getting their paper published in *Science,* the most prestigious platform for announcing their claims. The journal famously—and to the outrage of some scientists—insisted on exclusivity.

In Palo Alto, the ever-ebullient Zare found the ordeal of silence excruciating. When he was involved in something this exciting, he told the world. But, for once, he thought the stakes were high enough to justify a different strategy.

He did occasionally try to communicate some of his feelings about being part of all of this to a few trusted confidants. He tried to talk it out with his wife and daughters one night over dinner. He sputtered on for a while. "Okay, Dad," sighed young Bethany, unimpressed. "Could you just pass the ketchup."

For the group in Building 31 in Houston, their clandestine behavior was getting so obvious it was straining their relations with coworkers. Things were getting awkward. At this point, they had shared the information with only the minimum number of people they deemed necessary and politic. A year earlier, they had briefed Carolyn Huntoon, director of Johnson Space Center, and two or three others in the complex. Huntoon was a biochemist, and after seeing some of the images, she thought the whole idea was really "out there." Knowing McKay and Gibson to be men of integrity, she told them to keep working—but she also directed them to seek outside expertise. After all, she knew, anybody could overlook something, add two plus two and get five.

But the group had not told even colleagues who worked essentially next door. They tried not to make a big deal out of it. When developments warranted, the four of them—McKay, Gibson, Thomas, and Romanek—would meet in one of their offices to talk. If someone wandered in, they would simply change the subject. They kept track of their work in lab notebooks and, when nearing publication, tried to make sure not to forget telltale papers in the copy machines or anywhere else. When leaving their desks, they would turn those papers over. There had always been an atmosphere of trust along these corridors.

Outside the cone of silence, coworkers around them at the space center began to get suspicious and resentful. There was a mounting buzz around the project. Everybody knew that there were signs of water flow in the rock, and that the rock was extraordinarily old and had other special properties. As a visiting scientist working near the group would say later, he considered it "the worst-kept secret at the space center." As you walked by Kathie Thomas's desk on any given day, there would be a stack of papers lying there on topics such as biological interactions with minerals, or microbial life. And some colleagues could not help noticing the visit by Bill Schopf, god of the Precambrian, and the huddles with Vali, an expert on microbial life. So there was gossip that the group might be poised to make some kind of public gesture toward the possibility of biology on Mars. Almost no one anticipated that it would be the dominant thrust of their analysis.

An odd breach almost occurred in the spring of 1996. At the Kinko's a few blocks from the space center, employee Erin O'Shey, twenty-six, was jarred out of her sleepy routine one day when a man and a woman came into the shop with a half dozen photographs they wanted to photocopy in color. They seemed nervous and unusually concerned about the quality of the reproductions and kept all the rejects in hand, instead of throwing them away.

As O'Shey stole glances at the pictures, she thought they seemed pretty routine—enlargements of some kind of microscopic organisms. Then O'Shey noticed the words printed below the images: "Search for Life on Mars." She asked the woman if somebody had found Martians. The woman said, "We don't know," and wanted to know why O'Shey had asked. O'Shey pointed out the caption under the pictures and the woman replied, "Oh, yeah"—as if she had messed up. The woman then asked O'Shey to keep all this to herself. O'Shey agreed—but managed to make some extra copies of the images on the sly.

The space center workforce always provided a steady stream of customers, and these were not the first NASA pictures O'Shey had pinched for herself, thinking they might be something "cool." Through friends, the Kinko's worker soon mailed copies of the pilfered pictures to a leading advocate of the theory that an alien civilization had sculpted a giant face on the surface of Mars and that there had been other signs of extraterrestrials covered up by the government, in the mode of Fox's *The X-Files.* The recipient—the conspiracy theorist—would ignore O'Shey's pictures until the day the news broke worldwide. Then O'Shey would learn that her woman customer was Kathie Thomas. Once word of O'Shey's shenanigans got around NASA, via the Web, Kinko's fired her—collateral comeuppance from a rock that had augered into the Antarctic 13,000 years earlier.

The tensions between the clandestine operatives of the McKay group and their broader community of interest crackled and thumped like distant lightning strikes all through the first half of 1996, most visibly at a series of tribal events.

In the run-up to a March conference in Houston, for example—a major annual gathering of lunar and planetary scientists—the McKay team initially indicated that they would make a presentation on the magnetic crystals. They belatedly decided they'd better not. Gibson would say later that the thinking was "Someone might ask a question that would force us to give away our major finding. . . . If you were on to the biggest scoop of your life, would you tell people?" They withdrew the paper.

At a gathering on meteorites in Berlin, no representative from the McKay group showed up. This offended some who were there, who had heard the buzz and looked forward to seeing the team stand up before an audience of their peers who could give the claims a fair evaluation and a dose of constructive criticism. They wondered why the team was hiding.

Later that summer, the Lunar and Planetary Institute near the space center was the scene of a lunch seminar that brought together in a cramped room some sixty or so researchers to hear three colleagues describe their work on the Mars rock. One of the speakers was Chris Romanek. The scientists took turns describing what they had found and debating their conflicts. Soon, attendees began to catch on that Chris Romanek (back in town on sabbatical from his new job) knew things he was refusing to tell. He kept saying, "I can't answer that." He was withholding evidence! The lunch turned abruptly into a free-for-all, and people started to bash Romanek in "come clean, you dirty rat" tones: "Why can't you tell us?" "This is so unprofessional!" The crowd eventually calmed down, but the incident, again, left a residue of bitterness toward the McKay group.

What no outsiders realized at the time was that *Science* magazine had just accepted the McKay Nine's paper for publication. The magazine's rules strictly prohibited them from saying a public word about it before the formal announcement. Some scientists vehemently objected to this prohibition, in general, as infringing on the free flow of scientific information, but the McKay group believed it helped keep the process disciplined and responsible, preventing a free-for-all in the popular media, where inaccuracies could proliferate.

Throughout the first half of 1996, the McKay group was consumed with the preparations for going public. They started by thrashing through a series of disagreements over how to word the paper they would submit to *Science.* Gibson wrote the initial draft, but none of them was satisfied with it. As Gibson was the first to admit, it was a little too bold.

At some point, the group decided that McKay, who was not only cautious but also the senior scientist, would be the lead author.

In their line of work, being the first to publish a paper reporting a new advance, when it led to a succession of independent reports, was a source of prestige and authority—the bigger the discovery, the greater the rewards. And when there were multiple authors, the first name on the published work was generally assumed to have made the largest contribution to the project. Sometimes this was a close call, and other factors (such as seniority or reputation) would weigh on the scales.

The principals in the McKay group liked to say that they were not much for hierarchy. They were equals.

• • •

They reworked the manuscript. Messages and drafts flew between Houston and Palo Alto. Zare led a contingent from Stanford to spend a weekend in Houston. They went over the points they needed to clarify, the revisions they had to make. And all of this was just the first round of revisions.

In the abstract, or short summary, written for the *Science* manuscript, for example, one Stanford editing notation said of the final sentence: "Too strong! Suggest softening." On the other hand, it suggested strengthening the role of the organics discovered at Zarelab: "Isn't statement that these are the first organic molecules discovered from Mars important enough to put in abstract?" The group finally hammered out a written summary of the data and a hypothesis (what a trial lawyer would call the theory of the case) that they could all live with.

On April 4, 1996, the McKay team shipped the paper to *Science*, with the bland title "Search for Life on Mars: A Study of Martian Meteorites." The journal's editors treated it like the hot potato it was. They limited knowledge of it to as few people as possible, while subjecting it to intense scrutiny by authoritative outside experts—including the ill but alert Carl Sagan.

Suffering from myelodysplasia, a disease of the bone marrow, Sagan had just returned home from an operation when he learned that NASA scientists might have detected fossils in the unusual rock from Mars. Two years earlier, Sagan had written about the failure to find evidence of microbes in any of the Martian meteorites, adding pointedly and optimistically: "so far."

Sagan was the most famous and charismatic scientific personality since Einstein. He had mastered physics, genetics, chemistry, and geology, as well as astronomy, in the service of his lifelong investigation into the potential for life beyond Earth. He was the romantic bard of the cosmos—but he was also an aggressive champion of healthy skepticism, of the scientific method, and of the need for scientists to communicate more clearly with the public. It is easy to imagine these alter egos collectively relishing the news about the Mars rock and the range of possibilities, even as the man wrestled with his own mortality.

All through the summer, the McKay team responded to questions and provided rewrites. The process took an unusually long time.

Most of the assessments by the referees reflected the complexity of the topic and the number of specialties involved. The reviewers would come back with comments such as "In my area of technical expertise, with the following little corrections, I find the work to be reasonable. But there are areas in the paper I am not an expert in." The editor of *Science* made his own additional suggestions for changes. Among other things, the initial paper was much too long for the magazine, at eighteen to twenty pages of text plus another eighteen tables and images. It had to be cut to about one-third of that.

The McKay team made all the requested revisions. They knew from the comments of the reviewers that the tiny wormlike and egg-shaped features proposed as microfossils would likely be the weak link in their skein of evidence.

On July 16, 1996, McKay got word that the paper had passed muster. *Science* would publish it in August. This was it. The team was about to take the leap into the light. They thought they were firmly braced for what would come.

Whenever he paused to consider the probable public reaction, McKay concentrated on the tribe: his peers, experts who would pore over the details in the paper. He had evaluated the weak and strong points of his argument from that perspective, tried to guess who would attack and with what kind of firepower, as well as who would support him. He figured the story would probably make the newspapers but not be a huge deal.

McKay had read in a self-help book that when taking a risk, it is useful to have the support of a sponsor or colleague—someone with the stature to defend and support you. To fill that role, McKay came to regard Dick Zare as ideal—someone whose reputation was impeccable, who (it was said) was short-listed for the Nobel. McKay felt pleased and somewhat comforted to have him as a coauthor.

Once *Science* accepted the paper, life around Building 31 got even stranger. McKay and Gibson disappeared from the premises for days and came back wearing little beepers and looking drained. Word was they had been to NASA headquarters in Washington, talking to the brass and getting some kind of naked-lightbulb treatment over their impending publication.

The paper's acceptance triggered another round of dueling rewrites, this time between public affairs people at the Houston space center (and later at NASA headquarters) and Stanford University.

James Hartsfield, Johnson Space Center's point man on the Mars rock story, considered his involvement very likely the most important thing he would do in his life. When he got the assignment, he worked virtually night and day for three days preparing an initial draft of the press release that would herald the paper's public debut. His role as he saw it was to tell the story in simple terms that a broad audience could understand. The editing suggestions that came back from David Salisbury, his counterpart at Stanford, seemed to him overly academic and complicated.

Salisbury, for his part, thought the NASA folks tended to sensationalize the research a bit, overemphasize the microscope images, and underplay the role of Stanford. In an early draft of the press release, NASA called the evidence of possible primitive life on Mars "compelling." Stanford wanted to soften the language to say the evidence "strongly suggests." By July 29, Salisbury had replaced that with "compelling circumstantial evidence."

NASA wanted to single out Zarelab's detection of the PAHs as "most important." Stanford preferred to emphasize that all the lines of evidence fit together to make a strong case. NASA proposed that they say: "Some remains of the microscopic organisms were fossilized in the carbonate, in a fashion similar to the formation of fossils in limestone on Earth." Stanford objected: "Too positive about the fossils, need to be consistent in describing these forms as possible fossils." And so it went.

One torpid July day, Chris Romanek found himself lying on a Galveston beach drinking beer with friends from his days at Texas A&M—structural geologists working for the oil industry. He showed them the secret paper that would shortly become public. They were incredulous. "Chris, do you know what you're saying here? I can't believe you're going to publish this." The attitude was: "My buddy Chris is writing a paper saying he discovered life on Mars?!" Romanek was not exactly one of the world's premier specialists in the field. Still, neither Romanek nor his friends thought anybody would make a big deal of it.

Back at the space center, the tensions between the anxious collaborators and their excluded, suspicious coworkers sparked again on one of the dog days of early August.

On this steamy high noon, a group of scientists and students from Building 31, including two from the McKay group, were heading out for lunch. Visiting scientist Ralph Harvey noticed a sign explaining that NASA was planning to fumigate the building on such and such a date because of a persistent cockroach problem.

Harvey turned to Kathie Thomas and said, with mock alarm, "Oh, my God, is this your fault? Did you let something get out?"

The whole crowd laughed, except the McKay duo. Seeing the tense looks on their faces, Harvey wondered fleetingly whether they were peeved—or scared.

CHAPTER SEVEN

THE GRAND INQUISITOR

DAVID MCKAY AND Everett Gibson sat in leather wing chairs big enough to make almost anybody feel small. It was Wednesday, July 31, 1996, and they were on the top floor of NASA's Washington, D.C., headquarters, a few blocks southwest of the U.S. Capitol dome.

Dan Goldin, the NASA boss, glared across his desk as he pounded them with yet another question in his muscular South Bronx baritone. They had originally been scheduled to meet with him for thirty minutes. Instead, he had been grilling them for three punishing hours. He was the Grand Inquisitor of Space, his eyes glinting with a sinister light—or so he wanted them to believe. Goldin had spent the previous day at the White House, going over the implications of the claims about the meteorite from Mars with the president and vice president. The reactions there had excited Goldin, made him feel in his gut what he already knew in his head to be the true dimensions of the looming revelation.

From the moment his lieutenants had informed him that NASA civil servants in Houston were proposing a hypothesis about possible life on ancient Mars, Goldin had been mindful of its explosive aspects, its "giggle factor" potential, and the fact that this was happening in an election year. He had decided it was crucial for him to hear the whole story from the instigators themselves, face-to-face and in detail.

In addition to understanding a thing or two about rocket science, Goldin had studied management techniques and prided himself on his ability to shove people out of their psychological comfort zone. He would subject his targets to a flamethrower blast of intimidation—the Goldin Grill—designed to test human mettle while demonstrating who was in charge. Most people caved or, as he put it, "wiltered."

On this occasion, Goldin had taken the scientists from Houston on a methodical intellectual odyssey back to first principles; he pushed to know their thinking, their logic, all the steps that had led to their astonishing claim. He would end up with twenty-seven pages of scribbled notes. And in his head, he was silently grading the answers they gave him—their delivery and tone as well as the quality of their evidence. Goldin had been "staring into them," as he would say, trying to penetrate their souls to see what stuff they were made of.

Goldin was aware that he might soon be sharing a stage with them as they confronted a global audience and the clamoring media. His reputation and his credibility, as well as NASA's, would be on the line along with theirs. He had to know who they were, and how they had reached this point.

Now, after three hours of the Grill, he was satisfied. These folks were holding up.

Goldin was especially taken with the magnetic crystals. He had done a lot of work with magnetic materials in his earlier days, and this was where he thought the team demonstrated real strength. He probed intensely on that one, giving that argument an "A plus."

As the room emptied, he took more time to let it all sink in. If McKay and the others were wrong, he reminded himself one more time, a misstep could shatter the space agency's credibility (and his), almost as severely as an exploding space shuttle.

And if they were on to something real? The implications were almost beyond imagining; so were the opportunities. Almost.

For Goldin, the issue on the table that day went deeper than policy, politics, or personal ambition. McKay and Gibson had touched a theme that triggered something akin to music in his romantic soul. At age fifty-six, he was feeling the resurgence of a longing that had its roots back in his childhood, in one of those indelible emotional attachments to a waking dream.

Daniel Saul Goldin grew up in a house across the street from Public School 93 in the South Bronx, in a working-class neighborhood of Jewish and Italian families. By the time he walked into first grade there, Flash Gordon and Buck Rogers had captured his heart. At the Wood Theater on Westchester Avenue, he sat in the Saturday dark and watched Flash (Buster Crabbe) struggle against the alliterative Ming the Merciless on the planets Mongo and Mars, and he followed the adventures as Buck (again Buster Crabbe) blasted off to Saturn for reinforcements in his fight against the tyrant Killer Kane.

His course through life hinged upon a flaw in his own eyeballs, an elongation. He had progressive myopia, and a doctor had told his mother that if Daniel played contact sports, he could go blind. At times, Goldin felt he must have been born wearing eyeglasses. But however devastating the affliction had seemed for a rambunctious boy, his grown-up self concluded that it had been a lucky break. Goldin felt certain that his life was richer as a result. While the other kids were shooting baskets across the street on the school court, he was indoors reading, or listening to Milton Cross and the Saturday afternoon broadcast of the Metropolitan Opera, or going to museums. He had also found a talent for building model airplanes.

As the Cold War lit a fire under the American space program, the myopic boy from the Bronx went on to study mechanical engineering at Hunter College, part of the City University of New York. In 1962, when he was close to graduating, a recruiter from NASA's Lewis Research Center in Cleveland came to the campus and hooked Goldin by displaying some pictures of an advanced spacecraft design intended to go to Mars. To go to Mars! It was his childhood fantasy coming to life. Equipped with a bachelor's degree (the highest degree he would formally earn), he went to work for NASA at about the time John Glenn made the first U.S. manned orbital flight.

The young engineer, with his penchant for long hours and an exhaustive pace, was soon enticed away from NASA by the aerospace company TRW, near Los Angeles. By that time, Goldin and his wife, Judy, had two young daughters.

As Goldin rose through the company ranks over the next twenty-five years, he disappeared into top secret weapons programs. (He wasn't all work and no play, however: he kept a surfboard in the TRW lab for lunch-hour outings to the beach.) Along the way, he built a reputation as an efficiency expert. As his company worked on NASA projects, he developed passionate opinions about what was wrong with the American space program and how to fix it. For example, his team planned to make a public presentation on TRW's concept for a new line of small, lightweight spacecraft. Before the event, one of his superiors got an angry call from a NASA official who warned that the TRW proposal threatened vested interests—builders of the massive, instrument-packed billion-dollar babies then favored at the space agency, big-ticket research spacecraft to which Goldin and others referred, derisively, as Battlestar Galacticas. These galoots could take a decade to build and launch, eating up whole careers before a payoff arrived in the form of data flowing back from space. Galacticas were so huge and expensive they could not afford to fail—yet fail they sometimes did.

Goldin let it be known that he was infuriated by that NASA reaction. He was certain the agency's approach was bad for the space program and bad for the country. It was one more sign that, during the 1970s and 1980s, the once agile outfit that had sent men to the moon had hunkered into the hidebound and defensive mentality of a bureaucratic fortress.

For Goldin, this decline was about more than just rockets and Buck Rogers. To him, the American effort to explore space was a kind of mine canary for the national character, and he saw both of them faltering, becoming fainthearted, risk averse, mired in the drive to "survive and consume" instead of imagining, creating, and building.

His prescriptions for change—delivered sometimes in fiery public lectures—impressed key officials in the White House–led National Space Council. The first President Bush fired Goldin's predecessor and, in April 1992, appointed the fiery reformer as the ninth NASA administrator. To Goldin, it felt like poetic justice.

What happened next created the atmospherics, the context, into which McKay and Gibson would be lobbing their geochemical bombshell.

Many people thought Goldin was wading into professional quicksand at NASA. In 1991, the Soviet empire had dissolved, and along with it the last of the Cold War rationale that had sustained the struggling space agency. At the same time, space leaders were limited in their ability to push for new projects because of a cap on federal spending.

These external factors aggravated NASA's managerial shortcomings and contributed to the succession of problems and failures, which included the flaw built into the Hubble telescope, the loss of a billion-dollar spacecraft at Mars, the political fumbling of Bush's proposed human-exploration initiative, and the congressional threat to kill the planned space station. NASA had become a political embarrassment.

NASA was unusual in government, in that its charge from Congress and the White House was to enhance American prestige in the world by doing dazzling, difficult, and dangerous things—particularly difficult and dangerous things that no one had ever done before—and on top of that to do this high-wire act in full public view. This agency's job, in short, was the antithesis of the bureaucratic impulse. In addition, since the Apollo glory days, NASA had been expected to do this work on a budget that was at once both shrinking and unpredictable.

The new man arrived in Washington with a strong mandate from the White

House space council to get the mess in hand, to make the space bureaucracy responsive and effective again.

Goldin's appeal was not limited to the Republican White House. During his Senate confirmation hearings, then-senator Al Gore, Democrat of Tennessee, said, "I detect a backbone in this nominee." In 1992, after Bill Clinton and Gore took the White House, they would keep Goldin on.

Announcing himself as a man on a mission and an idealist, Goldin walked into the morass with "so much brass he clanked." Goldin didn't look like NASA. He wore bespoke suits, with flashy modern tailoring, expensive ties, and, sometimes, cowboy boots. Goldin didn't *sound* like NASA. The agency's previous leaders had tended to communicate in jargon, often with the leisurely accents of the NASA "crescent," the sun-belt swath where the major NASA facilities were hammocked between the Atlantic and the Pacific.

Not Goldin. Not only did he have the Bronx-Pacino patois (with a half cup of actor Paul Reiser in the delivery); he showed his emotions and pushed hot buttons. He got right up in people's faces. His public talks could veer from chilling warnings aimed at balky contractors to ruminations about the "neo-Malthusians" who saw forward momentum on some key space project as nothing but "pork," to flights of fancy about a future in which we could "turn over the keys to a real Starship Enterprise" to our grandchildren.

At the beginning, as the *National Journal* put it, the "crew-cut, pocket-protected old cranks" at headquarters snickered indulgently. Here came this wide-eyed outsider with a briefcase full of trendy management theory ("If you can't measure it, you can't manage it"). How far could this guy get?

Driven by his fine-honed sense of personal outrage, the new chief proved to be both riveting and frightening. He made decisions, took risks, picked a direction, and charged off. He stood out in the Washington ecosystem for his absolute willingness to make enemies. "Leadership is not being loved," he would say. "Leadership is doing the right thing."

At the outset, Goldin adopted a warp-speed schedule. He would often begin at four-thirty A.M. or so with a workout on his eighteen-speed ultralight Klein bicycle over the bike paths that curved along the Potomac near the Watergate, where he had taken an apartment. He would show up in his office in the predawn darkness and work until eleven P.M., arriving just as early the next morning, with notes or instructions for his staff that he had prepared overnight.

Goldin plunged into a total-immersion education on his new empire. On Saturdays and Sundays, he would summon experts to brief him on its various arcane activities. In public outreach, he came on like the Elmer Gantry of space, visiting sixty cities in two years, holding town meetings, and talking regularly to schoolchildren.

Now, after four years on the job—to the relief of some and the consternation of others—this volatile personality had conquered the NASA bureaucracy in a way that Washington had seldom seen. Goldin made the new NASA a model for the Gore-led campaign to "reinvent" government, and Goldin became a poster boy for can-do management.

But many people despised him for what he had done, or for the *way* he had done it, especially people whose work took them inside what might be called the Goldin "blast danger zone." Goldin believed that a measured ration of chaos was a useful tool for bonding a team as well as for discombobulating potential adversaries whose fiefdoms—"rice bowls" or "sandboxes" in NASA-speak—were to be dismantled or shaken up under the new order. Captain Chaos, some called him. He introduced a system of documenting procedures that, some said, made employees' lives miserable and was too rigid for an innovative agency. Some who got the Goldin push out of their jobs, and others as well, maintained that the administrator could be brutal, and that he lied, or misled, sometimes doing the direct opposite of what he had promised. Wes Huntress, years later, would compare him to both the good guys and bad guys in the *Star Wars* movies, saying, "Dan the man was a Jedi. Dan the administrator was a Sith."

Goldin had waded into the bureaucratic equivalent of street brawls over issues ranging from engineering philosophy ("lose the fat, keep the muscle") to a crusade for hiring women and black people. He asserted that the culture of space had become "too stale, male and pale." He sought out risk takers, "non-linear thinkers," those in whom he found intellectual brilliance and a "fire in the eyes."

He dismantled the main headquarters science office, got rid of about half the directors at NASA's centers across the country, and put NASA's clubby aerospace contractors on notice that the tradition of cost overruns (typically about 75 percent higher than contracted for in those days) would no longer be tolerated. He railed against the "iron triangle," a phrase used to describe the Washington-centered alliance of federal bureaucrats, aerospace lobbyists, and congressional staff who conspired self-servingly to thwart the desires of the White House and the tax-paying public.

Goldin and his top lieutenants cut the space shuttle workforce by thousands of jobs while maintaining the same rate of launches. This prompted charges that he was jeopardizing safety, even though he was personally avid in his advocacy of safety and prudence. As for the embattled space station project (the supposed destination for the shuttle), Goldin had shaken the aerospace world to its struts by staking his reputation on a radical overhaul that he vowed would finally break the project free of bureaucratic gravity and into orbit at an acceptable cost.

Veterans at Johnson Space Center in Houston rolled their eyes when Goldin began to "meddle" with their preparations for a daunting shuttle mission in which astronauts were to install a complicated set of instruments to fix the Hubble Space Telescope as it orbited Earth. Goldin ordered up unprecedented numbers of equipment tests, imposed record hours of crew training, and marshaled layers of technical reviews by outside experts. Some of the same NASA veterans who criticized him for these "distractions" later expressed grudging respect when the 1993 repair mission turned out to be virtually flawless. It yielded what was arguably the agency's most spectacular success since Apollo—a restored Hubble that would keep churning out revelations and iconic images of the universe for more than a decade.

In August 1993, the $1 billion Mars Observer (in the pipeline before Goldin's advent) vanished just as it arrived at the red planet. The loss provided fodder for Goldin's anti–Battlestar Gallactica approach, which others had termed "faster, better, cheaper" (or a common variant, "smaller, faster, cheaper"). Instead of pinning the agency's hopes on those huge-but-rare missions, Goldin decreed that program designers must take advantage of the latest developments in the rapidly evolving microelectronics industry. NASA would launch more spacecraft, more often. There was to be a flotilla of innovative missions that would study Earth as well as the rest of the solar system, with tightly constrained budgets and disciplined managers.

Under Goldin, taking risks was back in style; failure *was* an option—as long as it didn't involve incompetence or fraud. Failure henceforth (under the Goldin doctrine) should reflect a well-executed effort to push the envelope. If one of these little suckers failed because it was groundbreaking, well, it would be a learning experience, and that's what NASA was supposed to be all about.

At the time of Goldin's meeting with McKay and Gibson, the agency had a series of Goldin-style missions in development, many of them part of a revamped Mars-exploration program. One was an experimental robotic package scheduled to bounce onto the surface of the red planet on airbags—airbags!—a year later, on the Fourth of July 1997. Much of the buzz was about how it would probably become another NASA fizzle. Its name was Pathfinder.

Goldin acknowledged that in his initial zeal, he had sometimes hurt people unnecessarily, and he regretted that. "Guilty as charged," he would confess years later. "But you don't learn if you don't try."

Seemingly fearless on the surface, Goldin actually felt driven by some combination of fear and ambition. He felt his decisions would shape U.S. space exploration for decades to come. At times it seemed the whole program could go down—and it would have happened on his watch.

He had another motivation. A shameless romantic since boyhood in his enthusiasm for space adventure, the tough guy from the Bronx still considered the human drive to leave the home planet a "biological imperative—wired right into our DNA." This was where his heart was. His fondest dream, he would tell people, was for his grandson to walk on another world someday. Goldin wanted Mars for young Zachary Michael and his generation.

Goldin had kept up with the new picture of the natural world that was coming to light, the same revelations that had emboldened Chris Romanek to think his hush-hush radical thoughts about the carbonates in the rock: living things could thrive in extreme places previously thought to be barren; and astronomers had finally confirmed the existence of planets around stars other than the sun. Goldin's personal vision, outlined in numerous public appearances, heralded a day when NASA scientists would unveil for the citizens of Earth the first image of another blue planet like our own, around a star far beyond the sun, where mighty Earth instruments might detect the chemical signatures of life-forms harbored there.

As formerly top secret designs in optics and other Cold War technologies made their way to civilian hands, it appeared that a new golden age of astronomy had, in fact, blossomed. All of a sudden, mere human mortals seemed remarkably close to tracing the evolution of the known universe from its beginnings in a "big bang" forward through its manufacture of the first seeds of life, to the awakening of consciousness on at least one blue world, and beyond that all the way to the fate of the cosmos somewhere in infinity.

In these advances, Goldin and his science lieutenants glimpsed a worthy new mission for the agency, one that would surely revive dwindling public support—and funding—because it sought to answer some of the most profound questions ever posed about human existence. It would be a grand collective quest, summed up in the word *origins*. And for some time now, the perfect catalyst for this vision had been wending its obscure way—from Robbie Score to Duck Mittlefehldt to Chris Romanek to Gibson, McKay, Thomas, Clemett, Zare, and the other collaborators—and now to the upper reaches of the U.S. government.

Just as Goldin's passionate, brutal, charismatic, sensitive, *in*sensitive, infuriating, inquisitive personality had dominated and reshaped the space agency since his arrival, it now channeled the institutional waters in which the claims of possible ancient life on Mars were surfacing.

Given his years in classified military programs, Dan Goldin could appreciate the secrecy in which the investigation of the rock had proceeded. Only after they got their paper drafted, in the spring of 1996, did the McKay team slip word of their claims to key superiors at NASA's Washington headquarters.

Senior among these confidants was NASA's head of space science, the soft-spoken Wesley T. Huntress, Jr. He had worked at headquarters for eight years and had survived what he (and others) called the Great Purge in the science office soon after Goldin had taken over. Space science at NASA, despite its lustrous record of robotic explorations of the solar system and other achievements, was as low in the engineer-dominated agency pecking order as it had ever been. Having already gone through a drastic staff reduction, the program was facing an even worse budget drain. Huntress seized on the emerging "origins" concept as a way to rejuvenate the whole enterprise.

Huntress had worked with Goldin for four years now, long enough to know how to get along with him. He reacted to Goldin's onslaughts like the self-contained New Englander that he was: that is, he didn't react. If he felt off balance, he didn't let Goldin see it.

What made this effort worthwhile for Huntress was Goldin's remarkable vision, which Huntress heartily admired. Huntress felt that it was up to him and his staff to figure out the locus where Goldin's vision overlapped reality—

on the Origins program or any other undertaking. At that intersection, they could make something happen. There was a certain amount of fun in that, because it allowed them to think creatively.

So early one morning in April 1996, Huntress looked up to see several of the top people in his directorate standing around in his doorway. They were not on his schedule. "Wes," said one, "we've got to tell you about this paper." They started talking about signs of ancient microbes in a Martian meteorite. His eyes lit up. Not only was it a profound discovery, if valid, but it fit right in with his team's work on the new way of approaching the big, fundamental questions, all linked under the beguiling heading of "origins."

Wide-eyed with excitement, Huntress's group went over what was in the paper as relayed by McKay and Gibson in Houston. They included a series of cautionary notes: the evidence was not conclusive, the paper had not yet been reviewed by independent experts or accepted for publication, and so on. Huntress asked how long the process would take. At least a few weeks and probably a few months, they told him. He said, "Okay, I'm not going to worry about this yet. If the paper isn't accepted by *Science,* it's not credible. So get back to me when it's credible."

Some at headquarters had been hearing rumors about the McKay team's work, though not the details, since the 1995 planetary science conference in Houston, where they had discussed it publicly in sketchy, tentative, and preliminary form. The notion that they had found *organic material* on Mars—never mind biology—would be a huge deal if confirmed. The Viking landers had failed to find any such thing at levels as precise as a few parts per billion.

Huntress was not about to take such a flamboyant claim to Goldin until it had a stamp of approval. He frankly doubted that this would ever happen. The notion was completely wild. Huntress forgot about the rock and went about his business. Spring turned to summer.

On July 12, 1996, the same little crowd appeared at Huntress's door at around eight A.M. One of his deputies said, "Hey, Wes? You know that paper about life on Mars? It's going to get published." Huntress felt his jaw go slack. This time, for the first time, they had a copy of the actual paper to show him. They discussed the timing of the public announcement. The scheduled release date was August 16, just weeks away. Huntress thought, "Holy shit."

Huntress called Goldin's secretary. "Kelly, I've got to see Dan. It's a number one priority." Soon he was sitting in Goldin's top-floor office.

The news rocked Goldin back in his executive chair. He peppered Huntress with questions. Huntress answered them, and showed Goldin some microscopic images that would be appearing in the magazine.

Goldin activated a special team, which included representatives from public affairs, key political and policy people, and scientists. They launched a flurry of activity to plan the handling of this information grenade. Those in the know started referring to the object of their attention only as "the rock." Their operation became as secretive as the McKay team's had been.

From the time they got word that *Science* magazine would publish the McKay paper, the inner circle at headquarters was concerned that the story be "contained." If it leaked, they would blow their chance at what needed to be an orderly and carefully constructed story for the public. Huntress's worst fear was that the rock would become fodder for tabloid sensationalism and misinformation—"Killer Microbes from Mars!" Just as Orson Welles had once unwittingly terrorized listeners with an all-too-realistic fictional radio broadcast about a Martian invasion, the story of the rock, if mishandled, might alarm even rational people.

A key meeting took place on Friday, July 26, in Huntress's suite. Laurie Boeder, head of NASA public affairs (a Clinton administration political appointee) was there with other top scientists and public affairs people, with McKay and Gibson hooked in by phone. Boeder went over the strict rules *Science* had about releasing their information. One concern in the room was that NASA had paid for the research. So the question on the table was: Do we trust *Science* to be able to contain this so it doesn't leak? Or do we jump the gun on them and release the story before it can leak? The instant response around the table was: No, we've got to let the scientific process play out—but freakin' don't let this leak.

The group talked about the press conference that NASA would hold in conjunction with *Science*'s announcement and how they had to prepare at a relatively rapid clip. They decided to invite at least one prominent scientist to sit on the stage along with the McKay group and provide the appropriate caveats.

The other issue the group dealt with that day was the politics of the situation. The country was in the midst of a presidential election campaign. Boeder explained that the *Science* announcement would coincide with the news cycle in which Republican presidential candidate Bob Dole, at the San

Diego convention, would accept the party nomination. The Democratic Clinton administration might be accused of high-level orchestration, of trying to steal headlines from the rival party.

Goldin, when presented with the dilemma, told his team emphatically not to upset the magazine's normal process. As Huntress saw it, the administrator, to his credit, was determined from the get-go not to make this into a political thing, or a "NASA thing." Goldin called the editor at *Science* to discuss the issue with him, and they agreed on that point.

But the question of the release date and the conflict with the Republican convention was above even Goldin's pay grade. He had to take it to the White House. He set up an appointment with Leon Panetta, the White House chief of staff, and asked Wes Huntress to come along.

On the morning of Tuesday, July 30, Goldin and Huntress arrived at Panetta's office. When they walked in the door, Goldin laughed at himself as he said, "Leon, I'm here to talk to you about life on Mars."

In addition to his own personal zeal for the subject, Goldin was armed with talking points he had asked Huntress to provide: the McKay group's findings could turn out to be the most important scientific discovery of the twentieth century; additional work would need to be done to confirm the result; until that happened the result would be controversial; wide-scale media coverage could be expected when the story broke.

"NASA and the Administration will be thrust into the middle of this debate," Huntress had written, adding that the agency would have to take on the role of the "keenly interested skeptic. . . . We must avoid a frenzy of activity that could end up as a real embarrassment should this result not hold up under scrutiny." The notes listed logical follow-up activities, such as dispatching robots designed to look for more direct evidence on the surface of Mars, eventually trying to bring actual rock and soil samples back from Mars, and searching the solar system for other evidence of life.

Under the bullet regarding the search for planets around other stars, Huntress had written: "This program has resonated heavily with the public (show TIME cover) without any public announcements at all from the agency. The Federal Government should consider supporting the idea of 'Origins' as a principal (Dan—i.e. 'Kennedy-like?') goal for the agency. There are no large immediate resource requirements since the program is current technology-limited and investments in technology development will be required before any missions can be launched; sometime early in the next decade." In plain English, this meant that the Origins program could get started without a large infusion of bucks up front.

Under the heading "Political," Huntress had noted: "Dan, your call on this one." He'd added, "The president could play on the popularity of the space program, and the idea of life in outer space (as demonstrated by the current immense popularity of science fiction in print and movies), by recognizing this discovery as one of the most amazing results of his Administration. He could announce that his new Administration after the election will establish a program in NASA to search for life's origins in this and other solar systems. America will send small inexpensive robotic missions to Mars at every opportunity to follow up on this discovery."

The idea was to set the stage for the administration to propose starting up the new Origins program.

With guarded excitement, Goldin walked Panetta through the story, beginning with the rock's discovery in the Antarctic. He kept technical details to a minimum, turning to Huntress just a couple of times for elaboration. Panetta occasionally nodded and said, "Okay, I understand that." Goldin described NASA's official posture on the claims as one of "skeptical optimism."

When Panetta whipped out a yellow legal pad and started taking prodigious notes, Goldin said, "Leon, this is real 'classified' stuff. I don't want it to leak." He and Panetta agreed that the commitment to *Science* magazine must be kept. They also resolved, however, that if the story leaked to the "legitimate press," the government should call an immediate press conference because of the taxpayers' right to accurate knowledge of the results of tax-funded research.

As they were ready to leave, Goldin asked, "Do you think it would be appropriate for me to brief the president?" Panetta had found the story of the rock arresting enough that he turned to an aide and said, "I want the president to know about this right now." He asked where the president was. The staffer's answer was "We're not sure if he's awake yet, or if he's had breakfast."

Panetta told Goldin the president would speak to him right away.

Among White House concerns that day were a decision on the signing of a bill to overhaul welfare; questions surrounding a bomb detonated in the middle of the Olympics in Atlanta and the baffling explosion of TWA Flight

800 over Long Island; reported Democratic fears, as the party conventions were about to begin, that Clinton's supporters had grown overconfident about his ability to defeat Republican rival Bob Dole in the presidential race; and the need to cope with various legal proceedings stemming from the so-called Clinton scandals.

Whatever else was going on in the world that day, Huntress thought, this piece of news about a fundamental mystery of nature had become a priority.

Goldin and Huntress waited in a hallway for about fifteen minutes after their meeting with Panetta. At about ten A.M., Panetta reappeared and said, "C'mon. Follow me." He led the two NASA men into the Oval Office.

For Huntress, it was like being herded somewhere in a dream. If he hadn't been a scientist, Huntress had always imagined, he would have gone into some history-related profession—and now he felt he was witnessing a truly historic moment.

Bill Clinton stood alone in the Oval Office. Huntress had never met him before and was bowled over by his sheer physical presence. The president was huge. Huntress noticed that Clinton had a bit of that bleary "morning look" in his eyes. But he was impeccably dressed, and there was nothing else about his demeanor that indicated he was still waking up. The president seated himself in a rocking chair in front of the fireplace as Goldin, Huntress, Panetta, and Skip Johns, of the White House science office, took their places on couches.

As Goldin talked about the rock, repeating what he had told Panetta, Huntress could see the president waking up fast. The president's posture straightened, and his eyes opened wide. It seemed to be dawning on him that this could be big.

The NASA men were impressed by Clinton's apparent interest. He asked questions about how the discovery had been made, how they could know the rock was from Mars, and whether the primitive life-forms might have evolved into more complex creatures. The meeting went on for thirty minutes. Goldin thought, "My God, this is going into extra innings."

When they were ready to leave, Clinton said, "The vice president has got to hear this story. Leon, take 'em in to see the vice president." It was Huntress's impression that while the president had wanted to be informed about the story, he was delegating to Gore the decisions on what to *do* about it.

At some point during the discussion, Huntress had shown the president a copy of a magazine with a picture of an Antarctic meteorite in it. As they left the Oval Office, Clinton asked, "Can I keep this?" Huntress turned one last time on his way out the door and glanced back over his shoulder. The president was still standing in front of his desk, looking at the magazine.

Panetta ushered the men past a long line of people waiting to see Gore. Clearly some of them were getting impatient, and there was all sorts of hallway grumbling going on. Panetta spoke to Gore's secretary, and they went right into the vice president's office. There was no waiting around. Huntress thought this had to be unusual.

The vice president's West Wing office seemed more spacious than the historic oval. On one wall, the environmentalist Gore had installed a picture of the "full Earth," the iconic image of the fragile and isolated blue planet snapped by the *Apollo 17* astronauts (the last men on the moon) in 1972.

Gore greeted his visitors, and Goldin said, "I've got something to tell you about. We've just been to see the president and he asked us to come tell you." They crossed the room and sat on couches around Gore's fireplace. Goldin delivered his revelations for the third time that morning. "Wait a minute," Gore asked. "Our guys, government scientists, did this?" Their impromptu meeting lasted some forty-five minutes.

Gore peppered them with technical questions. Both NASA men were impressed by Gore's ease with the subject. The vice president (who was routinely accused of outwonking the wonks, and who some thought was better suited to a professor's life than a politician's) asked about the age of the meteorite, when the purported Martian microbes would have been alive, and how that related to the initial bombardment of both Mars and Earth. He asked about the relative conditions that would have existed on Earth and Mars at that time. Goldin answered those questions. Other questions he turned over to Huntress: whether this life-form, if it existed, could have evolved on Mars, how the McKay group's result related to the Viking experiments of the 1970s, and what this new evidence in combination with the Viking results might mean for the possibility of current life on Mars.

Gore then asked what actions might be taken, what all of this might mean for the future. Huntress was ready with the "origins" pitch. Gore seemed to know about the recent discoveries involving life in extreme environments and pushed for more information about how the evidence in the rock fit in with this emerging picture of "microbial diversity."

Huntress continued to be amazed by the appreciation both the president and the vice president exhibited for this kind of science and the overall struggle to approach fundamental questions. Goldin, years later, when the Clinton administration was out of power and he himself was on his way out as NASA administrator, would say of these exchanges at the White House, "I'm good. I know stuff. But I gotta tell you, these guys had a depth of knowledge. . . . They took me to deep places. It blew me away."

At some point in the meeting, Goldin and Panetta broached the politics. They told the vice president there were two issues. One involved the process by which the information would be released. They expressed their concern about what could happen if it leaked prematurely, and wondered "whether we should jump the gun on it ourselves to make sure there's no problems here and the right story gets out."

The vice president responded, instantly, "Follow *Science*'s process. Do not make this a political issue. Do not ever make this a political issue. This is a scientific discovery."

Goldin then told Gore about the timing conflict with the Republican convention. Imagine, he said, if the headlines the next morning are about possible life in a Mars rock instead of "Bob Dole Is Nominated." The administration could be accused of timing this to detract from Dole.

Gore said, "Don't even worry about it. Don't worry about it." And he repeated: "We cannot make this a political issue." The politics, it seemed, had all been dealt with—and dismissed in a flash.

Goldin and Huntress exchanged looks of relief. They were impressed. They hadn't been sure they would get the right answer. This was the right answer.

As soon as he had finished his presentation to the vice president, Goldin tried to reach McKay and Gibson. Neither man was in.

McKay was at home, taking out the garbage, when he got a call from his secretary. The White House was on the line; was it okay to give out his home number? McKay soon found himself talking to the NASA administrator, flanked by Wes Huntress. In the course of this call and a later one the same day, Goldin instructed McKay to get himself and Gibson to Washington right away.

Huntress had met Gibson once but didn't know McKay. The book on them at this point, his deputies told him, was that they were well-respected members of the geochemists' tribe, they had done a lot of work with meteorites, and they had one of the world's best instruments to work with. The only real knock against them was that they were at a NASA center instead of a university. The NASA officials themselves were aware of a widespread perception in scientific circles that if you were at a NASA center, you must not be as good a scientist as somebody at a university. But McKay et al. had university collaborators coauthoring their *Science* paper. They even had Dick Zare from Stanford, a member of the National Academy of Sciences, a heavy hitter.

The next morning, McKay and Gibson met with Huntress and other brass and went over the *Science* manuscript, the arrangements for the public announcement, and other issues. When they threw their images on the table, a commotion erupted. "Oh, my God . . . !"

The visiting Texans had been scheduled for a thirty-minute meeting with Goldin, but Goldin sent word that he would need them longer. He would clear his schedule. At two-thirty that afternoon, the visitors took the elevator to the top floor, carrying their written material and pictures with them. Huntress and Laurie Boeder positioned themselves in the back of the room. Huntress felt like a mentor whose students were explaining their research to a Ph.D. examiner.

Goldin allowed McKay and Gibson roughly fifteen minutes to summarize their findings their own way. Then Goldin went to work with his psychological dentist's drill. The inquisition ended at around five-thirty with the visiting team limping exhaustedly out of the room. Huntress remarked to McKay and Gibson, "You know, he put you two guys through a Ph.D. orals examination."

Gibson, for one, emerged from the three-hour grilling with a much higher opinion of the feared NASA administrator than the second- and thirdhand impressions he had gone in with. As the session ended, Goldin made a gesture that was turning into something of a personal trademark, an impetuous act that expressed his emotional, sentimental side in stark (and perhaps usefully disorienting) contrast to his harsh, aggressive side. He startled McKay, as he had other visitors on occasion, by asking, "Can I give you a hug?"

Chris Romanek, on his sabbatical from the South Carolina lab, was still at the Houston space center, helping to put the finishing touches on the *Science* paper. That day in Building 31 he went to find Gibson only to discover he was missing, along with McKay. "Where'd they go?" he asked an assistant, who replied, "Uh, they flew up to Washington, D.C., for, like, a double-secret meeting."

When McKay and Gibson got back, Romanek thought they each looked like they'd lost fifteen pounds. And he noticed they were wearing pagers.

Goldin felt satisfied that the claims—and the claimants—were ready for prime time. Since he had become NASA chief, Goldin had made a point of criticizing the tendency of some at NASA to wallow in Apollo-era nostalgia. The workforce seemed to break down into two groups: those who clung to the past, yearning for the "good old days," and those who wanted to "start writing history instead of reading it." It wasn't gray hair he objected to, Goldin liked to say: "It's gray minds."

Now the Houston guys had presented Goldin with a glimmering of evidence that could launch a big revival of public interest in space exploration. It could help the agency start writing history once again.

Immediately after Goldin and Huntress had made their rounds at the White House, someone passed the news on to President Clinton's close friend and political adviser Dick Morris. Morris had engineered a comeback for Clinton years earlier, as a young Arkansas governor, and in 1994, once again, the president had turned to him for help in rehabilitating his image after humiliating losses in Congress at the hands of the Newt Gingrich–led Republicans.

Morris was the resident Rasputin, a shadowy and influential political mystic with a certain ideological flexibility. He had worked for some of Clinton's staunchest opponents in Congress. Surfacing back in the Clinton camp, and initially known to the White House staff only by the code name "Charlie," he secretly helped the president devise a "family values" campaign to take back from the Republicans an issue they had owned for decades.

The political adviser was riding high these days over the historic victory he anticipated soon, when Clinton would become the first Democrat in fifty years to win a second term. Morris saw rich political possibilities for Clinton in the news of possible extraterrestrial life. He immediately started pushing for the president to go beyond even NASA's most ambitious blueprint and announce a manned mission to Mars on an accelerated timetable. It wasn't exactly the glory days of Apollo, but here was a space-related achievement that could stir the souls of the masses.

Morris's plan was wildly unrealistic. The idea of a hasty "Mars initiative" brought shudders to the White House science staff, which considered it ill-informed and naive. The tug-of-war fell to Goldin for resolution.

Not long after his visit to the White House, Goldin took a call that linked him into a meeting in progress there, with members of the White House science office and others. Morris got on the line and asked Goldin about the prospects for a major Mars initiative. The NASA administrator hesitated. The question put him in an awkward spot. For all his hopes and dreams about somehow getting NASA on track for Mars, he had to be honest. He told them that a grand foray to the red planet was simply not feasible under current circumstances—not technically or financially or in any other way. That ended that.

But Morris had shared the secret of the rock with one outsider. While working at the White House, the adviser—a married man—had been renting a $420-a-day suite in the Jefferson Hotel a few blocks north. There, he had been paying $200 an hour for the services of a long-legged call girl named Sherry Rowlands.

On August 2, Rowlands made a diary entry about a hush-hush item Morris had passed on to her after drinks and dinner. He'd whispered in her ear that scientists had discovered life on another planet. Something like a "vegetable in a rock" was how she recalled his description. "He said they found proof of life on Pluto!"

CHAPTER EIGHT

"KLAATU BARADA NIKTO"

OVER MILLIONS OF years, warm, shallow seas deposited limestone, shale, marl, siltstone, and an overlay of more limestone along the shifting shores where dinosaurs had once roamed in great herds and left their footprints in the sands. Now the inverted reflections of the ancient stone cliffs festoon the clear streams of the hill country around Garner State Park, about ninety miles west of San Antonio in Uvalde County.

The McKays had come up here on a long-planned escape from the steamy cauldron that was Houston in August. They joined other summer visitors who camped out or rented cabins and hiked in the shade of majestic bald cypresses under the limestone bluffs cut by the Frio River. The waters of the Frio were clear and cool—even in the hottest summers. And the visitors could hope for a glimpse of skittery white-tailed deer, Rio Grande turkeys, and waddling armadillos.

David McKay was confident that the big announcement was still more than a week away—timed for August 15, to coincide with publication of the paper in *Science*. Accordingly, three days ago (on Saturday, August 3), he had gone AWOL in direct violation of an edict from NASA headquarters. This pilgrimage was a decades-long McKay family tradition. Besides, after the tensions and labors of the last few months, capped by the Goldin grilling in Washington days earlier, David McKay welcomed a chance to catch his breath, enjoy the spring-fed waters, and collect his thoughts.

Ordinarily, McKay found it a pleasant bonus that modern communications had a hard time penetrating here in the backcountry. This time was different. As a precaution, he brought along his government-issue SkyPager. He expected to be back in plenty of time for the scheduled press conference, and in case anything went wrong—such as the word leaking out—he had the pager. But the device had registered no messages since he'd left Houston, and he found the isolation increasingly uncomfortable.

On this day—Tuesday, August 6—McKay's innate prudence finally prompted him to trek to a general store where there was a pay phone, to check in with his office. Standing at the phone, McKay was informed that, in effect, the schist had hit the fan and that his coworkers had been trying frantically to reach him. He and Gibson would laugh about the irony of it later. The pager, it turned out, was an embodiment not of the high-tech, cutting-edge, cosmic-virtuoso NASA but of the low-bid, behind-the-curve, bureaucratic NASA. The device did not work outside the Houston metropolitan area.

Dismayed, McKay absorbed the instruction from on high: he must fly to Washington immediately. NASA was throwing together a press conference. The nightmare scenario was a reality. The story was breaking loose on both sides of the Atlantic, and he had stuck himself out in the boonies with a pager that operated more like a magic decoder from a box of Cracker Jacks than a device favored by the designated agents of the high-tech, high-frontier future.

McKay was obliged to leave the quiet eye of the storm and enter the maelstrom. He and his collaborators were headed for total immersion in an experiment quite unfamiliar to them. They would become Exhibit A in a modern case study of the real-life consequences, as opposed to the well-worn Hollywood scenarios, when somebody ventures into an area that smacks of extraterrestrials. This would be the first case to demonstrate what can happen when such an event whets the appetites of the uncontrollable, everywhere-all-the-time, instant-news monster that had lately begun to take over the planet.

The McKay Martians were both hypothetical and microscopic, to be sure. They lacked the clarity of invading seven-foot Cyclops-eyed creatures with oily, fungoid skin, or of an oversized robot stationed at the ramp of a flying saucer on a greensward in the nation's capital, responding only to cryptic intergalactic idiom like "Klaatu barada nikto." Still, McKay and his partners would feel the very real and wrenching forces that can roil the intersection of media, science, and politics.

The story had been pushing at several portals, trying to escape.

Sherry Rowlands, Dick Morris's "paid lady," as she called herself, had been dickering with Richard Gooding, a reporter for the tabloid *Star*, over the sale of her diary. He wanted proof of her assertions that Clinton adviser Morris was transgressing beyond mere adultery—no big deal—and betraying the president of the United States by sharing state secrets with a call girl. Now *that* would be a story. In response, Rowlands offered the anecdote about the rock. She told Gooding about the evening when, following drinks and dinner, Morris had said to her, "Can I trust you? I have a military secret only seven people in the world know. We found life on one of the planets."

For all Gooding knew, it was a complete fabrication. (Actually, Morris or Rowlands or both of them had bobbled the facts: she wasn't sure which planet was involved; the information was not actually classified; many more than seven people knew about it; and the evidence of life on another planet was not conclusive.) But what she told Gooding was rooted in fact and would be something of a global scoop for the *Star*—if only the reporter could be sure.

Gooding pressed her about which planet.

"Well, what are we exploring now?" she asked Gooding.

"I don't know—Jupiter?" he suggested.

Gooding would remain a skeptic until August 6, when the story of the rock reverberated around the world and he realized Rowlands had been telling the truth.

At the same time, the tale of the "secret" made its way to the British media.

Everett Gibson, back in Houston, felt these first cracks opening up over the same weekend (August 3–4) that McKay left to play hooky. Gibson got a call from his friend Colin Pillinger in Britain. "I have an abstract with your initials on it," Pillinger told him. Gibson had left a copy of the galley proofs—bearing his initials—with Goldin in Washington, who had passed it to the White House. Apparently, somebody had faxed a copy to British media outlets, and they'd turned to experts like Pillinger for comment.

Gibson would later surmise that this, too, was the work of Rowlands, trying to sell the information after getting it from Morris. During the transcontinental phone call, Gibson asked Pillinger to maintain silence a little longer. "Just bear with me for a little while." Gibson thought he could stall the rush of events. He was wrong.

For one thing, the politically excited Morris had been promoting the story quietly (and with the understanding that the information was embargoed) to members of the White House press corps, who were pressing to break the news.

But the first person to put the story on the public record was Leonard David, a weathered freelance reporter writing for the industry weekly *Space*

News and also editing the magazine *Final Frontier.* In appearance, he was a blend of balding, long-haired hippie and cowboy casual. As a kid, he had been known to break up his eyeglasses and hook the lenses to toilet-paper tubes to make telescopes. An unabashed space enthusiast since well before Sputnik, he had spent almost four decades reporting on space-exploration issues for a variety of bosses and was still scrambling to string together enough checks to pay the rent. He had, on occasion, gone over to the other side to work for the government, including stints as a consultant to NASA and research director for a commission on space exploration set up jointly by Congress and the White House. He was a familiar presence in aerospace circles, closely attuned to the players, the gossip, and incipient news.

For months now, as he made his rounds at the scientific gatherings in Houston and other places, David had been picking up "hall talk"—the same hints and rumors that had been circulating with increasing energy in the corridors of Building 31. David had first begun to focus on the rock as a potentially big story in March 1995, in Houston, when the McKay group had discussed their preliminary description of Zarelab's detection of the organics—the PAHs—at a gathering of planetary scientists.

A few months after that, John Kerridge, a cosmochemist and meteorite specialist at NASA headquarters working on studies about how to find life on Mars, had slipped David a tip to watch the rock for a big development. "I'm telling you about this now, because when you hear about it, don't be too carried away," Kerridge had told his friend. "You know, it'll seem extremely spectacular when it appears, but you need to approach it from a very skeptical frame of mind, because there are a lot of pitfalls here, and it's not clear that all the pitfalls will have been gotten out of the way."

Kerridge and others at NASA headquarters had heard rumors that the McKay team was anxious to publish. Kerridge, for one, went from feeling intrigued with the evidence of the organics—the PAHs detected by Zarelab—to a horrible sense that "Oh, my God, this could be a disaster," because the Houston group might actually claim to have found life. Kerridge, who was at NASA on detached duty from the University of California, would argue later—after he became one of the most outspoken critics of the McKay group's claims—that he had not prejudged the question but was "just very concerned about the potential for, if you like, negative publicity if somebody comes up with something and it turns out to be incorrect."

In July 1996, Leonard David had attended a conference of space enthusiasts in Boulder, Colorado, on strategies and rationales for sending humans to explore and colonize Mars. Again he'd encountered a lot of buzz about this same rock. A couple of people had speculated casually about the f-word: fossil. From this point on, David had worked on the theory that the source of the buzz was some kind of fossil on Mars. He'd started "pinging" his sources at the Houston space center in earnest. It was when they started to clam up on the topic that he knew he had something big. During the months that he worked on the story, there was no single tipster but rather four or five different people who provided various small pieces of the puzzle that he could then use to leverage even more information out of sources. His dogged pursuit would later be widely referred to, somewhat misleadingly, as a "leak"—as if someone had kindly dumped a juicy scoop on his front stoop.

The final pieces fell into place at the end of July. A source left David a voice-mail message informing him of rumors that NASA was planning an important press conference in mid-August on an unspecified topic.

As July turned to August and David's next end-of-the-week deadline neared, he called Don Savage, a public affairs officer at NASA headquarters. He asked whether something was brewing that had to do with the Martian meteorite and a fossil.

Uh-oh, Savage thought.

Savage tried to walk the tightrope—to be as evasive as possible while responding honestly. He hoped (like Gibson) that he might be able to stall the rush of events. David didn't seem to have much in the way of hard facts. Savage told David that information about the story was restricted ("embargoed") and he couldn't talk about it. But that, of course, amounted to a confirmation of sorts. Now David knew there *was* a story coming to a head.

David then decided to take "a little bit of a gamble," as he would put it later. Nobody had ever laid it all out for him. But he wrote down all that he thought he had learned. He ran his draft past a "deep throat"—an informed source who assured him his story was accurate. He called his editor at *Space News* from his home base on Capitol Hill in Washington and told him he had an important item he'd like to get in right away. He kept saying, "I think this is really gonna be big." But the weekly deadline was looming, and work on the issue was almost closed out. When he quit work that Friday, David wasn't sure whether his exclusive would get in.

Over the weekend, he attended a party where he spotted Andrew Lawler, a reporter for *Science* magazine. David walked into the kitchen and, taking a wild shot, said to Lawler, "Hey, how about that Mars meteorite story." Lawler fired back, "You're not supposed to know about that!" That confirmed in David's mind that *Science* was the journal with the hot rock, and that his story was right on.

The following Monday, August 5, as David McKay was relaxing in the hill country, the short piece by Leonard David appeared in *Space News.* It was all but buried. It ran on page 2, the second in a column of eight little items. Only five paragraphs long, it began: "The prospect that life once existed on Mars is being raised following analysis of a meteorite recovered on Earth." At another point, David wrote that the Martian meteorite ALH84001 contained "indications of past biological activity on Mars."

The *Space News* item, in both its brevity and its placement, was so unassuming that it did not make a big, sudden splash. But it entered the circulatory system of the media beast. Like a virus, the word spread. The beast began to quiver and hum. People started calling Savage and his coworkers in NASA public affairs to inquire about the rock. By seven A.M. the next day—Tuesday—NASA had issued the emergency call for McKay, Gibson, and the other authors, wherever they were, to scramble to Washington for a press conference.

The NASA headquarters group had known that as soon as Goldin and Huntress paid their late-July visit to the White House ("one of the leakiest places on the planet," as they reminded one another), the risks would shoot up, given the magnitude of the story. They were aware that a leak could spring from anywhere—from their own ranks, from Building 31 in Houston, from the scientists scattered across the land who had reviewed the paper for the journal, from anyone in the expanding circle of those who knew.

The NASA group had made extensive preparations for such an escape, assembling charts, graphics, videos, written guidance for President Clinton in case he decided to comment, and everything else they might need for going public. But there were some things that could not be precooked and ready to serve. For example, the NASA team had laid out a by-the-book schedule of private briefings for relevant agency chiefs, key members of Congress, and other interested VIPs—to be implemented a few days ahead of the announcement.

Compounding the difficulties of that day was the fact that the headquarters civil service staff had been cut in half since Goldin had taken over in 1992.

The item in *Space News* opened a crack to the waiting flood. And for those closest to the story, it triggered a near-panicky transcontinental scramble.

Nan Broadbent, communications director for *Science* magazine, had been out sick on Friday with the stomach flu. Now, on the morning of Tuesday, August 6, 1996, though still in a fragile state, she decided to come in to the office, because she was (a) bored and (b) able to stand and walk. She and her staff went into a routine meeting, oblivious to the brewing media storm.

When the meeting broke up, the participants were startled to find a stack of messages waiting. Reporters from CBS and the BBC were among the early callers. "Is the embargo still in force?" they wanted to know. They were threatening to go public on this Martian meteorite thing.

Science magazine, established in 1880 by Thomas A. Edison, is one of the world's leading general science journals, rivaled only by *Nature*, an international publication with roots in Britain. *Science* is published by the American Association for the Advancement of Science (AAAS), the largest society of scientists in all fields. The magazine receives some seven thousand manuscripts every year, of which only about 10 percent make it to publication.

Science and the other journals provide one of the main platforms that scientists used to announce important advances. Journalists consider these publications important sources of original news, in much the same way that political writers cover a major policy speech by a candidate, or a court reporter relates a judge's decree.

Because of the meteorite story's importance, *Science* editors planned to distribute some five hundred copies of the McKay group paper a few days ahead of the publication date. In accordance with usual procedure, participating journalists and scientists—anyone who got the advance paper—would agree to keep any such embargoed information secret from the broader population until the ordained time. In turn, this would give the journalists several days to alert their editors and evaluate the importance of the story against the constant blizzard of competing developments in the world. It would enable the journalists to call around for independent opinions from experts and generally equip themselves to present more coherent and complete accounts than if they had to assemble the story in a mad deadline spurt. And it would give designated scientist-commentators—those independent of the McKay team but expert in the fields the paper dealt with—a chance to inform themselves in preparation for media inquiries, rather than being blindsided on announcement day.

This embargo system is controversial. Some people assail it on grounds that the journals use it to promote themselves, enhancing their own reputation and fortunes through manipulation of both scientists and the media. Some argue that a reporter sacrifices a degree of independence by accepting the arrangement and that the secrecy imposed on scientists impedes the free flow of information to others in their field and to reporters trying to conduct enterprising investigations.

Ordinarily, the orchestrated delay in the release of scientific information does not directly affect the public good. The issue of *timing*, for both scientists and journalists, usually matters more in terms of narrow effects such as who gets to claim credit. In certain cases where human health or lives might be jeopardized and timing matters a great deal (such as the discovery that a widely used drug is harmful), the process can be short-circuited accordingly.

In any case, journalists are free to go after stories independent of the embargo constraints, as Leonard David did, by not signing up or by getting the story well before an embargo went into effect. In fact, the advent of the Internet has made it easier for interested parties to get their hands on all sorts of preliminary reports and data—long before it reaches the embargo stage but also, often, well before it passes through peer review and other reliability-enhancing wickets.

On this Tuesday, the week before the planned announcement, no embargoed information had been sent out. The restrictions were not yet in force. The story at this point presumably was fair game for anybody who could ferret it out.

Even after the news began to surface, however, the embargo system cast a shadow. Dick Zare, for example, felt that the rules prohibited him, as an author of the paper, from revealing its import in advance to colleagues at the National Science Foundation—even though he was head of the foundation's National Science Board and even though the foundation had funded the Antarctic research program that made possible Robbie Score's discovery of the rock in 1984. Some at the foundation would express annoyance over this general state of affairs. For critics of the embargo system, it was useful fodder. And the ghost of the embargo (embodying the larger issue of who gets to control the information) would haunt the increasingly strident exchanges between *Science* magazine and NASA as Terrible Tuesday unfolded.

Broadbent and her coworkers saw only two choices in front of them: either release material describing the McKay group's paper and thereby help the universe of reporters get the story right or stick by the embargo and let a few journalists run with incomplete or inaccurate reports. From the day the McKay paper arrived, the staff had known that this one would be a big deal. They had taken extraordinary measures at every turn in its handling.

Like the scientists in Building 31, the magazine cloaked the topic of the rock in unusual secrecy. Unlike the typical submission, the McKay paper was not mentioned at the weekly meetings where *Science* editors discussed such things. At first, the knowledge was confined to three or four of the most senior people at the magazine. Gradually, additional members of the AAAS or *Science* staff were informed on a strictly "need to know" basis, and always behind closed doors.

One morning a few months earlier, not long after the McKay paper arrived in their hands, *Science* editor Brooks Hansen strolled into the communications suite to inform Nan Broadbent about the claims of possible Martian organisms. Broadbent and her staff had a "need to know" because they had to be involved in discussions with the authors of the paper and with NASA public affairs, among others. They were responsible for preparing the material that would be distributed to reporters.

In early summer (well before Goldin and Huntress made their visit to the White House with the news), she heard a rumor that, on a trip to Russia, Goldin had told Vice President Gore about the McKay findings. Concerned about a leak, she contacted her friend Rick Borchelt, a public affairs officer for the White House science adviser, to find out. Over dinner, she told him the story of the rock—in confidence—and asked about the rumor. Word came back that Goldin had told John Gibbons, the president's science adviser, in general terms to expect something stunning. Goldin had said nothing to Gore at that point. But now, by early July, at least one person at the White House knew the secret.

Broadbent and her assistant also confided in the ailing Carl Sagan, sending him a copy of the paper. Although he would be counted as one of the paper's reviewers, Broadbent had another motivation: she asked if he'd be willing to write about the Mars rock in his weekly column for *Parade* magazine, as a way of lending credibility to the story.

As the summer passed, Broadbent went through several exchanges with NASA about when to publish McKay's paper. NASA wanted the date set for as early as possible, August 9, in order to forestall leaks. But the logistics—the give-and-take with the reviewers and other aspects of preparing the paper for publication—dictated the later date, August 16. From as early as July 17, however, NASA was planning for the possibility of "a press conference on August 5, 6 or 7."

Because of the sensational nature of this particular paper, *Science* adjusted its own procedures. With encouragement from NASA, the magazine agreed not to send out advance copies of the Mars rock story until just three days ahead of the publication date instead of the usual six—to limit the leak opportunities. Also, the magazine planned to lift the embargo two hours early on August 15, the day of NASA's planned press conference, so that journalists in Asia and Australia could make their deadlines.

On the subject of leaks, NASA's Don Savage passed along to the *Science* staff the stated policy of his boss, Laurie Boeder, NASA's top public affairs official: "If we [NASA] received a call from a bona fide major media [representative] who had the story cold, wanted confirmation of it from NASA, and was going to break the embargo, we would have to give it to them and hold a press conference ASAP." The magazine staff balked at this, telling Savage that if there were a leak, it would likely come *from* NASA.

When things later went sour between NASA and *Science*, some participants would attribute it to simple frustration that everybody's hard work and planning—or, as some outsiders would say, their efforts to manipulate the media—were falling into a shambles. Each side was venting and sniping against the other. For some, though, the events of the frenzied forty-eight-hour span would leave abiding resentments.

Early on Terrible Tuesday, Broadbent went to tell her boss, Richard S. Nicholson, the publisher of *Science* and CEO of the AAAS, what was going on. He was in his office but in a meeting with the door closed. Broadbent plastered Post-it notes all over the door, all saying some variation of "Call me."

Around ten A.M., he called her. They discussed their options.

A communications officer on Broadbent's staff handled the calls to NASA—at first. But that staffer eventually came to Broadbent frustrated and upset, saying she had "had it" with NASA. That Laurie Boeder, the staffer said, was unbelievably rude. She wouldn't let you finish your sentences. She would raise her voice and say things like "Who are you to tell NASA or Dan

Goldin what to do?" The staffer refused to have further dealings with NASA or Boeder. Broadbent took over the negotiations and soon found herself having the same reaction.

The magazine was now in a full-blown head-butting conflict with NASA about how to handle the erupting media frenzy. Even though the space agency had for weeks urged the earliest possible release of the information in order to head off a leak, now that the breach had occurred, it was the NASA team that wanted to postpone a full-bore response. Boeder argued adamantly that the magazine should hold on to the bulk of official information about the rock until the next day, Wednesday, by which time NASA could gather the scientists for a press conference and release the news in an orderly fashion. Everybody should stay collected, calm, and cool and not get stampeded.

Broadbent decided to go the other way and release the material to the media in the interests of fairness and to ensure accurate and complete reports. She told Boeder so. But then, to her almost immediate regret, she "blinked." Boeder talked her into reversing herself and holding off on a press release. Broadbent's staff was aghast. Broadbent *re*-rethought her position. The NASA approach would work only if no information got out. This would require the newspeople who had been calling to hold off on running the story. And was there any reason to believe they would do that? Only that NASA had assured her they would. Broadbent thought of President Reagan's Cold War dictum "Trust but verify." So she called the reporters back and, sure enough, they told her no way; they were not waiting. They were going public this same day with whatever facts they could ferret out. NASA, they indicated, had even given them the go-ahead.

Around the offices of *Science* magazine and AAAS, the dispute was colored by a perception that NASA was deviously trying to take control of the story and maximize the agency's own press attention. Now *Science* staffers learned that NASA officials were essentially confirming the story—for those few reporters who called and asked—while requesting that the magazine withhold information from the rest.

Broadbent got Boeder on the phone again and said, "We're not playing." Broadbent had decided to revert to her original plan to release the material to everybody as soon as possible. Letting just a couple of outlets release the story, she said, would put *Science* in the awkward position of seeming unfair or dishonest toward the universe of other reporters.

Boeder was irate. She accused Broadbent of flip-flopping. "Yeah," Broadbent answered. "Our job is to release information in a proper manner. . . . Giving it out in spurts is not in the tradition of AAAS."

The hard-charging Boeder was a Clinton political appointee who for three years had been in the blast danger zone as Goldin's chief lieutenant for communications with the public. She understood that the story of the rock was extraordinarily high profile, not to mention potentially profound. But it was one of many high-visibility events parading through her calendar, such as a comet colliding with Jupiter and a historic agreement to cooperate with Russia in space.

The NASA hierarchy felt a strong sense of ownership of the Martian meteorite story. The agency's DNA carried the 1958 instruction from its congressional creators that it should "provide for the widest practicable and appropriate dissemination of information concerning its activities and the results thereof." And in this case, not only had the agency paid for the main research, but there was the added factor of Goldin's intense personal interest. At the same time, the NASA people were acutely aware that there was not much of a story without the imprimatur of *Science* magazine to give the claims a degree of credibility NASA alone could not bestow. Accordingly, NASA managers gave the magazine what—in their view, at least—was prominent and fair credit in their press statements about the rock.

Boeder's battle cry in every policy battle over the years had been "Credibility!" NASA's credibility. She had made the care and maintenance of that precious commodity her top priority. "If you don't have that," she would say to her staff, "you have nothing."

Regarding the problem at hand, her first thought had been how unfair it would be if David McKay's results were released without him being there. She felt the pressure to fill the ravening maw of journalism, to "feed the beast," but there were times, she thought, when you just had to say no.

The next thing Nan Broadbent knew, NASA administrator Dan Goldin was on the phone asking for her. Although he was no particular fan of the embargo system, Goldin felt humiliated, frustrated, and incensed by the story's early escape. It could undermine both his and the agency's credibility. And the worst thing of all would be if the "leak" had come from NASA. That concern was the source of much of the consternation and anger at NASA in those hours.

Broadbent did not take Goldin's call but passed it to her boss, Nicholson. At that point, Broadbent figured, Nicholson must have been convinced she was about to quit. He didn't realize she was still recovering from the flu. And she knew she *had* gone quite pale.

The two executives had an animated conversation. After a time, Nicholson came back down to Broadbent's office and remarked warmly of Goldin, "Well, that's a very pleasant man."

The magazine was releasing the Mars rock paper. The journalists had forced the issue. The AP's Paul Recer transmitted a wire story at 1:51 P.M. eastern time, saying, "A meteorite that fell to Earth after possibly being ejected from Mars may bear chemical evidence that life once existed on that planet, NASA officials said Tuesday." Less than ten minutes later, Britain's BBC1 television channel announced the news, and at 2:27 P.M. eastern time, CNN reported it. None of the early stories quoted the scientists making the claims.

The *Science* staff was appalled at some initial media missteps, notably some slapdash initial coverage by CNN, in which an old tape was recycled in a way that made it seem to be a new comment on the Mars rock, and a reporter calling the fossil-like shapes in the rock "something like maggots." The CNN coverage got better. Then there was a late-afternoon AP dispatch quoting an unidentified scientist "familiar with the study" (but not on the McKay team) as saying the discovery was "unequivocal."

Both NASA and *Science* were abruptly engulfed in a barrage of phone calls. A horde of relentless TV producers, reporters, editors, and other foot soldiers of the communications age converged on the story.

Broadbent's staff scrambled to distribute a four-page press release. They ended up having to release the McKay group manuscript itself with handwritten editors' notes still in the margins, because the final version of the paper had not been typeset.

Written text was not enough, of course. Reporters had questions. They needed give-and-take with actual humans. Clamoring for access to the paper's authors, journalists were dismayed to learn that NASA (though it had issued two press releases) still had the scientists under orders not to talk to the media before the big announcement at headquarters the next day.

But NASA could not control all of those who knew the story.

• • •

Keeping the secret of the rock had become almost unbearable for Richard Zare. He spent many a night sleepless from the combination of excitement and sheer terror of making such an amazing claim. He was more aware than ever of the ridicule and contempt that had plagued predecessors along this same path. He couldn't escape the feeling that he might be next.

Though Zare's public image was that of a big shot whose career had been crowned with success, he had never defined himself by his accomplishments. He was always driven to seek a more demanding and worthy test. He saw himself as driven not by ego but by the lack of it. This was why, he would conclude later, he was so unprepared for what was about to happen.

On August 5, while attending a conference in Woods Hole, Massachusetts, Zare had gotten a call from a distraught Everett Gibson in Houston. Bits and pieces of the story seemed to be getting out, Gibson warned him. Zare stuck to his schedule and flew back to Palo Alto the next day, Terrible Tuesday. Waiting for him was a series of messages Gibson had left on his answering machine, each one increasingly frantic. Zare was not to speak to the press under any circumstances, Gibson warned, before the press conference in Washington the next day. NASA had already contacted Zare's trusty postdoc Simon Clemett, along with David Salisbury, a seasoned Stanford University information officer, and they were ready to fly east. Zare had no choice but to join them and head right back to the airport.

A film crew from CNN awaited Zare and company at the departure gate at San Francisco airport with a volley of questions. Salisbury fended them off. The Stanford men boarded their plane, which meandered around the tarmac for half an hour before the captain pronounced the flight canceled due to mechanical problems.

The three retreated to the frequent-flier lounge to wait for the next plane to Washington—the red-eye, scheduled to depart five hours later. The TV monitor was tuned to CNN (which had always taken a special interest in space-related news). Zare, focusing his bleary eyes on the screen, was stunned to see NASA people on the air talking about the rock. He was also appalled to hear a CNN comment that compared the possible fossilized bacteria to maggots. This was seriously misleading on a number of levels, Zare thought. For starters, there were significant differences between bacteria and worms!

At around three-thirty P.M. West Coast time, *Science* magazine's Nan

Broadbent reached Zare's party in the lounge. She told them that the magazine was canceling the embargo, and—contrary to the instructions from NASA—she encouraged Zare and Clemett to start talking to the press immediately.

The phone next to Zare's chair in the lounge began to ring. The *New York Times, Washington Post, Philadelphia Inquirer, Boston Globe,* and half a dozen other papers had been directed to his location by the *Science* staff or the Stanford news office.

While NASA officials still had McKay and the Houston contingent under a gag order, they lacked the authority to muzzle the Stanford contingent. (Zare, among his many roles, was a member of the editorial board of *Science.*)

Zare and Clemett spent more than three straight hours on the phone helping a succession of journalists understand the complexities of the story. (Yes, there was evidence that microscopic life *might* have existed on Mars, billions of years ago, but it was far from conclusive; no, there was no indication that anything was still alive on the planet today—and so on.)

While Zare was talking to a *Los Angeles Times* writer, a film crew from ABC's *Nightline* came clanging in with their equipment to tape an interview. When the flight to the East was called at last, a CNN crew chased the three to the gate, where Zare gave them a few more sound bites. Network crews would be lined up to greet them at their destination as well.

During the flight, Zare had an attack of nerves. His head was telling him to treat this story as he would any other piece of research, never mind the Warhol factor. Was the hypothesis sound? Had he and his coworkers followed the proper protocols? But his heart was pounding audibly, overwhelming rational thought. For the first time, he allowed himself to feel the true dimensions of what he and the others were claiming. If their interpretation of the evidence was correct, the human species was on the brink of a turning point of staggering proportions. Then, once again, that stab of apprehension: What if they were wrong?

Either way, he realized, it was out of his control.

A few weeks earlier, Bill Schopf of UCLA, the noted paleobiologist and one of the experts the McKay group had consulted, had taken a call from NASA headquarters. It was an invitation to appear at the planned press conference to provide an independent evaluation of the McKay group's evidence.

Goldin and his top lieutenants were mindful that many scientists considered "press conference science" wrongheaded. Just publish in journals and let the work speak for itself, the argument went. But the work in this case *had* passed peer review. And by recruiting a heavyweight like Schopf to point out any weaknesses, the NASA people reasoned, they could ensure that the announcement would be seen as properly balanced and cautious.

Schopf soon realized that this was the same rock whose innards he had examined more than a year earlier when he'd visited Houston. He had told the group about his reservations back then, and he was surprised to learn their work had reached this stage now.

Before accepting NASA's invitation, Schopf asked the agency to send a copy of the McKay paper overnight to his office. After studying it, Schopf had so many reservations he tried to get Goldin's team to invite somebody else. But NASA persisted. Dan Goldin himself made the request. Schopf admired Goldin, and besides, he agreed with Goldin's strategy. The claim *did* need an assessment by "a hard-nosed outsider," and he certainly fit the bill. He reluctantly agreed to do it.

Schopf was in the process of preparing his arguments and some accompanying display charts when Terrible Tuesday arrived. He, too, got the summons from NASA headquarters: "Bill, get on the one-thirty afternoon flight. There's been a news leak—the press conference has been moved up."

Arriving at NASA headquarters the next morning, Schopf was escorted to NASA's lower-level studio, where McKay, Gibson, and the other principals were seated. Boeder and Savage and other NASA brass were there. The new arrival was a man of moderate build with short dark hair that had receded up his expansive forehead and a generous mouth with prominent teeth set in a jutting jawline.

This was a "prep session"—a dress rehearsal for the press conference.

Memories would differ about exactly who said what to whom in the course of this particular dress rehearsal. But the event embodied the conflicts that often emerged when scientists had to explain themselves to the public, with dashes of ego and institutional rivalry to spice up the fray. The importance of the topic intensified the differences. This session would create an impression in the minds of some that would lead to charges later on that NASA officials tried to hype the story of the rock for their own bureaucratic purposes.

First thing in that session, Schopf pulled out a bunch of transparencies showing typed text. To astrophysicist Edward Weiler, then NASA's chief scientist for the Hubble Space Telescope, it looked as if Schopf thought he was at a university symposium rather than a televised presentation to the global village. The two immediately got crosswise.

Weiler spoke up: "You're not going to show those at the press conference, are you?"

Schopf said, "Oh, yes I am."

"Have you ever done television before? That stuff won't show up. It'll be a disaster."

Events had led Weiler to become quite an enthusiastic student of the art of packaging science news for the media, especially television. He was a key player in overhauling the way NASA presented its most important research results, including a series of news updates that included video and graphics, dissenting points of view, context, and a tilt toward English over jargon. The perceived success of this approach in attracting media coverage fed a current of indignation among those who considered such efforts unseemly.

Weiler was convinced that the approach was an effective way to make valid NASA science news available to the people who paid for it. Reporters were always free to do their own assessment of the news offering, call their own independent experts, and blow the whole thing off if it didn't hold up.

Schopf was not the only one with unwanted last-minute additions that day. The McKay team had brought about twenty new graphics with them, to be shown on the TV monitors that would flank the stage at the press conference. The graphics showed various features in the rock and supporting data. The team insisted they could not make their presentation unless they showed all of these—in addition to about a dozen of their best transparencies, which had been sent in earlier. To the public affairs people, this was turning into something of a nightmare.

A scientist always wanted to present as much ammunition as possible, to make the point clearly. But this depth of detail would quickly lose an audience. Even *Science* and other top research journals required a lot more shortening and simplifying than many authors were comfortable with. The work of science was necessarily slow, precise, careful, detailed, complicated, and cautious. The journalist's enterprise was fast, short, often imprecise, and hungry for drama and certitude. It was hardly surprising that there was a clash whenever the two cultures tried to marry up.

In the end, NASA's Savage backed McKay. He decided it was more important that the researchers felt comfortable and were able to tell their story their way. He and his staff knew the press wouldn't use most of it. (They didn't.) But it was important for the scientists' state of mind.

Some of those on hand had witnessed McKay and Gibson's beaming, ebullient, high-energy performance under Goldin's interrogation just days earlier. The onlookers could hardly recognize the people in front of them now. Their delivery seemed plodding, flat.

Finally, McKay and his group completed their dry run.

Schopf gave his rebuttal. By this time, Goldin was among the growing crowd of listeners who had wandered in. When Schopf finished speaking, there was a marked silence. Then (as Schopf would remember it) Laurie Boeder addressed the McKay group: "Schopf has just demolished you. Can't you guys be more positive?"

Boeder could be tough. She had been arguing all along that the scientists who'd done the work should be the ones who put it out there. It was her policy never to meddle with the *substance* of the science on this or other stories. But she was also a political appointee who had spent her career in public service advising people how to communicate with the public about what they were doing. And she knew that anybody who didn't practice for an important announcement was not serving well either themselves or the people they were talking to. Scientists needed to communicate clearly. If you couldn't tell your story, you didn't get support—that is, money. You needed to be able to tell the people who paid for the work—the taxpayers—why it was worth their money. And that went to the issue of NASA's credibility—her mantra.

It was no secret that she was an advocate for the agency. That was not only her job, as she saw it, but something she felt on a personal level. She believed in the agency's role. "When I look at those Hubble pictures, I hear music," she would tell people. She felt the same way about the images from the rock. Now she had a modest supporting role in an investigation into one of the great, eternal questions. She didn't intend to blow it.

What concerned her at this moment were the personal dynamics in the room. Schopf was intimidating the others, and that was affecting the way McKay and his team were talking about their results.

So now, as Schopf concluded his arguments, Boeder prodded McKay and the others about whether there might not be a better way to talk about what they had found. As she would recall it, she said something like "That's a pretty good argument. He's making a really good point there. What's your response?" (She would take issue with Schopf's recollection, insisting that she would never have used the word "demolished.")

As Weiler would recall the exchange, he, Boeder, and some others advised the McKay group this way: "Do you believe in your result? Yes? Well, then say it! You know, don't be wimps up there. The reporters are going to pick up on your body language. If you come across tentative, it means you are tentative about your results. If that's the way you feel, fine. Be tentative. But if you feel these results are solid, show it."

After the prep session, as several of them rode the elevator to the first floor, Goldin gave McKay a final locker-room spur: "Don't wimp out."

Weiler wandered outside for a cigarette. There was Schopf—also a smoker. He was trying to relax before the press conference. They struck up a conversation. Weiler soon realized that Schopf had been under the impression that Weiler was just some dumb bureaucrat, trying to boss around the great scientist from California. Once he informed Schopf he was actually a fellow scientist who had done a lot of TV, Schopf seemed to lighten up. The two men called a truce. They stood in the steamy August heat and enjoyed their smokes. The fun was about to start.

CHAPTER NINE

IN THE BEAM

CHRIS ROMANEK WAS on a road trip. The previous year, when Everett Gibson had been unable to offer him a NASA job, he had moved to the University of Georgia's Savannah River ecology lab in South Carolina. But he'd spent most of the summer of 1996 back in Houston on a fellowship and had helped his comrades in Building 31 with final preparations for the big "coming out." On August 1, with a full two weeks until the scheduled press conference, he and his wife decided to pack up their dog, cats, and all the stuff they had brought with them for the summer and take a leisurely drive back to South Carolina.

When they reached Mobile, Alabama, they joined family members vacationing at a condo on the east side of Mobile Bay, with a postcard view of the rolling surf. Romanek went out to eat with his sister and his mother. They got home about eleven P.M. that Tuesday, August 6, 1996, and turned on the television set. The first thing Romanek saw was a picture of the carbonate globules—*his* carbonate globules, and then he heard somebody talking about a press conference. The next day!

He called his erstwhile collaborators in Houston, one after another. No answer. No answer. He finally reached Kathie Thomas's husband, who said, "Oh, yeah, they're all up at NASA [headquarters] right now."

Romanek felt sick. It seemed clear that he had been left out of the loop. He threw up his hands in frustration. No way he could make it. He was at the beach in shorts and sandals.

His family jerked him out of his funk. "Are you going to be a weenie here, or what? We can get you to Washington, D.C., in time if you really want to be in on this."

"Can you really do that?" he said.

"Of course!" His brother-in-law started calling and making plane reservations. His sister rushed home to collect all of her husband's suits. Her husband went six feet three, about 230 pounds. Romanek, a lean six-one, 180 pounds, found himself throwing on jackets, pants, shoes, and shirts, trying to find something that he could at least walk in without tripping.

But he couldn't shake the hollow feeling. "I'm being excluded," he thought. "I'm no longer a part of this."

He caught a flight out of Pensacola and made it to Washington by about ten-thirty A.M. the day of the one P.M. press conference. He got off the plane and headed straight for the nearest newspaper stand. There it was—front page of *The New York Times, The Washington Post,* and *USA Today.* There was his name. He couldn't believe it. He jumped into a taxi and said, "Take me to NASA headquarters."

Already in an anxious state, Romanek soon noticed they were driving away from the downtown area. "I think NASA headquarters is downtown," he protested. The cabbie said, "No, I know exactly where headquarters is. We have to go out this way."

They were headed toward Baltimore! Romanek saw an exit sign that said NASA GODDARD. (Goddard Space Flight Center is a major NASA facility in the Maryland suburbs northeast of Washington.) Romanek noted that there were few cars going in their direction, but that traffic was bumper to bumper going the other way—*into* the city. "What are you doing?!" Romanek yelled. "I've just broken my neck to get here just in time. If this is wrong—it's all for nothing. We'll never make it!"

Finally, Romanek persuaded the headstrong driver to pull over and ordered, "Wait for me." He ran into a building, found a phone, called information, and asked for the address of NASA headquarters. The operator gave it to him.

He got back in the cab and gave the cabbie the location. "Take me there as fast as you can," he instructed. "I've got a press conference in one hour!" Romanek was frantic, nauseated, feeling a cold sweat on his face. The cabbie, chastened, got back on the Baltimore-Washington Parkway, veered out of the slow lane, and went rumbling along the shoulder in order to shove past the slow-moving morning rush. He ran red lights.

They pulled up to the stone-and-glass facade of NASA headquarters with about twenty minutes to spare.

"How much do I owe you?" Romanek asked the cabbie.

"Nothing," the glum, shaken driver told him. "Please, just get out of my cab."

Romanek walked into the big, first-floor auditorium in his giant suit. "Where's the research team that's going to have the press conference today?" he asked somebody.

"Well, what do you want to know for?"

"I'm part of the research team. . . . No, really. I am."

Somebody finally helped Romanek catch up with his comrades. To his great relief, they greeted him warmly. "Chris! Glad you made it!" They gathered in Wes Huntress's office upstairs, and somebody came in with about ten big pizzas. They took five minutes to eat and then headed down to the auditorium.

A pumped Dan Goldin, on his way to the podium, told a companion, "I've been waiting for this my entire life."

They walked into pandemonium. For Romanek, such scenes had existed only on TV. Now he was in the middle of one: some fifty cameramen snapping and clacking away, lights flashing, people pushing and shoving and tugging him this way and that, reporters throwing questions, microphones stuck in his face.

There was no room on the stage for Romanek, and besides, he had not been prepped. He found himself sitting in the audience with Simon Clemett behind a phalanx of VIPs. The momentousness of the event would crystallize for Romanek shortly, when he heard Everett Gibson, up on the stage, say that this moment was more exciting for him than working on the Apollo missions.

"It was beyond my wildest dreams," Romanek would say later. He was somewhere high above, looking down at himself. He finally knew what it meant to have an out-of-body experience.

The countdown was at zero. The speakers sat on the stage in NASA's big main auditorium. McKay, Gibson, and Zare were exhausted. Zare had not slept in the last thirty hours. Kathie Thomas-Keprta (she had taken her husband's name when she'd signed the paper, as thanks for his patience with the recent craziness) looked calm enough, her mane of softly curled blond hair tamed by a dark headband. But she felt nauseous. She didn't know it yet, but she was pregnant.

Sweaty-palmed and dry-mouthed in varying degrees, the scientists in the bright lights sensed the restless focus of the peripatetic media horde settling abruptly on them like one of Zare's laser beams. This was their moment.

McKay, wearing his Apollo tie, and Gibson were struck by the contrast between the smaller, more relaxed, and virtually all-male press pack of the 1960s and the major-league circus in front of them now. Out beyond the stage, milling and swarming like cattle on the verge of stampede, was one of the biggest crowds drawn to a NASA press conference since the glory days of Apollo.

NASA had borrowed a chunk of the Allan Hills meteorite from the Smithsonian—the "ambassador" sample, which would be seen by the vice president and first lady, shown at a Senate hearing, and exhibited in the museum. It weighed 1.3 ounces (37 grams). Now it rested like a stolid punctuation mark in a glass case on a velvet swatch on the blue-draped table, right in front of McKay. It glowed in the blaze of light. Photographers hunched and squatted around it, searching for the best angle on the ignoble-looking lump.

The press conference and related proceedings, which would last two and a half hours, were being carried live on CNN, ABC, NBC, and CBS.

But first, the electronic eyes swiveled abruptly toward the other end of Pennsylvania Avenue. At 1:13 P.M., President Clinton stepped to the microphones to address the story of the rock from the Rose Garden (using some of the wording suggested by Huntress and NASA): "Today, rock eight four oh oh one speaks to us across all those billions of years and millions of miles. It speaks of the possibility of life. If this discovery is confirmed, it will surely be one of the most stunning insights into our universe that science has ever uncovered."

When the cameras refocused on the stage at NASA headquarters and the waiting press conference, Boeder introduced Goldin. As he took the podium and peered into the limelight haze, the NASA administrator was so excited that he sometimes tripped over his tongue as he went through the requisite niceties, introducing VIPs and thanking various institutions and people, including his father, who was on his deathbed.

Goldin had already emphasized in a written statement that nobody here would be talking about "little green men" and that the evidence was "exciting, even compelling, but not conclusive," involving at most "extremely small, single-cell creatures that somewhat resemble bacteria on Earth." Now he took care to repeat the caveats, adding that the scientists were "not here to say they found ultimate proof or evidence." The discovery was "certain to create lively scientific debate and controversy. . . . We [NASA] will be driven by scientific process and not a rush to go to Mars."

At the same time, he said, the team's work had "brought us to a day that may well go down in history." The scientists had come here to "to tell a fascinating detective story." He added, "We're now at the doorstep of the heavens. What a time to be alive."

The scientists on the stage faced an uncomfortable challenge. The average citizen could readily grasp the concept of life on Mars, but this story was highly nuanced. To many in the wider audience, the very *fact* of the press conference, added to the headlines and TV images of the last day or so, implied some sort of finality, of certitude. The bulk of the audience, including some journalists, would have little, if any, familiarity with the arcane meat of the evidence under discussion.

Wes Huntress, Goldin's chief space scientist, set the scene, reminding everybody about the paper soon to be published in *Science*, and handed the microphone to McKay, seated behind the meteorite chunk in its velvet nest. McKay's eyes behind his glasses were huge brown-satin buttons in his pale face. He cleared his throat, managed a fleeting smile, and began his story of "why we think we have found evidence for past life on Mars." He repeated Goldin's cautions, saying, "This is a controversial story, and there will be a lot of disagreement."

McKay briefly outlined the four lines of evidence: (1) the meteorite came from Mars and contained carbonates; (2) the carbonates' mineralogy and chemistry were compatible with biological origins; (3) the rock contained organic compounds (Zare's PAHs) sometimes associated with life; and (4) images of the rock showed fossil-like forms.

McKay noted that the team's interpretation of those forms as fossils was "perhaps the most controversial part of our presentation." He also said there were "alternative explanations for each of the lines of evidence that we see. . . . But," he concluded, "when you look at them all together collectively, particularly in view that they all occur within a very small volume [in the rock sample]—every sand-size chip has most of these kinds of evidence in it—we conclude that taken together this is evidence for early life on Mars."

He passed the mike to Gibson for an overview of the story, accompanied by the NASA animated video that would be replayed on TV around the world. It showed an artist's conception of the team's proposed scenario. It began with the rock's birth on Mars some 4.5 billion years ago, and followed with violent bombardment by cosmic debris that fractured the planet's surface. The "camera" zoomed in for a close-up of blue water flowing into the fractures in the rock, carrying specklings of minerals, and finally showed shiny bluish wormlike organisms (depicted at 100,000 times magnification) getting trapped in the gradual buildup of the carbonate deposits. The on-screen story, with Gibson narrating, jumped forward across billions of years to the frigid, dried-up Mars of about 16 million years ago, where the audience saw another large object slam into the surface, kick up a cloud of debris, and send the rock spinning off into space. The animation showed the rock plummeting onto the Antarctic ice sheet 13,000 years ago. As Gibson described its discovery by a National Science Foundation field team, there was film of a technician processing the meteorite in a Johnson Space Center laboratory.

Gibson's West Texas drawl accelerated with energy as he and the images took the audience further into the crevasses of the rock, and he briefly described the mineral composition. He ended up with an arresting color photograph of the carbonate pancakes, girdled by their Oreo rims, taken by Monica Grady of the Natural History Museum in London. "These are five times the diameter of a human hair," Gibson said of the carbonate globules, adding that they were unusual in a meteorite.

He passed the mike to Kathie Thomas-Keprta. She showed, in a cartoon cross section of a carbonate globule, how the black portions of the concentric Oreo rims were actually made up of fine mineral grains—magnetite (iron with oxygen) and pyrrhotite (iron with sulfur). In another region inside the orangey carbonate, she showed a patch of dark magnetite mixed with supposed greigite (also iron with sulfur). She noted that, on Earth, it is quite common for greigite to be produced by bacteria, and that, on Earth, this particular type of pyrrhotite can be produced either inorganically or by certain types of microorganisms. "In sum," she concluded, "although we feel there could be very complicated inorganic explanations for the presence of these mineral grains, the simplest explanation is that these are products of microorganisms on Mars."

Zare spoke next, explaining how his lab had found the minuscule amounts of organic chemicals—the greasy PAHs—deep inside the rock. Crisply, with the barely contained delight familiar to his students, he explained what exactly his Stanford team had found. Though they could have come from either biological or nonbiological sources, he noted, these PAHs were different from those found in other meteorites, and their distribution in the rock was much simpler. "It very much resembles what you'd expect when you have simple organic-matter decay," he said.

Also, he added, his team had mapped the PAHs in the rock and found that they were more densely concentrated toward the rock's core and correlated with the carbonate globules—the opposite of the pattern to be expected if the PAHs had seeped into the rock after it had arrived at Earth. Aided by the animation, Zare described his Stanford team's immensely sensitive laser technique, which enabled them to look at targets as small as a few thousand molecules.

Zare concluded his talk by stating what some knowledgeable listeners would describe as the most indisputably important news from the rock: "By this means, we've been able to look and see the first organic molecules that we believe come from Mars." Here were the long-sought signs, which the Viking landers had failed to detect, that once upon a time Mars had held crucial building blocks of life—and might still.

Next, McKay brought forth what he and others expected would be at the same time the most arresting and the most controversial portion of the evidence: the images the team thought might be fossils of ancient Martian organisms. (Only one of the fossil-like images had been reviewed by outside experts and approved for publication the following week in the journal *Science*—drawing criticism later on.)

The images on the auditorium screens zoomed in to a highly magnified view of a section of the iron-rich Oreo rim in the Grady image of a carbonate moon. The next image showed a grainy field of what looked like varieties of beads in different sizes cemented together in mounds.

McKay noted that "the biggest object in the picture is about one hundred nanometers across—that's about one-five-hundredth the size of a human hair. . . . You can see that there's one rod shape, there's one with a dark line up the center of it." Some of the shapes were probably the magnetic mineral grains Thomas-Keprta had shown earlier, "but some of these don't look like either magnetite or iron sulfide," he said. "They may be something else. And we're not quite sure what that is."

McKay then showed a structure that resembled a rough wall or reef. In the lower half of the image was a school of jelly bean–shaped objects that seemed to be "swimming" across it. "These are elongated forms, structural forms. We think that matrix they appear to be eroding out of is probably a clay mineral. We're [in the process of] confirming that. . . . The features that you see may be any number of things. For example, they could be dried-up parts of that clay. Or they could be microfossils from Antarctica. Or microfossils from Mars. It is our interpretation, the one that we favor, . . . that these are in fact microfossil forms from Mars," McKay said.

Once again, he added a caution. "But keep in mind that is an interpretation. We have no independent data that these are fossils. We don't have pictures showing cell walls or internal material characteristic of cells. It is simply an interpretation at this point."

He showed more grainy material with rodlike or wormy shapes seemingly caught in mid-squirm over the rounded surfaces and draped over the mounds, mostly in the left half of the image. "Are these strange crystals? Are they, uh, are they dried-up mud? We believe, we interpret that these are microfossils from Mars. They are extremely tiny. The longest one is about two hundred nanometers. This is very high magnification. . . . We are looking at rocks and minerals at a scale that has really not been used before."

At this point, for comparison, McKay showed an image of possible terrestrial nanobacteria on an Earth rock—features found by a completely different group and interpreted as organisms in a 1994 paper. (The things looked something like white jelly beans and candy bits strewn on a crumbled pile of dark cake.) They were found "on calcite, calcium carbonate, the same kind of material we are looking at on Mars," he said. At about 500 nanometers, unusually small for most known bacteria, he said, "these things are the same size and shape as many of the forms we're seeing in the Mars sample."

At the next image, the audience let out an audible gasp. It was the feature that would become known as "the worm," a segmented shape resembling a stretched-out Tootsie Roll, reclining on a grainy slope. "As we move on, we see a few of these elongate forms which appear to be segmented," McKay said. This one was about one-hundredth the diameter of a human hair, "again very tiny but now we're getting up into the size range of common terrestrial microbes and bacteria, and whether this is a microfossil or whether it's a dried-up mud crack we can't really say because we have no data other than what you see. . . . But again, it is our interpretation that this and similar features have a high probability of being Martian microfossils."

McKay showed another image of the rock produced by a different technique, taken by team member Hojatollah Vali at his lab in Canada. This time, the scene resembled a bas-relief of snakes slithering across a frieze. "We don't have chemistry on these," he said, "but one possible interpretation is that they are similar kinds of Martian microfossils" to the ones seen under the microscope in Houston.

McKay concluded with a slide of "real bacteria . . . , which turn out to be about the same size and about the same shape as the things that I've been showing you in the Mars sample." These organisms had been found in a sample drilled from volcanic rock more than a mile underground in the Columbia River area of Washington State. The native Martian environment from which the Allan Hills rock had been blasted might once have resembled the modern microbial habitat in the Columbia River formation: that is, a subsurface fracture in basalt, through which water flowed.

During the question period, McKay and others would elaborate further on the evolution of the Martian environment—how at some point in the planet's history "things went bad." The atmosphere mostly disappeared, either into space or into carbonate rocks below the surface, where it might remain locked up today. And the water dried up, some of it wafting into space, some possibly still on the planet in the form of permafrost or "even as a groundwater system." But what had happened to any life-forms that had existed there before this sorry turn of events? They could not live on the surface, but it was possible that some of them might survive underground—like the Columbia River organisms on Earth, drawing their energy from hot springs, and hydrothermal areas, and interactions between the atmosphere and the rock.

McKay ended with another recitation of caveats. "We have a number of forms which it is very tempting for us to interpret as Martian microfossils—*but* we have no confirming evidence, and you'll hear more about the pitfalls of identifying such things based on appearance alone. We don't have the chemistry of these. We'll find out if they have cell walls or not. . . . That will be part of our future work. But for now, we have to use these images and interpret them the best way we can. So I want to finish up here by simply saying that we have these lines of evidence and none of them in itself is definitive. But taken together, the simplest explanation to us is that they are the remains of Martian life."

For David McKay, the past forty-eight hours had been a wrenching journey that had carried him light-years, at warp speed, away from the clear waters, patient cliffs, and peace of the Texas hill country. Now, despite the dislocation and fatigue, he had told his story at last.

Huntress gave the last word to Schopf, the designated skeptic.

While the other scientists were talking, it seemed to Zare, Schopf had sat there wearing the sour face of someone sniffing a carton of milk to see if it's gone bad. A few years later, writing about the news conference in a book, Schopf would portray himself as having been sandbagged by the superior preparation and graphics of the McKay group—a version that Zare would disagree with.

Schopf began by saying that he preferred to render his comments "as part of a discussion rather than a debate. . . . I do think this is a fine piece of work. And this is not easy science." He moved quickly to his main point by reading a quote attributed to the absent Carl Sagan, as the words flashed up on the auditorium screens: "Extraordinary claims require extraordinary evidence."

Speaking with the assurance of the seasoned lecturer and teacher, Schopf mentioned his long history of searching for ancient life on Earth, and outlined criteria he used to test such claims on Earth. "And those criteria in my opinion must be met on Mars as well," he said.

He showed a slide depicting his own crowning achievement: microfossils from the Apex chert (a type of quartz) in Australia. At 3.465 billion years in age, they were the oldest evidence of life on this planet, he noted. Magnified on-screen, they looked like long, sinuous filaments clearly divided into segments, again like skinny, stretched-out Tootsie Rolls. "They are demonstrably cellular, as you can see, and they are composed of organic material. Their cell walls are made of organic matter."

On the next slide appeared another set of fossils from the same deposit: "They have conical end cells, they have rounded end cells, they have demonstrable cells, and all that. These are demonstrably fossils." He drew attention to a minuscule strand one-half micron in thickness. "This is a bacterial strand, it's 3.5 billion years in age, it comes from this Earth, and it is *one hundred times larger* than these microscopic objects that we have just seen from Mars. And that is one of the smallest—shown in this slide—one of the smallest fossils that has been found on Earth."

Schopf's final slide showed a confidence-rating chart that used seven criteria to compare the evidence of life on Earth to the evidence of life on Mars as just presented. Schopf punched his words. "I want to emphasize this is *subjective.* It says *subjective.* It is italicized *subjective.* It is my *o-pinion.*"

Noting that the Olympics had just ended, he said he was using a similar scoring technique—a scale of one to ten. He gave the probability that the meteorite was from Mars a nine. The age of the carbonate globules, too, seemed "pretty well established" at around 3.6 billion years. "I give that a confidence rating eight."

Evidence of the rock's environment and history was the subject of debate, he pointed out. Another team of scientists had recently argued in a published paper that the carbonates had been formed at 450 degrees Celsius (842 degrees Fahrenheit), much too hot for biological activity. They had also suggested that the rock had been fractured during the impact that threw it off the Martian surface, rather than well before, as the McKay team maintained. "I'm not taking sides in the matter. I'm simply saying that this is not a resolved issue as yet in the minds of some people," Schopf said, adding with a smile, ". . . although I think the guys here at NASA make pretty good arguments."

Schopf found it likely that the organic matter and the fossil-like objects, like the rock that contained them, were also from Mars. "I think it's very likely that even though they occur in fractures where groundwater [on Earth] can introduce things, I think the data are good," he said. "I give it an eight or nine rating that in fact those things are as old as the fractures in that rock" and therefore are from Mars.

However, addressing the possibility that the PAHs might have come from a *biological* source, Schopf turned feisty: "I take a rather different view. With regard to the polycyclic aromatic hydrocarbons, I *note* that such compounds are found in interstellar dust grains. I *note* that PAHs are found in interplanetary carbon grains. I *note* that PAHs are found in other sorts of meteorites, like carbonaceous chondrites. In none of those cases have they ever been interpreted as being biological. This is, after all, a meteorite." Therefore, assuming they're not contamination from industrial pollution on Earth, he said, "I'd say the first guess would be that they are probably nonbiological, just like PAHs that occur in other meteorites. The burden of proof is on those who claim that they're biological."

As for the fossil-like objects, he said, "I note that they are one hundred times smaller than such fossils that have been found on this planet. I note that there is no evidence of their composition. The best guess at this point would be that they are made of mineralic material. At least, there are no data—and it's because they are so small there are no techniques at present to analyze their chemical composition. But there's no evidence that they're made out of carbonaceous [i.e., organic] material. We don't know that yet. Thirdly, there's no evidence that there is a cavity within them, a compartment, a cell. Why do you need that? Well, that is where the juices of a living organism reside. That's where the chemistry that makes things live works. You've gotta look inside these things, see if they have cell walls, see if they're compartmentalized, see if they're cellular, see if they're composed of organic material. There is no good evidence as yet of life cycles, or of cell division—the tests that we also apply to the fossil record [on Earth]."

In his opinion, he said, the biological interpretation was probably unlikely. "I finally come back to Carl Sagan's quotation, which I think is applicable. . . . Personally, I think that this is exciting. I think it is very interesting. I think they are pointing in the right direction. But I think a lot more, or a certain amount of, additional work needs to be done before we can have firm confidence that this report is of life on Mars."

In the months and years to come, McKay would hear the Sagan quote repeated constantly by Schopf and by others. And he would confess now and then that it annoyed the hell out of him, given that Sagan had helped referee the very *Science* paper that was at issue.

There followed an hour-long barrage of questions from the assembled journalists as well as from reporters at other NASA facilities across the country, linked in by satellite. These exchanges focused heavily on what the findings might mean for public policy in space- and earth-based research. But there were a few queries about specific problems with the data, triggering a brief point-counterpoint between Schopf and Gibson. Someone asked what it would take to convince a doubter like Schopf. What was the "smoking gun," and was it obtainable?

"Absolutely," Schopf said. "And I'd like to see data on a couple thousand individuals. . . . You say, oh, that sounds like a lot. In the Apex chert, the oldest fossils on this planet, I have personally measured nineteen hundred, nineteen hundred and fifty cells. That is what is required to nail this thing.

"Okay, give me five hundred. That's enough. That's enough." The audience was laughing now.

When Goldin declared the press conference over, reporters surged toward the scientists at the front of the room to press for still more information.

By the end of it, the sleep-deprived Zare felt a growing desire to escape from what he considered the "surreal grandiosity of the moment," the relentless, burning attention of the TV lights and staring reporters.

Instead, as the day wound down, he scrambled to his hotel to finish an opinion piece he had been working on for *The New York Times*, an article that was suddenly in urgent demand. Then he was hustled off to a public television studio across the river in the Virginia suburbs, to tape an interview with PBS's Jim Lehrer for that evening's news. Zare had expected to be paired with McKay for that appearance, but NASA decreed that Huntress should do this one instead.

When Lehrer asked Zare why all this should matter to the rest of us, the scientist responded that he thought everyone must be a little curious about the fundamental nature of life, how it starts, how rare it is, and the possibility that biological material, on occasion, might have traveled from Mars to Earth on a meteorite and might even have provided the first seeds of the living world billions of years ago. Zare posed the question: "Is it possible that actually we're all Martians?"

None of them was prepared for the magnitude of the public response.

By early on August 7, the day of the press conference, *Science* magazine and the AAAS had posted the McKay paper on two Web sites (its fledgling EurekAlert and the online version of the magazine). Within days, almost a million people had seen the paper by that means.

NASA's Web site scored more than half a million hits the first day. By the end of that week, the NASA public affairs staff had counted more than a thousand stories about the rock broadcast in the top thirty-nine U.S. TV markets alone, in addition to vast coverage in newspapers and magazines. The story eclipsed former space news events such as the first landing on the moon, the first flight of a space shuttle, and the explosion of the shuttle *Challenger*, the staff reported.

"*Des Traces de Vie sur Mars,*" trumpeted the French newspaper *Le Figaro*. "In many ways . . . the headline mankind has been waiting for since the first human eyes looked into the heavens and saw God, fear or some amazing future journey," said a cover story in *USA Today* about the potential implications of the news.

Despite the unmannerly way in which the story hatched out of its protective shell, the predominant opinion among a sampling of scientific and journalistic evaluators was that the bulk of the initial reports were balanced and accurate—though inevitably lacking the amount of detail many scientists would always prefer. A few outlets were deemed guilty of some degree of exaggeration, most often an overstatement of the degree of certainty that Mars had once had microbes. And there were isolated irritants, such as that fleeting invocation of maggots, which so upset Zare and others. Regardless of the content of the reporting, Schopf, for one, suspected that the sheer volume of headlines and other coverage created the false public impression that the McKay team had already made a persuasive case for primitive life on Mars.

Congress scheduled hearings on the rock, with testimony from scientists and policy makers. The Clinton administration scheduled a "space summit," to discuss the appropriate government follow-up. An auction house in New York ran an ad in *The New York Times* proclaiming that the twelve known "meteorites from Mars have been thrust into the position of being perhaps the most precious natural objects in the world." It offered for sale the only collection of SNC fragments in private hands.

The iconic seventy-five-year-old Ray Bradbury, author of *The Martian Chronicles* and other science-fiction classics, gave the rock a curmudgeonly nose thumbing. He refused Ted Koppel's invitation to appear on *Nightline* to talk about it, because he didn't believe the claims and "didn't want to be a grouch."

Microsoft's chief strategic daydreamer, group vice president Nathan Myhrvold, wrote an essay, which *Time* magazine excerpted from *Slate* online, cheerfully calling the McKay group's announcement "the biggest insult to the human species in almost 500 years." In the short span of human existence, the revelations of science had managed to demote the home planet from the center of the universe, where Ptolemy had put it in the second century, to insignificance somewhere in a cosmic outback. And now it seemed the planet was losing its status as the lone oasis of life. But, Myhrvold and others suggested, location is not always everything. Humans still had the only known *intellects* in all creation, didn't they?

Back at Building 31 in Houston, the contemplative calm was shattered. Employees arriving for work might find as many as twenty news crews lined up in the lobby, waiting their turn to go into the meteorite lab. For days and even weeks, the inhabitants felt as if they had been invaded.

Everybody connected with the McKay team was inundated with phone calls, requests for interviews and autographs, and other kinds of attention. Collectively, over that fall and winter, the group would do some 260 TV, radio, print press, and lecture appearances around the world. In Australia alone, McKay would give five lectures in five cities in five days.

Gibson would do more traveling in the nine-month stretch following the press conference than he had in all his twenty-eight years at NASA combined. He appeared before the Royal Geographic Society in London, at the same podium where Stanley and Livingston and Amundsen had been welcomed.

Robbie Score, the rock's discoverer, had quite recently left NASA and Houston behind to answer the lure of the Antarctic. Just as "her" rock was becoming world famous, she had taken a job that would enable her to spend the austral summers back in the forbidding, enchanting place where she had found it. Still, with the help of the National Science Foundation, the media troops tracked her down. In numerous newspaper and magazine articles she became the face of the Antarctic meteorite search teams.

Chris Romanek, also off at a different job, would get sporadic calls from NASA officials trying to fill the public demand for speakers familiar with the story of the rock.

The rock also helped force the resignation of presidential adviser Dick Morris. In one of its odder ripple effects, the NASA press conference lent credibility to call girl Sherry Rowlands's tales of presidential eavesdropping, convincing the *Star* reporter that she was worth paying $50,000 for her revelations. The tabloid splashed the story across its front page on August 29. Morris quit the same day.

The reaction that McKay and his team cared most about came, of course, from fellow scientists. The group knew they had dared to stake out a preliminary claim to one of the so-called holy grails of science. There would be only one chance in history to discover the first proof of life on another planet. If they were right, they would also displace Schopf, in the textbooks, for having found the oldest known life on any planet, including Earth. If they were right, their discovery would be as unsettling to existing models of reality as the Copernican revolution had been. And their names would be in the history books forever.

As Schopf had made plain, such a coveted trophy would not be retired without a fight. But no one on the McKay team had foreseen the ferocity of it. The response began like a temblor, a subtle rumble of something beginning deep down and building toward release at the surface.

Then it rose up in their faces like a screaming, scouring Martian dust storm.

CHAPTER TEN

SCHOPF SHOCK

When he arrived at the Friendship Hotel in Beijing, Bill Schopf could tell at once that the type of limestone used to tile the registration desk and lobby floor had been constructed 850 million years ago by slimy microorganisms. The distinct curvilinear layers of light and dark that graced the decor were laid in cycles of long-ago life and death.

In the summer of 1982, fourteen years before the NASA press conference, Schopf and his wife were guests of the Chinese Academy of Sciences, doing research at the Institute of Botany. The forty-year-old visiting professor took a picture of that lobby, with his wife in the foreground for scale. He recognized the rock formations because he had just been working near the village of Jixian, east of Beijing, at a site where they occurred in nature.

The Chinese called them "flower ring rocks." To geologists like Schopf, they were known as stromatolites (pronounced stroh-MAT-oh-lites, a word derived from the Greek for "stony carpet"). The Chinese had been using the material in their buildings for more than two thousand years. Schopf saw examples in the Forbidden City, at the Temple of Heaven, and in the columns of the Great Hall of the People on Tiananmen Square.

Just as the Chinese had constructed their edifices of flower ring rocks, Bill Schopf was building a career on them. Sometimes it seemed his entire world, everywhere he looked, was layered in the ancient husks of microscopic builders.

One day, Schopf and his wife arranged to visit a vast factory where Chinese workers cut and polished the rocks. Schopf noted the variety of colors and designs the microbes had devised—"light gray with jet-black layers, dark gray set off by beige or light gray layers, dull red to pink and purple flecked with whitish zones."

Schopf's hosts even gave him a bargain deal on a damaged but still spectacular slab of Precambrian stromatolites arrayed in columns of bright red crossed by veins of sparkling white calcite. The Schopfs later arranged to have the piece embedded in a place of honor in the entryway of their Los Angeles home, which was strewn throughout with assorted stromatolite paperweights and doorstops.

Over tea and cakes in the factory managers' offices, Schopf tried to explain why he had become an aficionado of stromatolites in all their manifestations: he thought they might just be the mother lode of clues to the elusive history of Earth's earliest living things. His hosts could not have known that Schopf's quirky preoccupation with their construction material would help alter human perceptions of how they (humans) came to exist and, in the bargain, make him the "god of the Precambrian."

Before the year was out, Schopf would have in hand the specimens from which flowed the discovery that would bring him international acclaim. This was the path that would lead Schopf to sit on that brightly lighted NASA headquarters stage with David McKay and the others on August 7, 1996.

The encounter would accelerate the interweaving of both knowledge and doubt from two fields that had been segregated: studies of the most ancient rockbound mysteries of life on Earth and the search for their counterparts on Mars. And it would touch off a bitter personal antagonism between the fiery academic and the reserved NASA civil servant.

Schopf approached his public comments about the meteorite with the gimlet eye of a man who, after all, had devoted his life to the daunting task of finding ancient signs of life entombed in rock. He knew this kind of work involved bad odds piled on top of evil odds—even on his own home planet, where he could pick his shots, and where there was no doubt that life abounded. But a single specimen from Mars, selected randomly in a primordial bombardment—?

Schopf knew what it took to beat back the critics in such cases. This was his turf. It was where he had spent his life, and it was a point of pride that he was among the few to acquire the unique blend of expertise the task required, in chemistry, microbiology, geology, paleontology, physics—the list went on.

In 1960, as an eighteen-year-old sophomore at Oberlin College in Ohio, Schopf had developed a keen interest in finding the missing record of Earth's earliest life—even though it was a singular mystery whose resolution seemed frustratingly beyond reach.

Scientists had found a stunning diversity of multicellular life forms revealed in the remarkable fossil record of the "Cambrian explosion," beginning about 550 million years ago. But if Darwin's theory of evolution was correct, there ought to be some sign of the long buildup to this rich biological harvest. And yet, the first four billion years of the planet's history remained, for the most part, a stubborn blank.

This mystery period—the Precambrian Eon—encompassed some 90 percent of Earth's prehistory and almost a third of the age of the known universe. The sun and planets had condensed from a cloud of dust and gas, and the Earth formed out of the collisions of smaller bits of debris. Organic molecules appeared. At some point, the planet's primordial chemical soup spawned what might be considered the mother cell of all known life. That entity diversified into the known domains—bacteria and their single-cell cousins on one branch (plus an entire branch, the Archaea, still undiscovered in 1960), and on another the more complex domain of animals, plants, and fungi.

Until the 1950s, pioneering souls who attempted to claim discoveries of Precambrian life signs met with either embarrassing reversal or skepticism and ridicule from influential and dogmatic scientists. Some people scoffed at the radical notion that anyone could find actual proof of bacteria in rock half a billion or more years old.

The skeptics had plenty of ammunition. On teeming Earth, all but a fraction of a percent of primitive living things got eaten or otherwise used by other living things. Very few biological scraps lingered long enough to be preserved (buried) as fossils. As for the rock that entombed this record, the planet recycled most of it. Most terrestrial rock had been heated to incandescence or altered by water or otherwise processed by nature in ways that wiped out any signature left by fragile life-forms of long ago. Geologists referred to this state with the useful acronym FUBAR ("fouled up beyond all recognition").

In 1961, just as Schopf's youthful interest in the subject was perking up, an expedition to a supersalty lagoon called Shark Bay, on Australia's west coast, found the first compelling evidence that stromatolites were a product of living creatures. Exhibit A was a community of stromatolite colonies currently alive and thriving—and building layered rock that looked remarkably like formations known to be billions of years old. These constructions, in other words, had not evolved in all that time and, as researchers would later show, neither had the microscopic builders themselves.

The discovery of live colonies ended a long-simmering controversy that had hobbled progress on the mystery of Earth's earliest life. Schopf was troubled by a phenomenon that had contributed significantly to the delay. That was the combative resistance of a smart, powerful, well-credentialed scientist, Sir Albert Charles Seward. "His aggressive skepticism, delivered from his throne of unquestioned authority, was a disservice to the field," Schopf would write years later. "His dogmatic pronouncement . . . stifled the hunt for the missing Precambrian fossil record for nearly 40 years."

Schopf himself, as he sat on stage with David McKay in 1996, would be in a similar position, that of powerful authority figure with the imprimatur of textbook wisdom. He, too, would be influential in hindering or accelerating the pursuit of an unconventional line of inquiry. Right or wrong, he knew, the stature of individuals affected the process. But Schopf, like Seward, had to rely on his ability to read the evidence put in front of him. The evidence—the data—was the final arbiter.

At least one mentor warned the young Schopf that his chosen line of inquiry, so tricky and contentious, could be a career dead end. But Schopf had grown up in a family of scientists, his mother educated in botany and math, his father a paleobotanist. He was loaded with outward self-confidence, he was comfortable with conflict, and he had, as he says, "the Precambrian fire in my belly." He believed Darwin was right and thought he, Bill Schopf, had as good a chance as anyone of proving it.

In 1963, after graduating from Oberlin, Schopf entered Harvard University as a student of Elso Barghoorn, the noted paleobotanist—a specialist in plant fossils and geology—who had helped give birth to the new field Schopf now embraced. Barghoorn had helped identify milestone fossils, the first ever found from the mysterious Precambrian time. A colleague had discovered the fossils in the Gunflint Formation, a band of rock along the northern shore of Lake Superior in Ontario that was rich in a glassy black rock called chert (ideal for producing sparks and used as flints for early muzzle-loaders). The fossils were embedded in wave-washed outcroppings of the rock, stacked like hotcakes in concentric rings up to three feet across. Stromatolites!

The Gunflint discovery represented the first known microscopic fossils and, at 1.9 billion years old, the most ancient fossils of living things yet found.

In the course of his work with Barghoorn, the young Schopf found himself a co-conspirator in a demonstration of just how cutthroat even the most respected scientists can be when the stakes are high. As in the Seward case, Schopf was struck again by the realization that the pursuit of truth can never be isolated from the character of the individuals pursuing it.

For years, Barghoorn had been putting off logical follow-up work needed on the breakthrough Gunflint discovery. He had intended to interpret, describe, and name the early life-forms represented in the fossils. In the summer of 1964, another scientist abruptly preempted Barghoorn by submitting a manuscript to *Science* magazine that accomplished the follow-up Barghoorn had neglected to do. The editor at *Science,* as part of the routine peer-review process, sent a copy of the paper to Barghoorn for an honest, independent assessment of its validity. That was not what the editor got.

"Barghoorn was livid" when he saw the other man's paper, Schopf would write years later. Barghoorn instructed his protégé, Schopf, to help him deliberately delay the rival paper for a couple of weeks, so that the two of them could use the time to get their own version of the work ready for publication.

This tactic led them to cut a significant corner. Because of their competitive rush, Barghoorn and Schopf did not take the time to examine the actual individual fossil specimens but, rather, relied on photographs for their analysis. "This, I am sorry to say, is not good science," Schopf would write. "It's true that we were forced into this by unusual circumstances, but this was not the proper way to do this job."

Apparently through his stature and sheer audacity, Barghoorn persuaded both the author of the rival paper and the editor of *Science* that Barghoorn's paper should be published first. And so it was.

Schopf acknowledged that, to an outsider, this kind of maneuvering over who got credit "must appear unseemly, even tawdry. And of course it is. But science is done by real people, and it's much more competitive than one might expect."

In the world of science, *credit* is prime currency.

Decades later, some of those feeling the sting of Schopf's criticisms would point—fairly or not—to this incident, or more particularly his attitude toward it, as a sign of his take-no-prisoners approach in dealing with perceived rivals and threats to his status. (This was a perception that Schopf would vigorously challenge.)

Meanwhile, a gathering weight of evidence soon satisfied the wider world that Precambrian fossil hunters had indeed found an effective strategy for ferreting out the hidden record of Earth's early biology. The strategy was this: the hunters would look for flower ring rocks—stromatolites—composed of cherty, coal-like rock (rich in organic carbon) with a fine-grained texture that showed it had not been subjected to the heat and pressure that rendered most fossils unrecognizable.

In 1982 (the same year as his tour of duty in China), Schopf's travels took him to a site in the rough, dusty outback of northern Western Australia. His destination was one of the oldest rock formations known on the planet—and more flower ring rocks.

On this trip, he and coworkers took samples from a rock unit 3.465 billion years old, known as the Apex chert. The collection would make his reputation—but not without about a decade of tedious and sometimes contentious labor by him and his student assistants. The real hunt took place not in the outback but back in the laboratory, during long, meticulous microscope sessions. The features under study were charred, shredded, cooked, fragmented, and generally difficult to make out—hardly worth his attention, he noted, if they weren't so ancient.

Two graduate students in succession studied the specimens and "found nothing." A third, in 1986, detected something interesting but moved on to other work. (These difficulties—the degraded state of the samples and the fact that the grad students had seen nothing significant in them—in the year 2002 would take on importance.)

Schopf himself took another look at the Apex chert samples in 1989, while on detached duty in London.

He probed for three kinds of evidence to prove he was seeing tiny cellular Precambrian fossils: the presence of layered formations built by microscopic organisms; the organic matter produced in biological processes; and the fossilized cells themselves. He wanted cellular shapes that had been preserved in three dimensions instead of squashed like roadkill.

After several months, Schopf identified a total of eleven different types of microorganisms from the chert samples. He described three of them in an article published in *Science* magazine in 1992, and reported eight more in 1993. Schopf wrote that he could see clearly in these ancient forms the outlines of rounded cell walls and other shapes familiar in modern-day microbes.

His interpretation of the evidence upset the conventional thinking. In the last decades of the twentieth century, Schopf and others painted a stunning new picture of Earth's evolutionary history. In this scenario, life began much earlier and evolved more rapidly than anyone had suspected.

The opening of a window on the mysterious Precambrian revealed that, as Schopf liked to put it, "for three billion years, this was a planet of pond scum!" And "the modern world—the familiar fauna and flora of air-breathers and oxygen producers, the eaters and the eatees—is merely a scaled up version of a microbial menagerie billions of years old."

With that insight, plus the revelations of "extremophiles" thriving in all sorts of toxic nooks all over the planet, at the approach of the millennium the evidence seemed overwhelming that the most conspicuous and enduring feature of terrestrial life was its stable microbial population. Just as Earth had been displaced from the center of the universe, the proudly big-brained human species was in some ways being dislodged from its position of dominance in the realm of living things.

When he took NASA's stage as devil's advocate that day in August 1996, Schopf had the authority born of those 26 years in the paleo trenches. He knew as well as anyone how titanically tough it was to identify life in ancient rock.

When challenged or annoyed, Schopf could be intimidatingly forceful, his voice rising with the roar of a freshening gale. But he was also capable of a folksy, professorial charm, and it was in this vein that he delivered his assessments of the McKay group's hypothesis that day.

Schopf had an undeniable advantage over McKay and his Houston-based collaborators (though not Zare) in terms of prestige, credentials, and funding. He was Goliath to McKay's David. As Schopf was well aware, history had shown that smart and powerful people could be wrong—even determinedly, stubbornly, persistently wrong—and that people lacking in influence could be right. The process of learning something new could be skewed in any direction by a player's status and ability to sway others.

But in the end, advances in knowledge did not rest for long on power, influence, charisma, or credentials, and they did not depend on majority rule. All of the players in the current drama shared the sacred creed: that in the end, the slow accumulation of evidence would reveal the truth.

Schopf's comments generated wide praise, and the heartfelt appreciation of Goldin, Huntress, and others in the top ranks of NASA. They knew he had been reluctant to get involved with the Mars rock, but he provided the balance they considered essential.

The photogenic "worms"—the putative microfossils—became the stars of the media coverage following the NASA press conference. In a society that was more and more visually oriented, a good picture was worth a thousand sound bites.

But Schopf stated clearly that a provocative shape was not nearly enough. To be convinced that the wiggly forms in the Mars rock were indeed fossil remnants of dead bacteria, he required nothing less than what he himself had found in the Apex chert of Australia, with his record-setting fossils. Among other things, he wanted to see rounded cell walls made of organic material. And yet, Schopf noted, the McKay team had included in their seven-page technical article only four sentences on the fossil-like features, with a single reference to their resemblance to terrestrial microorganisms.

The size problem was a deal breaker, in Schopf's view. He knew, of course, about the controversial evidence on Earth of so-called nanobacteria, which had influenced Chris Romanek's early explorations of the Mars rock. For many microbiologists, however, these entities were about as plausible as Bigfoot. Schopf was among many who questioned whether anything that size—Martian or terrestrial—could contain even the most rudimentary cellular machinery required for life as we know it.

The basic cell—the smallest unit of matter that scientists considered alive—had to house plenty of water, a set of instructions (genes), something to carry out life functions (proteins), and something to manufacture the proteins (ribosomes).

A cell was like a little room, or a capsule. If the Martian "fossils" had ever been alive, they would have needed some kind of wall to separate their chemistry from the environment in which they lived. This was a fundamental feature in all known forms of life. But the total breadth of some of the proposed Mars "organisms" was less than that of the simplest known bacterial cell *wall*. They were smaller than any bacteria whose existence had been confirmed—actually a million times smaller in volume than a typical bacterium and more like bits of cellular machinery (such as a ribosome).

Schopf's smiling but unsparing rebuttal set the tone for much of the reaction among scientists—as it was meant to do. Many repeated Schopf's unassailable admonition about the need, in a case like this, for "extraordinary proof."

But the aftermath was more complicated than that.

The McKay team was caught completely off guard by the sheer fury of the challenges from many colleagues. They were staggered by the lacing of bitterness, of sheer personal vitriol.

The more extraordinary the claim, the more certain it was that the combatants' full range of sensibilities would be engaged: the constellation of fears, resentments, and rivalries, the friendships and loyalties among individuals and institutions, the native ambitions, dogmatic beliefs, sense of cultural and financial pressures and tribal memories. A particular irritant was NASA's reputation—born of Apollo—for mixing science, politics, and public relations to a high and, in the view of many, unhealthy degree.

"This is half-baked work that should not have been published," declared Edward Anders, a leading authority on meteorites and the birth of the sun and planets and a veteran of the Apollo lunar-sample investigations. Anders had a lustrous track record on the very question at hand—prospecting for biological signs in meteorites.

In 1961, as students of meteorites were well aware, researchers in New York had announced a similar detection of possible life-forms, structures like "fossil algae," in a meteorite. Their samples had come from a shower of carbon-rich meteorites—fragments from an asteroid, not Mars—that had fallen in the previous century near Orgueil, France. In 1961, as now, the announcement had caused a great hubbub in the press. Then, as now, skeptics had conducted their own analyses and refuted the claim. Anders, then at the University of Chicago, had produced a series of critical papers that played a crucial role in sinking the claim. A consensus had developed that the alien "fossils" were actually Earth contamination—some of it likely plain old terrestrial ragweed pollen.

This and other discredited claims constituted a flashing warning for others working the same terrain. It had certainly pulsed in the heads of David McKay and his collaborators as they'd checked and rechecked their work on the Allan Hills rock for two years, in preparation for going public.

Yet Anders felt compelled to disrupt his retirement in Switzerland. In an e-mail to McKay soon after the paper came out, Anders began with congratulations on "outstanding aspects of your work, especially your superb data and techniques." Then he lambasted the team's conclusions, alluded to the proposed nanofossils as "turd-like shapes," and hammered away at the "distressingly biased" interpretations" and "illogic."

Science magazine published a toned-down version of this message. His words resonant with authority, Anders compared the McKay team's claims with the 1961 case. "The Orgueil meteorite was bad data and bad interpretation. Now [with the McKay claims] we have good data and bad interpretation." Summing up the McKay group's various lines of evidence, he concluded, "Five maybes don't make a certainty!"

Geologist Ralph Harvey, who had done his own studies of the Allan Hills rock, joined the ranks of the outspoken, lamenting that a premature claim of this nature could be a setback for space scientists of every stripe. "Every single person who has written a letter to the editor saying 'We don't need this dreck about space' will say they were proven right if this [McKay] paper turns out to be wrong."

Even if the McKay group turned out to be right, Harvey asserted, their claim had a political taint. "The fact that the president of the United States spoke out on their behalf in an election year, and that the head of NASA was known to be looking to build the case for Mars exploration, means this was science being used for a political end. It scares a lot of us, frankly."

Harvey had recently become the National Science Foundation's man in charge of the annual Antarctic meteorite hunts, the same program that put Robbie Score on the ice back in 1984 and made the current controversy possible. As Schopf mentioned at the NASA press conference in August, Harvey and colleagues had already published one paper at odds with McKay, indicating conditions on Mars too hot to accommodate life.

In the months after the August announcement, the debate got so acrimonious that *Newsweek* reported its descent to the level of "you ignorant slut!" (shorthand for the parody of supercilious talking heads made famous on NBC's *Saturday Night Live*). Anonymous quotes—among the least admirable or credible features of journalism—flew in all directions. The magazine quoted one unnamed meteorite researcher as saying that the McKay team consisted of "an inferior group of people [who] are setting the agenda for others who have real science to perform." Other anonymous missiles included:

- "He wouldn't know the truth if it bit him."
- "He wouldn't know a meteorite if it hit him."
- "I am appalled he is still in the business."

Gibson, ever the most outspoken member of the McKay group, countered at one point that one of his critics, whom he didn't name, "has been in this country for 32 years and hasn't held a permanent job."

NASA's reputation as a scientific "entitlement agency" with a special mandate (which critics often lumped together with the Department of Energy in this regard) fed the flames. Some people, in political as well as scientific circles, loathed the agency's approach to allocating scarce research funds—an approach established at its inception under the Cold War umbrella—because (despite recent efforts by administrator Goldin to change this) they so often saw NASA operating outside the true rigors of the merit-based, peer-review regimens of the scientific establishment.

A corollary concern wafted through the atmosphere in which the McKay team produced their claims. It flowed from the specter known as cold fusion—a "discovery" that had made big news in March 1989. Two chemists claimed they had invented a tabletop device that produced more energy than it consumed, by means of nuclear fusion occurring at "cold" room temperatures rather than at the multimillion-degree temperatures that exist where it goes on inside stars and hydrogen bombs. If valid, the work could have led to a form of cheap, abundant energy that would transform civilization. But the chemists had announced preliminary results of their experiment before any review by independent scientists—or before they even managed to reproduce their own results.

The event, staged in Utah, turned infamous when numerous outside scientists were likewise unable to duplicate the experiment. While a few true believers would press on with the research, the wider scientific community rapidly concluded that the chemists had made serious mistakes in the conduct and reporting of their experiment and had never seen the "amazing phenomenon" they'd claimed.

In that case, the process of testing and knocking down had worked—and worked fast. But the *way* it worked—the initial burst of high-profile news reports (albeit in most cases with appropriate caveats), including the covers of *Time* and *Newsweek*, followed by a highly visible reversal—felt embarrassing to many scientists. The higher the stakes, the greater the sting of collective humiliation. Spectacular blunders of the cold-fusion sort could lead the public to question the credibility of unrelated projects.

Now, amid the furor over the Mars rock, a lot of people mentioned their desire to avoid "another cold fusion"—even though the McKay group (unlike the cold-fusion chemists) had done appropriate checks on their own work and passed the tests of peer review. One meteorite expert told *Newsweek* that the intense bitterness stemmed from a "profound fear" by meteorite scientists "of what this might do to our field. We're at the bottom of the pecking order in NASA's budget, and people are concerned that if this turns out to be as stupid as cold fusion we'll be out on the street."

With the shades of cold fusion, the Orgueil "fossils," and other misadventures, as well as NASA's waywardness, haunting them, and with the added aggravation of having been caught off guard when the Mars rock announcement had escaped untimely into the public domain, many scientists were not in a mood to be charitable toward McKay and company.

At the time of the McKay group's announcement, almost fifty laboratories around the world had pieces of the rock from Allan Hills. Quite apart from any possible signs of Martian life, the rock's other unusual properties (the carbonate deposits, the fact that it was the oldest rock known from any planet, and so on) made it a hot item.

Once the McKay claims surfaced, many of these researchers, from San Diego to New Mexico to London, took their samples out for another look. One researcher was moved to open up his freezer, where he had stored a large amount of ice from Allan Hills, hack off a big chunk, and melt it down for testing as well.

McKay and his eight collaborators had spent more than two years refining and checking their analysis. Other scientists who had not already received samples of the rock were now starting from zero—at a time when getting a piece of the rock had suddenly become difficult. Some would have to wait months.

After the August announcement, the suddenly beleaguered interagency working group responsible for allocating pieces of the meteorite to interested scientists halted the distribution on grounds that the group needed time to consider the blizzard of requests and select the most promising ones. The group set out to develop a strategy for distributing the material while holding a significant portion in reserve for, say, ten years in the future when advanced instruments might clear up the mysteries.

As government committees worked out the new rules, several researchers charged that NASA was trying to limit studies of the rock to the "pro-life" side. McKay, who continued to seek outside experts to help with the group's work on the rock, inadvertently aggravated this sore spot by dispatching some of the team's own specimens to selected colleagues without going through proper channels. His NASA bosses reprimanded him.

Not long after the McKay team's paper came out, there was a party at Johnson Space Center to celebrate. Dave McKay was making the rounds, shaking hands, when he spotted colleague Allan Treiman, of the neighboring Lunar and Planetary Institute. A meteorite specialist, Treiman was one of those who had been interviewed on TV several times about the McKay team's findings. He and McKay had worked closely together a few years earlier and had become friendly. McKay started outlining ways he thought Treiman could help with the next phases of research on the meteorite. Treiman reacted harshly. He had not yet "bought into" the Mars rock claims, had not volunteered to work on the project, and he thought McKay was being a little presumptuous. "I'm just like a little kid," he would say later. "When someone tells me to do something, I sort of resist it. . . . Dave didn't like it when I didn't sign on with him, and that was sort of the beginning of a downward trend in our relationship."

Treiman had been nursing a resentment about work he had done on the Allan Hills rock himself a couple of years earlier, with Kathie Thomas-Keprta serving as his sherpa on the transmission electron microscope in Building 31. (It was one of her duties to make sure people who came in to use the microscope didn't push the wrong button and blow the system up.) Treiman was fascinated by the same carbonate globules that had first attracted Chris Romanek, and particularly the little magnetic crystals inside them, but he was interested in them from a purely inorganic standpoint. Like many people, he was interested in the history of water on Mars; he wanted to understand the compositions of the solutions that might have deposited those minerals, because they were the only traces of what water on Mars was like. As he and

Kathie Thomas-Keprta worked at the microscope together, he freely discussed his ideas about what they were seeing, and shared his results.

When he learned later from a third party that she had been secretly studying the same features but had never offered to collaborate, Treiman suspected that she had deviously used his work for her own purposes. He came to regard the McKay core group as ingrown and isolated. But he never confronted Thomas-Keprta about any of this.

Instead of acceding to McKay's wishes that he join the research, Treiman decided to start a Web site that took a neutral position on the controversy and provided running updates describing the main lines of argument. He was inspired by a stockbroker he knew who had said he didn't care if the market went up or down so long as people bought and sold stocks. As he would say later, "I figured, since I didn't know which way I was supposed to go on this, the way I could benefit me and the institute around me was to try to be in the middle."

After his 1996 announcement, McKay set up a series of weekly seminars to talk about Mars, astrobiology, the origins of life, and related topics. People were invited to come one weekday afternoon and give talks in the little conference room in Building 31. But the gatherings died out after a couple of months, Treiman said, and he thought it was because the McKay group was not sharing the inside information that their colleagues were intensely interested in—their ongoing work on *the* meteorite.

In early 1997, Treiman got an angry call from McKay. An intern had apparently told one of McKay's daughters about negative comments Treiman had made about the McKay group's claims. "I don't remember saying that stuff, Dave," Treiman answered. "You know I have problems with your work, but I don't remember saying any of that in particular."

McKay himself soon decided that he had gotten overly defensive in the combative months following the announcement, to the detriment of his real work. He regretted it and started trying to shift his energies back to the laboratory, to generating more data on the rock.

Technically, a person isn't supposed to get emotionally vested in a particular outcome. Science is an enterprise in which reasonable people can weigh their disagreements together as they seek truth. The textbook version of the deal is, you propose a hypothesis, which leads to a prediction, which is then put through "risky" tests by experiment. The outcome will confirm or refute your hypothesis. Or at least it will appear to, until an exception or complication crops up. Failure is always an option. Controversy and failure can be as valuable (to the process if not to the individual) as triumph. This is how humans, with their powerful, "self"-conscious minds, learn new things; it is how they sort reality from myth, rumor, lies, and speculation.

But scientists obviously do care, some passionately. To sort through the chaotic jumble of "data" and come up with a useful picture of nature takes imagination, and an investment of individual spirit. "About thirty years ago there was much talk that geologists ought only to observe and not theorize," Darwin wrote in an 1861 letter to one of his prominent defenders, a British politician and economist, "and I well remember someone saying that at this rate a man might as well go into a gravel-pit and count the pebbles and describe the colors. How odd it is that anyone should not see that all observation must be for or against some view if it is to be of any service!"

Some people thought that for cases such as the Mars rock controversy, another model for scientific inquiry was preferable to the textbook one. Carol Cleland of the University of Colorado, a specialist in the philosophy of science, argued that "some of the controversy over ALH84001 derives from a misguided insistence that all research conform to experimental practice." She decided to teach the case in her classroom.

"It just drove me nuts when the Allan Hills meteorite news broke," Cleland would say later, "and people just lashed out" at the McKay group on grounds "they hadn't 'proved' their hypothesis. Well, please. We haven't proved $E = MC^2$ [Einstein's famous equation], either. I mean there's no such thing as proof of this kind in science."

Cleland considers the traditional approach generally ill suited for certain fields, such as paleontology, archaeology, astrophysics, and planetary science. They require a different—and, in her view, equally effective—way of confirming a hypothesis, called the *historical* method. Science is not a monolith, but ideally employs different methods based on the different ways that nature is physically put together.

While the classic experimental approach looks *forward* in time (making predictions), historical research looks *backward* in time. Experimental researchers are interested in hypotheses about regularities, about the universal, the general: All copper expands when heated. (Not under all circumstances.) All swans are white. (Until you find the black ones.) $E = MC^2$. Historical researchers, by contrast, focus on a particular event, such as what caused the extinction of the dinosaurs on Earth. They aren't out to prove that all meteorite impacts cause extinctions, but to investigate whether a particular impact caused a particular extinction.

Instead of a single hypothesis to be tested, historical researchers propose multiple hypotheses to account for traces and clues from past events. They then hunt for a key piece of evidence—a "smoking gun"—that will point uniquely to one of the competing hypotheses as the best explanation. (A few decades ago, some fifteen hypotheses had been put forward to account for the dinosaur extinction, including disease, climate change, an exploding star, and the one that was finally accepted after many years, a cataclysmic meteorite impact.)

Cleland likens historical researchers to Sherlock Holmes, stationed in the present, trying to re-create events in the past. It is hard to erase all traces of an event once it has occurred; this is the basis of good detective work. It is difficult to commit the perfect crime. There are many traces of the past in the present, and you only need a tiny portion of those to infer the existence of a past event.

What initially happened in the case of the meteorite, Cleland argues, was that people hit the McKay group with criticisms from the vantage point of their own different disciplines. "The critics were complaining, 'Oh, you haven't tried to falsify it,' or 'you haven't tried to prove it.' All of this was to hit them with inappropriate methodology, . . . [with] claims that they should do something that I would argue would be irrational for them to do."

Some scientists were mortified by the whole spectacle. Others were convinced that the public had been exposed to the conflict at too early a stage. "Some [people] might draw the conclusion that scientists are morons who can't make up their minds," said the geologist Ralph Harvey. Some scientists, alarmed at the image this furor might be conveying to the public, thought the press reports exaggerated the "mudslinging" nature of the debate.

Others thought the case was a dandy demonstration of scientific give-and-take at its most passionate, and for stakes that could not be much higher. NASA funded not only the McKay group but most of their opposition, and no one was penalized for a given stand on the matter. For better or worse, the scientific process was on display, warts and all.

At the height of the controversy, a single pinprick-sized carbonate globule from the meteorite could easily have fetched $10,000, one meteorite specialist estimated. However, the curators took extraordinary precautions to make sure that no piece of the rock ever reached the market.

The controversy soon had as many snarling, snaking heads as Medusa. And, in the course of the protracted battle now begun, it would speed the process in which the search for life *beyond* Earth increasingly intertwined itself with the struggle to understand the origins of life *on* Earth.

The Mars rock would generate scores of papers and lead researchers around the globe on a years-long descent deeper and deeper into the microscopic recesses of the rock's ancient landscape. The investigators would mobilize an arsenal of the world's most advanced technologies to the task, and enlist humankind's accumulating knowledge in such varied specialties as how atomic nuclei decay, how plastics behave, the history of magnetism on Mars, how bacteria on Earth find nutrients in murky waters, bacterial immunology, properties of impurities in Antarctic ice melt, how various minerals precipitate out of water as acidity goes up or down, and the minimum requirements of a living cell. It would be difficult for any single individual to be truly expert in all the diverse precincts of knowledge that the puzzle in that one fist-sized rock brought into play.

Throughout the bloody aftermath, McKay and his group would take particular comfort in one thing: no one—not the most truculent of adversaries—ever questioned the validity of the data they had published. It was the audacity of their interpretation—not their evidence—that had opened the can of Martian worms.

CHAPTER ELEVEN

EXPLORATIONS

On the chill, drizzly afternoon of December 11, 1996, David McKay again found himself in Washington, D.C., this time seated at a big table with the vice president of the United States, several senior government officials, and almost two dozen prominent scientists, theologians, and educators.

The reserved scientist from Texas felt a mixture of pride and amazement. These accomplished people had gathered here at the White House as a result of his group's work on the rock. He would never have imagined such a thing two years earlier, when he, Gibson, Romanek, and Thomas-Keprta had agreed to begin their collaboration. In fact, they were all still feeling frankly stunned by the intensity and breadth of the public reaction.

Now McKay was trying to soak up every detail, so he could give his wife a full report that night when he got home to their woodsy cul-de-sac.

The belligerencies triggered four months ago, the day he'd sat in the bright lights on the NASA stage, showed no signs of abating. Nevertheless, the Clinton administration had decided to go ahead and stamp the White House imprimatur on the tricky topic of extraterrestrial life.

At last, it seemed that nature was pointing the way around the giggle factor. The McKay team, in tandem with the burgeoning insights about Earth's most extreme life-forms and the proliferating discoveries of planets around other stars, had put legitimate research questions on the table. In Washington, this meant that E.T., at least in a primitive, microscopic, and still hypothetical form, might be politically rehabilitated where it really counted—in the federal budget.

Many person-hours of planning had gone into today's closed gathering. A scientific workshop, university symposium, and other events had laid the groundwork. Officials at the White House science office, the National Academy of Sciences (and its National Research Council), and NASA had carefully pruned the invitation list. They had written and rewritten a suggested script for the vice president.

After introductions and pleasantries, Gore laid down a ground rule: "Let's assume for the sake of discussion that David McKay is correct."

That makes it nice, McKay thought. At least here, in this heady air, he didn't have to feel defensive. In fact, this was turning out to be fun. Somebody at the table did eventually ask McKay how sure he was about the claims, and what the team intended to do next. McKay responded that he was "reasonably sure," but he emphasized once again that there was "an awful lot of work yet to be done."

The group steered clear of budget talk, per se. They focused instead on the softer societal, cultural, and religious, as well as scientific, issues exposed by the claims of possible extraterrestrial life and other recent scientific developments, and on the research opportunities that might now be seized.

Seated at the head of the table, the vice president was flanked on his left by John Gibbons, the White House science adviser, and on his right by NASA administrator Dan Goldin.

The other men and women around the table included leading lights from the fields of atmospheric chemistry, origins of life, biogenesis of membranes, planetary systems, planetary geoscience, evolution of the universe, biological complexity, stars, and bacteria.

One of the better-known participants was the evolutionary biologist Stephen Jay Gould, the author of almost twenty books. His celebrity status was approaching that of Carl Sagan and the other leading scientist celebrity, Stephen Hawking. (In 1982, Gould had appeared on the cover of *Newsweek*, and in 1997, he would attain the ultimate in cachet with a cameo on TV's *The Simpsons*.)

Gore asked him what would be the minimum consequence if the McKay claim turned out to be true. Gould responded that if life had sprung up on only one world and been transported to the other, if Earth had "seeded" Mars or vice versa, that would say little about the potential abundance of life in the broader cosmos. It would mean scientists still had essentially only one case.

Gould was delighted with the news about the Mars rock. He suspected that the "fossils" were inorganic, not biological, though he wouldn't bet heavily either way. He found the McKay group's *Science* paper to be well written and properly cautious.

The news had broken about the time Gould's book *Full House: The Spread of Excellence from Plato to Darwin* was coming off the presses. In it, he argued that the human species was not some glorious peak in the evolutionary march from simplicity to complexity. Humanity was, rather, one accidental variation in a realm of variation. Bacteria and other microscopic life-forms had been the genesis life-form, and they had never gone away. The McKay group's line of inquiry, whether their interpretation held up or not, was in harmony with this theme.

Two years earlier, in a talk at George Washington University, Gould had summed up one of the credos of exobiology (soon to be subsumed in the new field of astrobiology) when he told the audience that all life on Earth was the product of one single experiment, and we can't fully understand how that experiment proceeded "until we find another experiment independent from Earthly life." He concluded, "That other experiment is as close to a Holy Grail for biology as anything else we could conceptualize or ever know or find."

Gould would echo that sentiment after the vice president's session, telling reporters that if life on Earth was found not to be unique, "the implications just cascade. They're just enormous."

Gore then asked astronomer-historian Steven Dick, who had just published *The Biological Universe*, about the other extreme—what would be the maximum consequences of life on Mars. Aside from the profound implication that we might not be alone in the universe, Dick said, what was at stake was humanity's worldview and the possibility of constructing a universal biology in the same way that Kepler, Galileo, Newton, and their successors had demonstrated a universal physics. "We are trying to determine whether the ultimate outcome of cosmic evolution is merely planets, stars, and galaxies, or life, mind, and intelligence."

The meeting planners had originally assumed the session would focus solely on the scientific questions, but President Clinton, because of his interest in religion, had asked specifically that the group address theological questions and the interplay of science and religion. How, for instance, would the discovery of life beyond Earth affect humanity's self-image, and its image of God? What were the views of the world's major religions regarding the possibility of extraterrestrial life? Was there a danger in this realm—as in the burgeoning research on the human genome—of offending vast taxpayer blocs?

Religious leaders had been invited, Goldin would say later, because "it's crucial that we . . . have broad consultation with the American people. When you have science—free-flying science—funded by tax dollars, you want to avoid crossing ethical boundaries."

His point was punctuated by the virulent reactions in some religious circles. Richard Zare, for one, after talking about the meteorite on ABC's *Nightline* and other TV programs, had heard from all manner of people, including UFOlogists, but was most unsettled by the angry fundamentalists. In these calls, including some from Europe, people screamed at him about the religious implications. His story did not match what was in the Bible. Internet chat rooms took up the banner. As a protective measure, Zarelab had taken down its Web site, along with information about how to contact Zare's office. Things stayed scary for about a month before the hubbub subsided.

But the public outpouring around the world also included more thoughtful musings, often by prominent people, on the theological and philosophical implications of extraterrestrial life. There was a sense that the world's religions were flexible on the issue.

The White House discussion mirrored the broader discourse. Catholic theologian John Minogue, president of De Paul University and an obstetrician-gynecologist, said there was considerable support for the notion that science and religion could coexist harmoniously. Religion was rooted and stable. Science by its very definition disturbed the status quo, pushed the envelope. "We need both," he said.

One absence was keenly felt in the room that day. Just before the vice president's session, Carl Sagan had informed Goldin that he would be unable to attend. His hair gone, his body withered, Sagan was losing his battle with the bone-marrow disease. He would die nine days later, on December 20, at a cancer research center in Seattle.

Sagan was arguably the most effective champion of the scientific hunt for extraterrestrial life and had galloped to its rescue on numerous occasions. Now he managed to express himself once again on the issue at hand. He had sent White House science adviser Gibbons and NASA chief space scientist Huntress a letter offering suggestions as to how NASA might best follow up on what he called its "extraordinary accomplishments" that year, which had brought the space agency "a degree of public support it has not enjoyed in years."

He alluded indirectly to the giggle factor that had fueled Congressional objections to SETI (the Search for Extraterrestrial Intelligence) and urged scientists to try again to revive the program. "The putative finding of fossil microorganisms on Mars and what is beginning to look like abundant planets around other stars surely enhances the plausibility of extraterrestrial intelligence," he wrote. "If such intelligence were found, it would be a turning point in human history. If after a comprehensive search program it were not found, it would calibrate something of the rarity and preciousness of intelligent life—information well worth having also."

In the public interest, Sagan urged that the report on these developments "be thoroughly de-jargonized. . . . Furthermore, we should be sensitive to arguments that strike the lay reader as implausible. For example, how could we possibly know that a rock found on the Antarctic ice comes from Mars? To the novice this seems the wildest guess. A paragraph that discusses the definition of an isotope and the Viking findings on the composition of the Martian atmosphere is an investment well spent."

Sagan had always believed that, in tandem with the human advance toward new knowledge, it was important to provide the public with a travelogue about the journey—and in clear, simple language. This business of popularizing science had made Sagan wealthy, famous, and admired in the general populace—but a pariah among many of his colleagues.

In his early years, Sagan had churned out a torrent of important research that interwove astronomy, biology, chemistry, and earth sciences. He eventually garnered eighteen honorary doctorates and over sixty awards or medals, including a Pulitzer and three Emmys. Yet Sagan had been denied tenure at Harvard, and in 1992, in a move Sagan partisans attributed to professional envy, the National Academy of Sciences, the nation's most prestigious scientific organization, had rebuffed an attempt to get Sagan voted in as a member.

Over the years, some people had criticized him for drifting further and further from active research as he gave in to the seductions of celebrity, or they'd murmured that he had become a prima donna with the ego of a rock star, or had tainted his career with his political activism. Some considered it frivolous and counterproductive for a scientist like Sagan to spend time trying to "dumb down" their way of speaking among themselves—an argot designed to be extremely detailed and precise—to benefit the lay public.

For citizens with a curiosity about science, his death would leave a void, and a long search for another approaching his rare mix of intellect, skills, charisma, and stature.

In one of his final essays, echoing Schopf's echo of his own words, Sagan cautioned that "the evidence for life on Mars is not yet extraordinary enough." But he added that the McKay team's discoveries in the rock opened up the field of Martian exobiology.

Arguably, on this wintry day in December 1996—as David McKay's recent experiences illustrated—science had reached a state where studies of the most remote corners of the galaxy, and of the most humble ribbons of gas in space, and of the most unassuming flecks of dust or rock all constituted research on the same fundamental and deeply fascinating topic: the story of life—life on Earth and wherever it might be found. It was this sense that moved the McKay team along the unlikely track they had followed into the depths of the rock. It was the intellectual sea in which they swam.

A couple of weeks after the Mars rock burst into the headlines, a team of genetic researchers led by Craig Venter and based in Rockville, Maryland, just outside Washington, published in the same journal—*Science*—word of a stunning development for the understanding of terrestrial life: in the culmination of some two decades of work, they had confirmed the existence of a third major domain of life.

They had done this by sequencing the genes of methane-belching microorganisms that dwell on the floor of the Pacific, tucked into the crevices of the hot-water vents that spew boiling-hot, mineral-rich fluids into the cold sea. The organisms thrive in the extreme heat without oxygen or direct sunlight and do not need organic carbon as an energy source.

The microbes called Archaea joined the other two major life domains: the bacteria (prokaryotes) and the more complex life-forms (eukaryotes), including plants, animals, and humans. "We were astounded to find that two-thirds of the [archaea] genes do not look like anything we've ever seen in biology before," Venter reported, adding that the research showed "how little we know about life on this planet."

The accumulating discoveries about extreme-loving microbes reinforced speculations that life might once have sprung up, might even exist today, in hidden pockets on Mars or in some other extraterrestrial oasis not yet dreamed of. And NASA's Galileo spacecraft, in orbit around Jupiter, had pumped up the excitement by taking provocative photographs of the Jovian moon Europa that showed evidence of slushy ice or liquid water in a vast subsurface ocean there—another potential haven for life.

All of this was in the minds of those gathered in the White House Treaty

Room. As the damp and wintry day darkened, the planned two-hour session stretched on to almost three hours because Gore had become so engrossed in the conversation.

He was unusual among elected politicians in his ease with scientific, as well as theological and philosophical, issues. In fact, for him this exchange was a genuine treat, not political theater. This quality had proved to be a mixed blessing for his political hopes, earning him praise for his intellectual heft but also feeding the stereotype of Gore as a wooden, condescending überwonk.

As a young congressman, he had been a precocious presence on science and technology committees. In 1992, inspired by the near death of his young son in a street accident, he published *Earth in the Balance,* a best-selling book about his environmental—and spiritual—concerns. In it, he quoted Erwin Schrödinger, a pioneer in quantum physics, on the topic of how a pattern of life can emerge from a formless cluster of molecules, "escaping the decay into atomic chaos."

Now, as the Treaty Room discussion continued, Gore departed from the prepared script and asked a question that echoed that passage from his book. "Okay, I've been listening to all of you on the possibility that life might actually be ubiquitous in the universe," he said. "How does that square with the second law of thermodynamics, which says that the universe tends everywhere towards *disorder*—but we all know that life is a highly *ordered* process?"

Jeez! The vice president's guests were caught with their answers down. They were also delighted. It was an excellent question. (If anyone but the vice president had asked, someone might have responded that the "law" he cited had to do with closed systems, and a living creature was not a closed system but one that would "drink orderliness" from its environment.)

"There was not a scientist in the room who wasn't impressed" with Gore's performance that day, astronomer Anneila Sargent of Caltech, one of those at the table, would remark afterward. McKay was no exception. He was struck by the fact that the vice president had taken time to give the issue so much thought, enough to develop his own ideas about it. "That guy is really smart," McKay would tell Mary Fae when he got home and described this remarkable day.

The discussion might have resembled a philosophers' salon. But McKay was aware that he was also witnessing the unavoidable dance of science and politics. Somebody had to try to harmonize the scientists' world of open-ended, unpredictable inquiry and the government world of election cycles and budget plans. The link was money. The scientists needed it; politicians dispensed it.

The rapid advance of discoveries about the nature of life and the cosmos had provided context and inspiration for the McKay team as they approached the rock. Now, the NASA headquarters team carried in their pockets a blueprint—part policy statement, part marketing tool—that would build on those developments and meld their disparate threads into an ambitious, long-term government assault on nature's deepest mysteries—profound questions previously limited to the provinces of philosophers, poets, and priests. The policy makers were eager to align themselves with the broader struggle to document the narrative that began with the genesis moment known as the big bang and continued as the thermonuclear furnaces of stars forged the elements that led to the chemistry by which the first life assembled itself.

The excitement surrounding the Mars rock seemed an ideal kick start for the strategy. From the moment NASA officials learned that *Science* had accepted the McKay paper, they maneuvered to make the most of the anticipated surge in public attention and among researchers. As Huntress and Goldin saw it, the new Origins program would give the struggling space agency a compelling purpose that would engage the public and, consequently, be of aid in the perpetual battle of the budget, which NASA had been losing badly.

President Clinton, in his comments about the claims of possible life on Mars the day of the August press conference, had announced that he would ask Gore to "convene at the White House before the end of the year a bipartisan space summit on the future of America's space program. A significant purpose of this summit will be to discuss how America should pursue answers to the scientific questions raised by this finding."

NASA and the other agencies organized the summit as a multistep process, with preliminary meetings on relevant topics leading up to a final rare bipartisan confab with the White House and key members of Congress. This session in the Treaty Room was supposed to be the next to last in the series.

In late October, at the request of the White House science office, NASA and the National Research Council (an arm of the National Academy of Sciences) had convened a workshop to lay the groundwork. The theme was (as Huntress had suggested) "The Search for Origins." The National Research

Council selected, and NASA approved, some three dozen biologists, planetary scientists, astronomers, and cosmologists to attend.

Among the group's written conclusions (which highlighted NASA's new key word in all caps):

- "For the first time in history, we have achieved the level of understanding and technical capability to press for answers to fundamental questions concerning our ORIGINS, our history and our context in the Universe."

- "The ORIGINS quest informs, excites, and inspires the public. Its outcome may well have as profound an effect on human thought as the Copernican and Darwinian revolutions."

Running throughout the activities was an obvious awareness of a political dilemma: how to demonstrate political commitment to tackling the big, fundamental questions when there was little hope of prying loose new money to do it.

As part of its "guidance" for the summit, NASA had prepared a memo for the White House science office: "Make it clear that NASA has successfully restructured its program to focus on research and development and has [relevance] to the American public." The agency's stated goal was to win support for a merely *stable*—not larger—budget from 1997 to 2001 "at the $13.6–$13.8 billion level." But a hand-scrawled notation on the page differed: "Not [illegible] *enuf!* 14 at least w/o Mars initiative!" The memo went on to ask, "If budget is held to FY 1996 (vs 97!?) outyear runout, what are policy decisions on what to cut?"

Though genuinely excited about the potential implications of the Martian meteorite findings, the White House was preoccupied with the deficit. The administration was grappling with what one space policy expert described as its "uncertainty over how best to deal with the firestorm of interest unleashed by the president's words [in connection with the Mars rock story], compared to its desire to avoid major new, expensive space commitments."

Hopes for another major, Kennedy-style push in *human* space exploration had been as persistent as they were unrealistic. The last time anybody had tried it was in 1989, when President Bush's proposal for a human mission to Mars had sunk without a ripple, largely because Congress found it so easy to laugh at NASA's stolid, self-serving proposal of a $400 billion sticker price.

Clinton and Gore were not about to follow that path.

But well-founded research on the compelling question of extraterrestrial life was another matter. The summit process served to underline the fact that, as Michael Meyer, the top NASA exobiology official, put it, "this is not crazy research but, in fact, has become cutting edge with mainstream scientists."

When the meeting finally broke up, waiting reporters wanted to know about the resulting plan for action. The media wanted news, not schmooze—but from their perspective, the latter was all they got. White House science adviser John Gibbons commented that the event had provided "a good deal of flavor and perspective" for the vice president, and should help administration officials be more definitive in their testimony before Congress. The group around the table had concluded that Goldin and NASA were on the right track, he said, "to do more with the same dollars."

The vice president issued a statement calling NASA's new Origins program "a vital contribution to our national and global pursuit of knowledge." He would prove to be a staunch supporter.

An ebullient Dan Goldin, his voice tight with emotion, called the event "a highlight of my career in Washington." He said it was in harmony with NASA's new approach: in the past, NASA had defined itself by the engineering temples it wanted to build and then picked the questions to fit. That was about to change, Goldin vowed. The overarching great questions about humanity's place in the cosmos would henceforth determine the engineering instruments to be built, not the other way around. Budgets and engineers and infrastructures were finally going to labor in the service of the science.

As 1997 approached, the American space program seemed to be moving out of its long purgatory. Now the United States, with other nations, was beginning a new thrust, to dispatch flotillas of spacecraft to Mars, to other planets, to comets, asteroids, and moons. Engineers were poised to finally get the long-planned and ever-controversial space station off the drawing board and into orbit—its ultimate purpose to serve as a springboard to human exploration of Mars and beyond. At this moment, it seemed that space exploration was becoming exciting and vital again.

Within two months, the "space summiteers" would lop off their final meeting, saying it was unnecessary. The goal of the event—to assure NASA of "stable and sustainable" funding—was already in hand. The White House and Congress would allow the agency to beat the Washington expectations game by getting if not an increase, more money than anticipated. The agency would halt its steep five-year slide downward with a budget pegged at around $13 billion running out to fiscal year 2001 or 2002—better than the budget targets of under $12 billion that the Office of Management and Budget had threatened for months.

And in the realm of space science (as opposed to human spaceflight and other categories), the budget would increase over the coming years, as NASA shifted funding internally.

Around Johnson Space Center, at least, McKay, Gibson, and Thomas-Keprta would be honored for their role in bringing public attention, an improved funding outlook, and other benefits to the realm of solar system research. They had never felt themselves motivated by such concerns. At first, when people accused them of exaggerating their claims about the rock in order to enhance NASA's bottom line, they felt surprised, and they would always feel stomach-knotting twinges of dismay and anger.

Those who accused NASA of trying to get political mileage out of the Mars rock were correct on the facts. But far from being defensive, Goldin, Huntress, and other agency officials expressed pride in their approach and their accomplishment. They felt they were doing just what they were paid to do—coming up with a set of projects that appealed to taxpayers, that would bring credit to the country and benefit the human race. If NASA did well in the bargain, well, so be it.

The administration's decision to make the new Origins initiative the centerpiece of its NASA funding request was the response many space scientists had hoped for, the right follow-up to recent discoveries. Buoyed by public excitement over the rock, Origins would make biological research an integral part of the space program in an unprecedented way. It would boost the number of biologists winning NASA grants and catalyze the ascendancy of an emerging field of study newly named astrobiology. It encompassed the interplay of life and planets and the universe and how they evolve together, and it required the melding of many different fields of expertise.

Matters of budget, policy, and metaphysics aside, for McKay, the Treaty Room session was a moment to be savored. It was a gestalt experience, he thought, enjoyable at the visceral as well as the intellectual level, and very heady. The mild-mannered geologist from Houston was sorry when the gathering broke up and it was time to head out into the chill and blustery dark.

Pathfinder bounced onto the Martian surface on the Fourth of July 1997, marking a triumphant return of American robots after a decades-long absence. In the pipeline well before the announcement of the Mars rock claims the previous summer, the mission was designed not to do science but to test technological capability. Pathfinder's assignment was basically to cruise the dusty surface and sniff rocks. It traveled no more than the length of a typical auditorium and would produce no significant change in the human view of Mars. And yet the mission was wildly popular, and succeeded far beyond its modest goals.

A vast worldwide audience watched the adventures of the loose-jointed pet-sized robot unfold live on the young Internet. The adventure reintroduced people to the forgotten pizzazz of "touching" and exploring another world.

With a view to the Goldin regime's emphasis on risk taking, groundbreaking research, and efficient, clever management, the space program had geared up for the second great assault on the red planet in the history of the space age. U.S. missions—often in collaboration with Russia and other space-faring nations—would be dispatched at every opportunity: roughly every twenty-six months when the planets were in favorable alignment. Pathfinder was the opening salvo.

Planners had to consider how the findings in the Mars rock affected those future missions. As the controversy over the rock waxed and waned, and as the McKay core group pressed its hunt for more data, some planetary scientists concentrated on the aspects of the rock's meaning that were not in dispute. What tales did it tell most clearly about the young Mars? The implications were huge even without claims about biology. The rock's fascinating geochemical signatures seemed to restore the rich potential for biology in the Martian past that the 1970s Viking landers had eviscerated. No matter who won the "pro-life"/"anti-life" battle over the rock, the allure of the extraterrestrial was back. And the arguments about the selection of landing spots and sample types for the Mars missions took on a new edge.

The people who studied Mars most intently, whose work it was to plan these interplanetary excursions, had adopted a single organizing principle: follow the water. The edict became the overarching theme of the struggle to understand the evolution, geology, atmosphere, potential for biology—every aspect of the planetary personality of the red planet. Data flowing from the arriving robot explorers would soon churn up baffling evidence concerning the riddle of the Martian waters, shaking up many of the old assumptions.

Even as they fought about the meaning of one tiny chunk of the other world, some of the combatants were finding a kind of cease-fire zone in this work directed toward the world itself—the pipeline of missions to Mars. No matter how violent their differences over the rock ALH84001, the warring factions agreed that many of their most vexing questions could be resolved only on Mars itself.

The ancient rock that divided them was pressing all of them back toward its cradle, toward Mars.

In 1997, in the aftermath of the rock fight, John Rummel held the title of "planetary protection officer." In casual moments, he would confide that, based on the reactions he got from kids and even some grown-ups, it seemed the next best thing to wearing Superman's *S* on his chest.

But he was no comic-book action hero. He was an agent of the U.S. government working in the real world, a former naval officer who had flown anti-submarine warfare missions and then become an ecologist and sometime microbiologist. On occasion, he would recall that he'd begun his scientific career sailing the Caribbean studying lizard population densities; then he would laugh to himself and wonder, "How did I end up here?"

One big reason was the shift toward biology within NASA. After the McKay team's 1996 announcement, administrator Goldin had expressed dismay at the dearth of biological expertise in the space agency culture (except for studies geared to astronaut health). He directed NASA to start hiring biologists, all kinds of biologists. Goldin would also soon set up the NASA Astrobiology Institute in California and hire a Nobel prizewinner to run it.

Rummel was NASA's man in charge of ensuring that we humans prevent contamination of our own planet by alien microbes, and also avoid the spread of our own ecological undesirables to Mars or other sites in our little corner of space. He was supposed to oversee the writing of the rules.

The main focus of the planetary protector's concerns was summed up in the term "sample return." This referred to scientists' cherished goal of dispatching a team of robots to gather and bring back rocks and soil from Mars. Simple as it might sound, it would be enormously tricky and expensive to design a robotic package that was (a) lightweight enough to be practical, (b) smart enough to do the job (make a sophisticated selection of desirable rocks and soils, for example), (c) able to land safely on the rugged Martian landscapes designated as most promising for biological clues, (d) able to take off again, and (e) able to deposit the treasure safely and cleanly back on Earth. And, again, there were related, and increasingly convoluted, issues of contamination.

In early 1997, a blue-ribbon science panel warned that the possibility of bringing back dangerous Martian life-forms, while small, was not zero. At a time when Ebola outbreaks, genetically altered agricultural produce, and bioterrorism had revved up global anxieties, Rummel and others felt unprecedented pressure to make sure the public was comfortable with the protective measures taken by their government.

The National Academy of Sciences mobilized some of the brightest minds in microbiology to figure out methods for complying with the Rummel Rules. They hammered out standards for biological signatures and ruminated on the challenges of differentiating dormant life-forms from active or dead ones, as well as distinguishing between life and nonlife—distinctions that remained frustratingly elusive. The very ambiguity of the evidence in the Mars rock became a catalyst for change in NASA's approach to this issue and the design of the sample-return mission.

The Mars rock dispute helped planners realize how far short they fell in terms of the level of understanding and technology required to justify the stunning costs ($2 billion or more) and the years of effort involved in landing equipment to pick up maybe two pounds of carefully targeted Martian soil and rock, then launching the cargo back to Earth.

The anti-contamination people decided they would have to assemble not only a much better knowledge base but an arsenal of new technologies and techniques, containment facilities, and tools that would constitute an effective twenty-first-century Maginot Line against interplanetary contamination in all its modes.

Engineers pondered such arcane questions as how to detect possible pathogen leaks on an Earthward-bound sample freighter in time to redirect it past the home planet. Higher-level officials debated whether, if the samples were allowed to parachute to Earth, the retrieval teams should be from NASA, the military, the Centers for Disease Control and Prevention, or some other group.

Rummel somehow had to make the imperatives of planetary protection mesh with the primary purpose of the mission: the widespread distribution of the Mars samples among expert Earthlings. The goal was to ensure sophisticated analysis of the samples by the best instruments and minds available on Earth. But those two designs were sometimes in direct conflict.

A central question for those working on the quarantine issue was: What criteria must be satisfied before the samples could be released to waiting scientists around the world for study? The answer involved such trivial issues as how to determine whether the samples had life in them and, if so, whether this life was a danger to Earth—the very questions that the Mars rock controversy had demonstrated would be infernally difficult to answer.

And beyond the sample-collection project, the notion of sending humans to Mars seemed to recede even further into the future. Even if such a mission were technically and financially feasible, there was the freshly exposed worry that "sending humans may alter the biological history of Mars," as astrobiologist Michael Meyer told a colleague. "You have a couple of problems. One is you're likely not to recognize something [alive] on Mars because you've overwhelmed it with your own biology, and the other is you might end up contaminating the planet."

In late 1999, with breathtaking suddenness, NASA's bold blueprint for the Mars investigations fell into ruins. Mission teams based in Pasadena, at Caltech and the Jet Propulsion Laboratory, as well as in a control room in Denver, watched in disbelief as first one and then another catastrophic failure wiped out an entire generation of U.S. missions as they arrived at the planet. An orbiter and, three months later, a lander carrying two small surface probes crashed before they could begin work. White-haired planetary geologists saw a significant chunk of their life's work, and their hopes, vanish in a digital blink.

The losses, eventually traced to avoidable errors, staggered NASA's planetary program and demonstrated that there was a practical limit to Goldin's "faster, better, cheaper" strategy. Overburdened managers in this case had failed to follow the rules designed to prevent such errors, and had stretched their teams too far.

Huntress, for one, was appalled as he watched Goldin shrink back from his bold push to encourage prudent risk taking throughout the space agency. Instead, in the wake of the Mars disasters, Huntress saw the policy transformed to "thou shalt not fail," and the culture tilted toward one of "inquisition" and fearfulness.

The failures of 1999, combined with the controversy over the rock and puzzlements cropping up in the scientific data from the orbiting Mars Global Surveyor already at the red planet, forced mission planners to rethink virtually their entire repertoire of techniques and strategies for exploring Mars.

NASA officials pushed the vaunted capture of Martian samples more than a decade further into the future, until at least 2014, mainly because of the lack of money in the wake of the failures. But among those privy to all the scientific uncertainties, there was also the sense that they had dodged a bullet—and a suspicion that even an extra decade might not be enough to lay all the necessary groundwork. They had realized they were simply not prepared.

The scientists' heady days of White House summitry and their dreams of ambitious exploration suddenly seemed at least as far away as the florid and treacherous sister planet. To many, the turn of events was like a warning echoing across interplanetary space: "You are not ready."

CHAPTER TWELVE

AT DAGGERS DRAWN

On August 7, 1996, Andrew Steele—"Steelie" to his friends—was lying in bed at home in Portsmouth, on the southern coast of England, when he heard the headline on the BBC: "Life on Mars?!" He had dislocated his knee playing football (soccer, in American parlance), and his leg was in a plaster cast. But his imagination was leaping.

With his chiseled features, aquiline nose, and flowing mane of light-brown hair, he could have been a rock star or an *Esquire* model, but he was, instead, a cheeky microbiologist with a droll sense of humor and a shiny new Ph.D. from the University of Portsmouth. He had focused his attentions so far on the interactions between bacterial biofilms and metal surfaces in marine and freshwater settings. His goal was to learn more about the corrosion caused by microbes, and especially the role of bacterial slimes in the corrosion process. His project was sponsored by the U.K. nuclear industry.

But Steele was something of a romantic, quietly idealistic and already a bit restive at the prospect of a career in what he and his friends sometimes referred to—with great affection, of course—as "stools and fuels." He had been an avid reader of science fiction all his life, and he knew something about the recent discoveries of organisms thriving in extreme environments. He found himself yearning to be part of that excitement. Now, as he nursed his knee, he heard the siren song of the red planet. He felt the surge of public fervor at the news. It got his heart to thumping, and agitated his mind.

Steele didn't have a phone at home, so he hobbled into the lab where he was continuing work as a postdoc. He dialed the international information service and asked for the number of Johnson Space Center. He called David McKay's office, spoke with an assistant, and explained that he was a specialist on a high-powered instrument called the atomic force microscope. He said, "I think I could probably get you a three-D image of the 'worm.' "

Following the assistant's instructions, Steele sent McKay a written "mini-proposal." For weeks, he heard nothing. His friends in the lab teased him:

"You're going to get a piece of Mars? Oh, yeah. Right."

His supervisor told him, "You're mad, mate. You're mental!"

Surely, they said, the McKay group had already subjected their rock to the atomic force microscope.

The McKay group had not.

In mid-September 1996, McKay was called to testify before a congressional committee, where he mentioned that he was sending a piece of the meteorite to the University of Portsmouth, in England, for expert microscopy. The young Brit, who had never seen a meteorite before in his life, couldn't believe his eyes when he read McKay's comments in *Science.*

After several more attempts, Steele finally got through on the phone to McKay himself, who told him that, indeed, the package would be arriving by FedEx. McKay explained that, as a result of Steele's first phone call, he had queried other experts about the feasibility of the atomic force microscope technique. He had sent a piece of the rock off to the maker of the microscope. The feedback was encouraging.

Steele told his friends this thing was really happening. But his sense of humor was familiar to them, and they figured he was just winding them up.

One chilly day in November, the package arrived. Steele and his chastened friends took pictures to record the event. They were half afraid to open the thing. When Steele finally did unwrap the contents—a plastic vial—he thought there had been some mistake, or possibly a prank. The vial was empty, wasn't it? They saw nothing at all in there.

Distressed, he called McKay. "Uhhhh, you can't see it in there. Are you sure you put the sample in?" he queried. Steele had a sudden vision of the man at the other end of the line weeping softly into his handkerchief and wondering, "What have I done?"

Together, they figured out the problem—and tracked down the missing item where it was hiding out. Steele's piece of Mars was a 300-micron particle that had happened to stick at just the worst possible place, so that he and his colleagues had mistaken it for a "full stop" (British lingo for the punctuation mark Americans call a period) at the end of the printed number that identified the sample.

Steele felt a flood of relief. Naturally, in celebration, he and his lab team took their sample, which they nicknamed Chip, to the place where, as Steele liked to say, "all the best British science gets done"—the neighborhood pub.

The word had spread throughout the kingdom—well, at least to Steele's far-flung circle of friends. One lad drove 250 miles from Preston to be there. The little band set the container holding their microscopic prize in the middle of the worn wooden tabletop and stared at it. "What do we do now?" they asked each other.

They worked out a plan. They didn't even know which way was "up," that is, which surface held the mysterious carbonate globules where all the interesting stuff was.

Their next step was to consult Monica Grady, the accomplished meteorite specialist at the Natural History Museum in London, who had taken detailed images of the meteorite. She and her husband, Ian Wright, had worked with Colin Pillinger, Everett Gibson's collaborator on the *Nature* paper years earlier.

Grady got Steele's "full stop" mounted properly, carbonate side up, and his group went to town on Chip. They worked on the meteorite through Christmas and New Year's. It was important to zero in on features similar to those that McKay had focused on, and to match as closely as possible the conditions under which he had operated.

One of the initial criticisms of the McKay group's claims held that the worm-fossil shapes had actually been created by accident in the laboratory, in the process of preparing samples for the microscope. McKay routinely coated his samples with a thin layer of electrically conducting gold-and-palladium alloy, and if the coating was not uniform, it could have created artificial shapes on the rock.

Using the atomic force microscope, Steele (with colleagues Dave Goddard and Dave Stapleton) applied the same coating and took images of the same areas of their meteorite fragments. While this microscope can magnify up to about 10 million times and distinguish individual atoms, the team used it to map the targeted microfeatures in the sample at the same level of magnification used by McKay—but this time in three dimensions and both before and after the gold-and-palladium coating had been applied. (Quite a trick, requiring considerable work, to find that same spot again: it was 2 microns by 2 microns on a 350-micron sample area. A fine human hair is about 50 microns across.) The device worked something like an old-fashioned phonograph turntable, with a needle arm tipped with a sharp point of silicon nitride that traveled lightly over the surface of the object under study, rising and falling with the tiny topography. The point of a laser beam, in turn, rode piggyback atop the probe, tracking these ups and downs. By measuring the beam's displacement, the researchers could build a three-dimensional map of the sample's topography.

Studying areas of the carbonate similar to those described by McKay, both coated and uncoated, they looked for any differences. They found none. Because of the way the images are displayed under the microscope, the globules looked luminous, almost wet, and mottled like a gilded alligator hide. Under the powerful magnification, Steele could see that the gold plating had formed only a slight crazing pattern of gold on the sample surface. Each dot was 7 nanometers across, or tens of thousands of times smaller than the diameter of a fine hair.

He concluded that any artifacts formed as part of this laboratory process were too small to account for what he called "the wormy guy," the McKay group's possible fossil. He had effectively ruled that argument out. Now he had a paper to publish, with new information to contribute to the Mars rock debate.

He decided to take his news across the sea to the annual Lunar and Planetary Science Conference in Houston. He had no idea what kind of spectacle he was in for.

The meeting, held in mid-March 1997, was the first chance for the warring camps to face off in one big room and assess the state of play on the Mars rock.

There was a palpable aura of tension and excitement among the more than eight hundred planetary scientists gathered at the Gilruth Center, a recreation complex on the Johnson Space Center campus. Up for discussion were thirty-four new papers concerning the rock. Duck Mittlefehldt was one of the presiding emcees or, some would say, referees. After each set of findings was described, there would be open discussion. At times, the proceeding sounded more like O. J. Simpson's criminal trial, with opposing arguments, conflicting witnesses, and disputes about contaminated evidence. The arguments ran through hour after hour of the formal sessions and rippled out across the cocktail get-togethers and dinners where so much of the real cross-pollination of science traditionally gets done.

Two members of the McKay team were temporarily out of combat.

In the frenetic and stomach-churning aftermath of the group's August announcement, David McKay had gone in for a physical, which revealed troubling signs of artery blockage. (David's father had died of heart failure.) His doctor told him to go get further tests. But he was so consumed with fending off attackers and responding to interview requests that he put the task off through the holidays and into February. He went in for an angiogram just before he and Mary Fae were set to take another pleasurable trip to Japan, where the National Institute of Polar Research had invited McKay to discuss his work.

The procedure, performed on a Friday, showed serious blockage. Three days later, surgeons at the Texas Heart Institute cut through McKay's breastbone and constructed bypasses for four arteries. As the face-off at the Gilruth Center commenced, he was still recovering.

Kathie Thomas-Keprta was on the verge of giving birth. This left Gibson as the group's main defender.

Gibson was energized, defiant. His relish for the fight was clear to anyone dropping by his jam-packed workspace during this period. Gibson had ordered custom polo shirts with the insignia "Mars Meteorite Exploration Team," which he wore proudly and distributed to coworkers.

Around his office—amid journals, tape cassettes, and papers stacked on the floor and tables—Gibson displayed a set of plastic action figures of Warner Bros. cartoon aliens, the Marvin the Martian line. Alongside old air-show posters and an airplane mobile, a tear sheet from the tabloid *Weekly World* blared: "Deadly Mars Rock Virus Infects 4 Researchers." Rocks were strewn on top of filing cabinets. There were globes of Mars and the moon. A big wall poster showed the comparative sizes of viruses, cells, and such. And, on a shelf, six big loose-leaf binders, titled "Martian Chronicles, Vols. 1–6," contained his notes and records pertaining to the team's work on the rock. As the team's official record keeper, he had in mind that he might write his own book after he left the government.

Over at the Gilruth Center, facing the critics, Gibson declared the evidence of past biological activity in the rock "much stronger now than when we wrote the paper." He added, "We believe the criticism that has been leveled at us can be answered. . . . This is science in action." But Gibson was no more combative than some of his assailants.

The main points of disagreement among the combatants were these:

TOO DARN HOT?

If the McKay team was right and the mysterious carbonate moons had incorporated the trappings of living organisms, that meant the carbonates with their distinctive Oreo rims must have formed at moderate temperatures. That was presumed to mean no more than about 140 to 170 degrees Celsius (284 to 338°F); the upper limits for heat-loving bacteria found in recent years on Earth.

But models developed by various groups put the ranges for the rock's formation at anywhere from well under 100 degrees Celsius, the boiling point of water at sea level, to 700 degrees or more (well over 1,000° Fahrenheit). To complicate things further, some people would agree that, yes, the temperatures were moderate but there was nevertheless no biology present.

In their *Nature* paper published two years earlier, Romanek, Gibson, and their British colleagues had proffered the initial evidence (based on the oxygen isotopic ratios) that the carbonates had formed from what amounted to a "Goldilocks" broth, neither too hot nor too cold. If the high-temperature model was correct, Gibson argued, the carbonates would have homogenized into an undistinguishable mass, and the organics inside would have disintegrated. Neither had happened.

Joseph L. Kirschvink, of Caltech, and others (including Hojatollah Vali of McGill University, one of the original McKay Nine), took an unusual approach and produced striking—if not conclusive—results.

Magnetic crystals—the iron sulfides that Kathie Thomas-Keprta had detected inside the carbonate globules—are known to retain the signature of any magnetic field present at the time they form, but will lose this telltale imprint at high temperatures, conforming again to the local field as they cool. Kirschvink and coworkers used an ultrasensitive superconducting magnetometer system to detect, in adjacent fragments from a fracture zone in the rock, signs of two distinct magnetic fields aligned at almost a right angle. Presumably the fragments had shifted against each other when a sudden impact fractured the rock.

The researchers estimated that the magnetite grains—and the carbonate around them—could not have been heated above roughly 325 degrees Celsius (617°F) and probably not above about 110 degrees C. (230°F) in at least the last 4 billion years. If they had been, they would have lost their memory of those opposing magnetic fields.

Arguments about biology aside, this was potentially significant work, showing for the first time that Mars once had a magnetic field. That field, like Earth's, might have shielded the planet and any emerging life-forms from the deadly "wind" of electrically charged particles that flows off the sun. The

Kirschvink discovery also fit with—although it did not prove—the McKay group's hypothesis that ancient Martian swimmers might have formed little internal compasses to help them navigate in a magnetic field, just as Earth organisms had done.

Modern Mars, scientists had shown, has no magnetic field.

Within four years, the Kirschvink finding would get a boost from NASA's Mars Global Surveyor, a robotic orbiter that in this spring of 1997 was well along on a three-hundred-day cruise to the red planet, where it would arrive in September. Once in operation, Surveyor would send back evidence of "remnant anomalies" in the Martian crust indicating that, in its early youth, Mars indeed had a global magnetic field.

"One thing is clear," Gibson asserted at the time of the meeting. "Any model for a high temperature origin for the carbonate globules is clearly incorrect and invalid."

Science magazine would report the data evenly divided between the high- and low-temp camps.

THOSE DEAR OLD PAHS

The tongue-twisting hydrocarbons represented a significant find as long as they came from Mars, whether or not they were by-products of living Martians. That was because they suggested an environment that *could have* developed and nourished life on the red planet. But were they from Mars? Jeffrey Bada of Scripps Institution of Oceanography and his coworkers argued that the organic molecules were not from Mars at all but represented contamination from Antarctic meltwater that had flushed through the rock during its 13,000-year entombment. As for the unexpected distribution of the PAHs—denser toward the center of the rock, and concentrated around the carbonate globules—Bada said chemistry, not biology, likely had done that too.

Simon Clemett, of Zarelab, countered that the Scripps experiments were flawed. The challengers, in his view, had failed to distinguish between soluble and insoluble types of the molecules. No other meteorites recovered from the Antarctic ice had the level of PAHs found in the celebrated rock from Allan Hills, and analysis of melted ice from Allan Hills showed no PAHs.

Nanobes We Never Knew

As the McKay team expected, the heaviest counterfire was aimed against those ovoid and tubular, sometimes undulant shapes they had presented as possible nanofossils—remnants of microbial Martians. One prominent microbiologist referred to the shapes as "nanothings." There were jokes about "no-no-fossils."

Steele reported that his Portsmouth group had ruled out the argument that the nanoforms had been created in the laboratory when the meteorite sample was coated in preparation for electron microscope viewing. (The shapes were still present when viewed with other techniques.) Others argued that the addition of the coating could enhance an illusion of segmented wormlike shapes that were actually nothing more than angled views of crystal layers and protruding ledges along fracture faces in the rock. (The McKay team would argue in rebuttal that layered minerals are much more orderly and widespread than the possible microfossils, and that they could tell the difference.)

Here, size *was* important. Bill Schopf had argued earlier that the wormy shapes were so small that they had no room for even the barest genetic essentials of a living cell. Romanek calculated that, in terms of relative size, if you could balloon one of the orange carbonate globules (the largest of them barely visible to the human eye) up to the size of a football stadium, the fossil-like shapes would be no bigger than hot dogs.

The McKay team pointed (again) to the work of Robert Folk, at the University of Texas, who had recently published images of what he claimed were fossils of previously unknown Earth bacteria approximately as small as those in the rock. "You can't tell which are my pictures and the pictures from Mars—they're absolutely identical in size, textures and shape," Folk commented in late 1996. He argued that they could be an intermediate form of life somewhere between viruses and bacteria that did not need the same amount of genetic equipment as typical Earth bacteria in order to function. Folk's interpretations, however, were well nigh as controversial as the work of the McKay team, while the conventional wisdom was all on Bill Schopf's side.

At least one avid critic concluded that it was the very unprecedented precision and power of their instruments that had led the McKay group astray. In this view, they were like Percival Lowell seeing artificial "canals" on Mars, and like other aggressive, possibly overreaching explorers on any frontier.

As if they had suddenly been transformed into insects in their own backyards, they (like everybody else) were seeing this realm for the first time at a scale—millionths of a meter of resolution—so radically small that they could—anyone could—easily misinterpret its unfamiliar warps and rills.

THE MYSTERY CRYSTALS

The magnetic crystals that Kathie Thomas-Keprta had found constituted the strongest—or at least the most intriguing—evidence offered by the McKay team.

Biologists had come to accept perfect, well-organized little formations of grains, or crystals, as persuasive evidence of biological activity on Earth, where certain bacteria were known to produce chains of them. The McKay group had described defect-free, elongated crystals in the Martian specimens that, as far as anyone had yet shown, could be produced only in biological processes.

But variants of the crystals could form without biological activity and were common in ordinary rocks. The peripatetic Ralph Harvey, with colleagues, reported seeing the small, defect-free grains, all right, but also "a zoo of other forms" rampant with defects. These included elongated "whiskers," which, on Earth, were known to form only around volcanic vents in settings with lethal temperatures of around 650 degrees Celsius. And they included crystals with defects that weighed against their having formed inside the safe haven of an organism, including something called a "screw" dislocation (in which atoms in one row of the crystal matrix were offset), a defect known to occur in crystals grown from vapor but not in those grown inside living cells.

Of all the challenges, McKay was most perplexed by the isotope studies. Because living things on Earth showed a marked preference for the lighter carbon isotope, carbon 12, as they built their cells, the carbon isotopes in the rock, in theory, should have provided a powerful test. But scientists were finding that the carbon signature was muddied and difficult to read. Even more dramatically, the sulfur isotopes seemed to show the clear opposite of the signature expected if Martian microbes had made the selection.

None of these disagreements, per se, troubled the young Dr. Steele. He had flown all the way from England in anticipation of a rational, earnest, dogma-free discussion aimed toward advancing knowledge. Instead, he felt that those on both sides went beyond the scope of the available data to states of unjustified certitude and unnecessary rancor.

As the meeting wore on, he grew increasingly dismayed as he watched seemingly intelligent grown-ups with rhetorical knives and daggers drawn, trying to shout each other down over isotopic ratios and crystal structure.

At one point he finally took the microphone. He rebuked the crowd for "acting like children brawling in a school yard." He expressed his disappointment. He wanted a consensus, and if there was not yet enough data, he wanted everybody to put their heads together and plan out some experiments that would provide it. He didn't want to see any more of this chorus of "my ball is bigger than your ball."

Mary Fae McKay was in the audience. At just about every such meeting she attended, she would pick out someone especially interesting, with whom she'd like to spend time in conversation. About this particular conference, she would say later, "Andy was my 'find.' I wasn't the only person who noticed him, but he stood up several times and made some cogent comments [including] this sort of Rodney King talk at the panel discussion, basically saying, you know, can't we just all get along?"

Some considered the debate just another useful demonstration of the tug and pull of high-stakes science at its most fervent. Douglas Blanchard, head of the solar system exploration division at Johnson Space Center and straw boss of the McKay regulars, thought it was natural that the process became adversarial. Individual passions about the work helped drive the scientific process. He was one of several in and outside of NASA who believed that the claims about possible signs of life in the Mars rock were generating an "explosion" of learning and new data that would be valuable whether or not the assertion turned out to be correct.

For better or worse, because of the importance of the question and the intensity of modern communications, it was inevitable that these skirmishes would be on display. And one of the few things people on all sides seemed to agree on was this, repeated frequently: "Well, it proves that scientists are human." It was never the saintly perfection of individual scientists that kept the process honest but the "community" working through the process—the checks and balances of hundreds, sometimes thousands of scientists—operating like some vast organism slumping toward the cheese at the end of the maze.

Still, that process could be painful for those within its coils. Some weeks after this meeting, Zare would issue a statement echoing Steele's sentiments, decrying the ongoing debate as "emotional and highly disruptive," as opposed to the skeptical but highly reasoned and dispassionate discourse he had anticipated. The quality of the commentary "has not been terribly impressive," he said. There seemed to be "a strong element of the blind men and the elephant at work here," in which "different researchers are looking at different parts of the meteorite and coming up with wildly different reports."

Like other members of the McKay team, Zare would be dismayed that the constant need to respond to the criticism disrupted his lab research. Zare's assistant Simon Clemett found himself reacting, working to combat the charges that the organics were mere Earth contamination, when he would rather have moved on to probe the rock for amino acids and other higher-order organic compounds. He got to the point where he was sleepless and depressed.

But here at the March melee in Houston, an astounding thing happened after Steele delivered his sermonette. A man named Gerald Soffen walked up and greeted the bold Brit. Soffen was something approaching a legend in the space program, the man who had led the Viking science team that had carried out the first remote experiments on the surface of Mars. More recently, he had helped Dan Goldin shape NASA's plan to search for life in the universe.

"I want you to come to work for NASA," Soffen said. "Come meet Mike Meyer, come meet . . ." And he began to introduce Steele around to NASA officials attending the meeting and to Gibson and others working with the ailing and absent McKay. Steele was stunned—but not too stunned to accept. No matter what bullshit went along with it, he thought, it would be an honor and privilege to be able to do this work.

He soon found himself ensconced in the sanctum sanctorum of the "life on Mars" debate: Building 31. The University of Portsmouth arranged funding for a Ph.D. student to assist Steele. He hired his future brother-in-law, Jan Toporski, an environmental geologist with a sideline in paleontology. They started working under NASA's auspices in October 1997 and moved shop to Johnson Space Center in November on a two-year postdoctoral grant.

Then a second thing happened to Steele during this already heady period, causing him to develop a serious emotional conflict. Katja Toporski was a captivating anesthesiologist he had met years earlier while working at a hospital in Ireland. Now they were expecting.

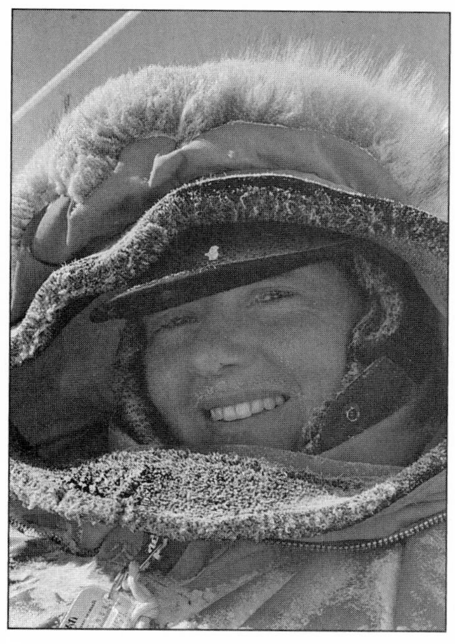

Robbie Score, shown during the 1984–85 Antarctic meteorite hunt, when she was credited with the discovery of the meteorite that would become famous. It was her first visit to "the ice." (Catherine King-Frazier)

The 1984–85 meteorite search team. Standing, from left: John Schutt, Carl Thompson, Scott Sandford, Bob Walker; sitting in front are Robbie Score and Catherine King-Frazier. (Robbie Score)

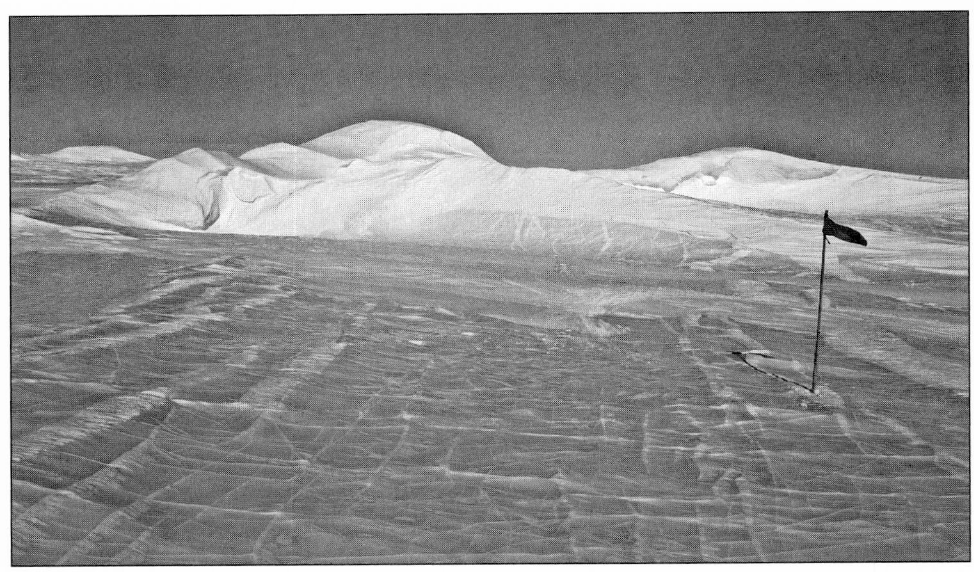

The desolate site at Allan Hills, on the Far Western Icefield, where Robbie Score found the unusual Mars rock some 16 million years after it was blasted off Mars, and 13,000 years after it landed on Earth. (Robbie Score)

Intrepid meteorite hunters in the Antarctic have collected some 30,000 specimens, including about three dozen from Mars, since scientists realized that special conditions were at work to preserve and isolate them on the polar ice. (Linda Martell, Antarctic Search for Meteorites [ANSMET])

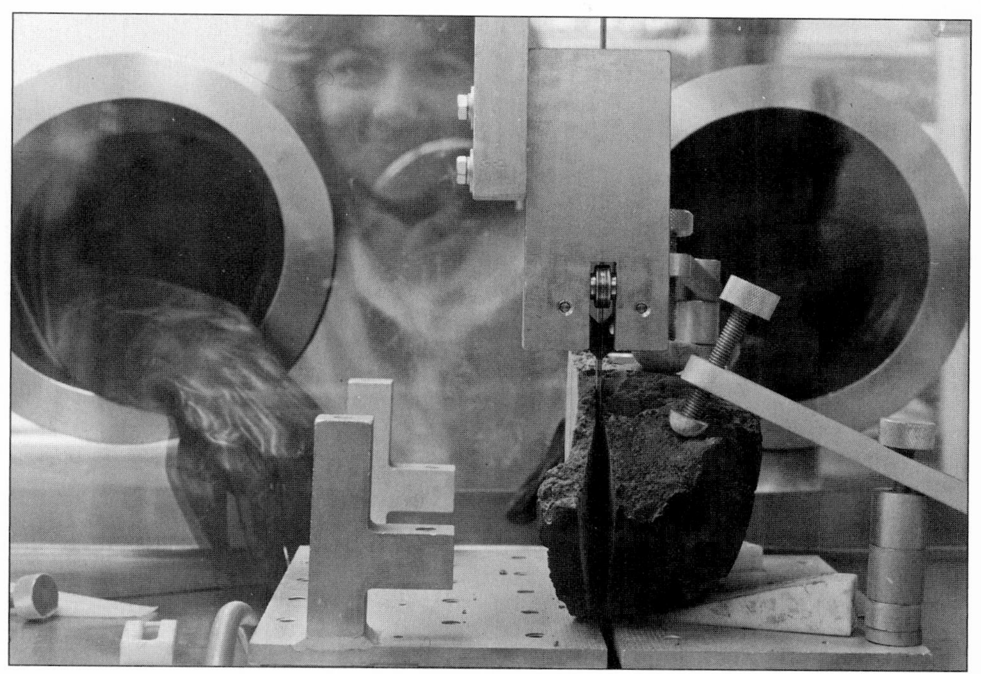

Robbie Score at work in the meteorite lab at NASA's Johnson Space Center in Houston, shown here sawing a Mars rock found in a part of Antarctica called Elephant Moraine. Other researchers determined that this specimen contained trapped gases that matched the atmosphere of Mars, as measured by the Viking landers. (NASA)

Apollo astronauts and geologists picnic among the trees during a February 1969 geology training trip to western Texas, near Sierra Blanca, about 80 miles southeast of El Paso. Astronauts Jim Lovell and Fred Haise are in the left foreground, and Neil Armstrong and Buzz Aldrin (back to the camera) are in the background. Among the others (mostly geologists from NASA and the U.S. Geological Survey) is NASA's David McKay, third from right. (NASA/Andrew "Pat" Patnesky)

David and Mary Fae McKay, circa 1975, at home in a Houston suburb near the NASA space center there. The McKays' romance had flowered in Japan. (Courtesy of the McKay family)

Duck Mittlefehldt's research on ordinary meteorites led him to an extraordinary rock. He discovered that meteorite ALH84001 was not a chip off an asteroid but a very unusual chunk of Mars. (Kathy Sawyer)

Chris Romanek, shown on a ship during a research expedition in the South Pacific in 1996, had specialized in studies of carbonate seashells and become fascinated by the presence of carbonate globules (also called rosettes, pancakes, or moons) in the unusual Mars rock. He was the first to consider possible similarities between structures in the ancient Mars rock and those seen in forms proposed as "nanofossils" found around hot springs on Earth. (Courtesy Chris Romanek)

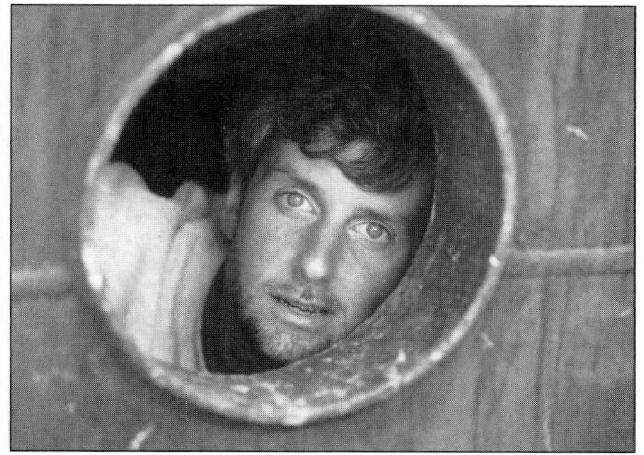

The Allan Hills meteorite, before it was sawn apart, weighed 4.25 pounds. A field team member described it in initial notes as "highly-shocked, grayish-green," and covered with a black fusion crust from its fiery descent. There was also the editorial comment "Yowza-Yowza." (NASA)

The rock, which Score designated ALH84001, is shown after it was cleaved at the NASA lab in Houston. It would turn out to be the oldest rock known from any planet, packed with clues not available in any known Earth rock about the early history of the planets. (NASA)

Close-up of carbonate globules (each about 100–200 microns across), where the most interesting and puzzling microscopic features in the rock are concentrated. These deposits suggested the rock had interacted with water, a prerequisite for life, on the young Mars. This photo originally appeared in the journal *Nature.* (NASA/Monica Grady)

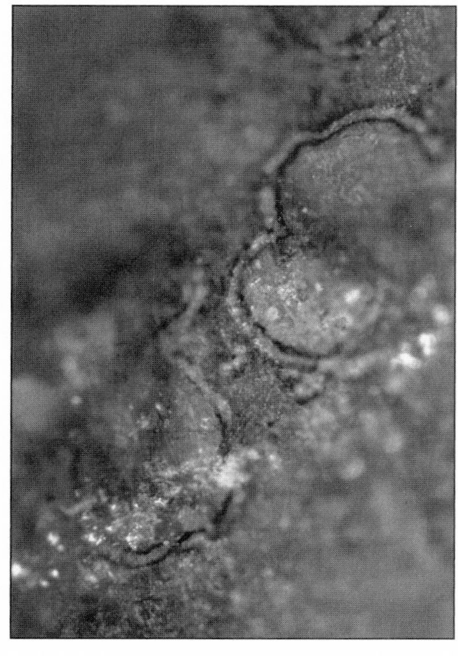

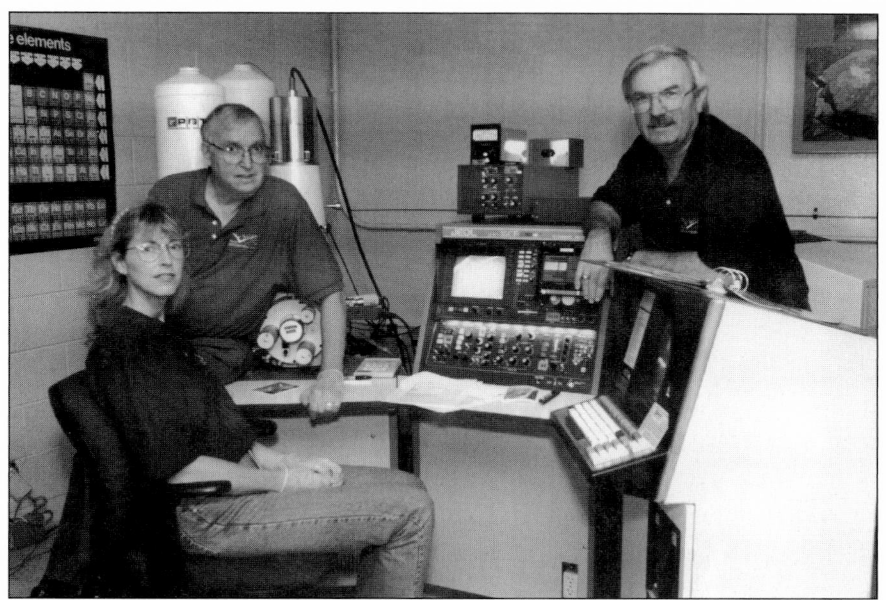

Secret collaborators Kathie Thomas-Keprta, David McKay, and Everett Gibson in the electron microscope lab in Building 31 at NASA's Johnson Space Center, Houston. (NASA)

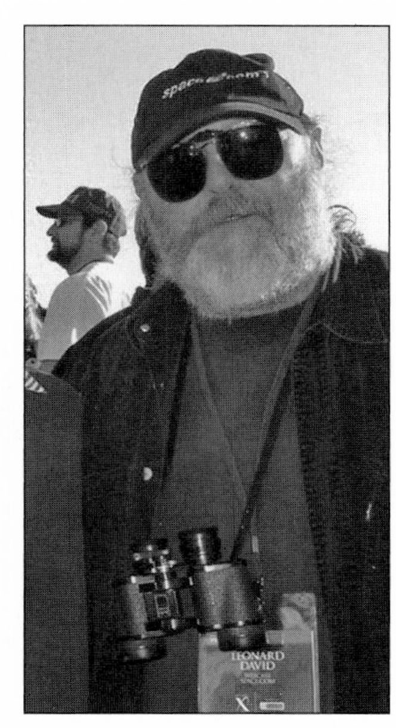

Veteran space reporter Leonard David got wind of the McKay group's work and, in August 1996, triggered a media feeding frenzy that forced officials to move up the announcement of possible evidence of biology in the rock from Mars. (Courtesy Leonard David)

WASHINGTON—
THE 1996 NASA HEADQUARTERS PRESS CONFERENCE

Wesley Huntress, Dan Goldin, David McKay, and Everett Gibson appear upbeat as they gather around the rock on the day of the press conference that launched global headlines and triggered a hostile blowback from many scientists. (NASA/Bill Ingalls)

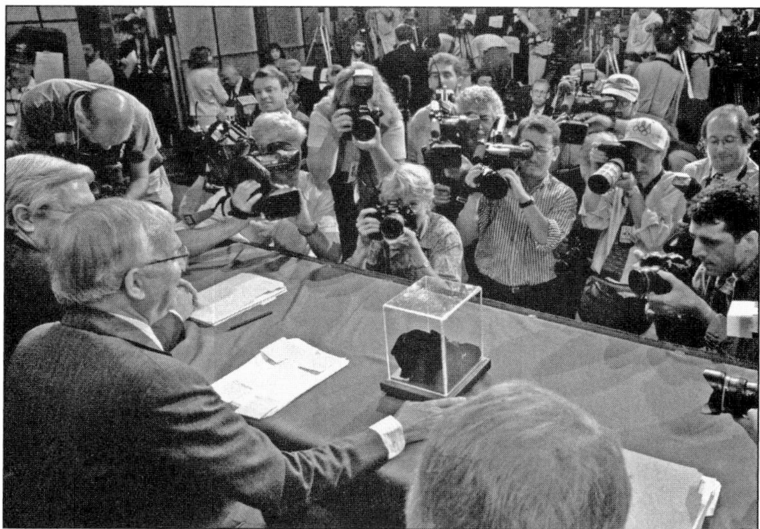

Reporters and photographers press toward the dais as McKay, with his hand on the glass case containing the Mars rock, looks out on the excited scene in the NASA headquarters auditorium just before the press conference is to begin. (NASA/Bill Ingalls)

Dan Goldin introduces the scientists, including J. William Schopf (second from right, between Hojatollah Vali, far right, and Richard Zare). Schopf would become a leading critic and antagonist of the McKay interpretation of the evidence in the Mars rock. (NASA/Bill Ingalls)

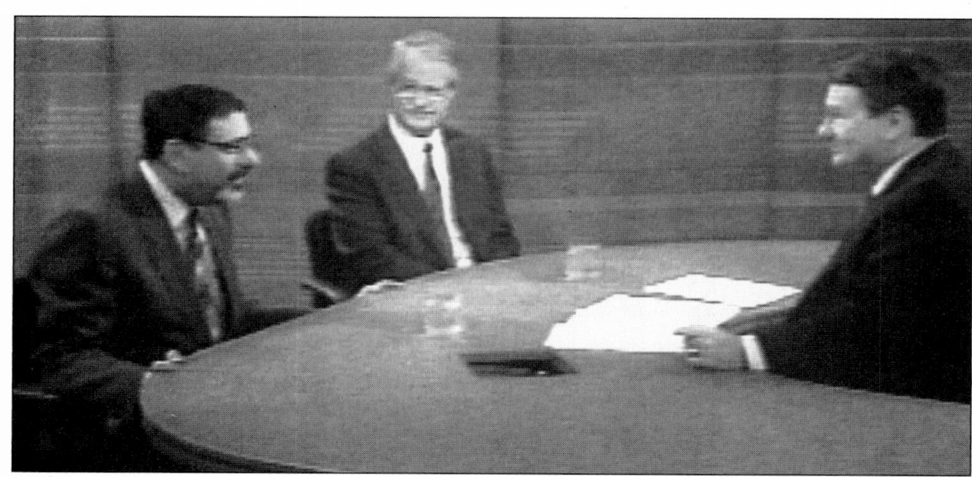

Hours after the press conference, a sleep-deprived Richard Zare and NASA official Wesley Huntress explain the Mars rock findings to Jim Lehrer of PBS. (Courtesy Stanford University)

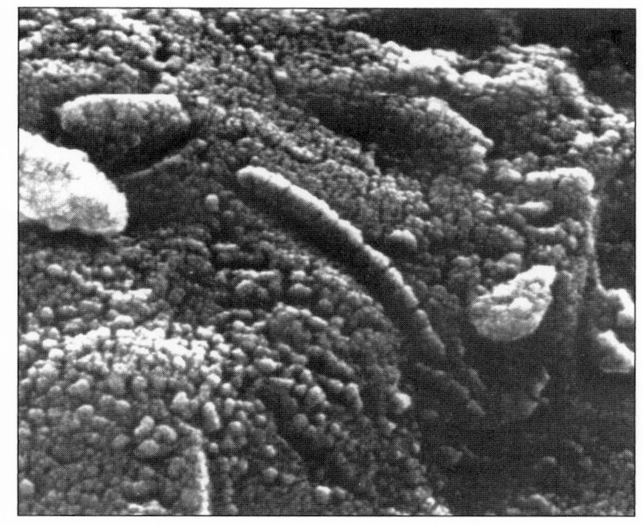

This high-resolution scanning electron microscope image of ALH84001 shows an unusual tubelike structural form found by the McKay group. This structure was not part of the report the group published in the August 16, 1996, issue of the journal *Science*, but they showed it at the August 1996 press conference at NASA headquarters and it appeared in media reports around the world. Numerous scientists quickly disputed the proposition that it might be evidence of primitive life on Mars 3.6 billion years ago. (NASA)

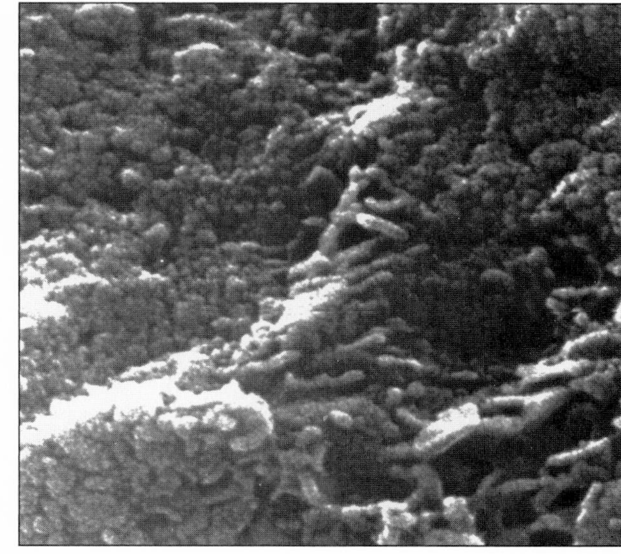

This electron microscope image shows tubular structures of likely Martian origin that are very similar in size and shape to extremely tiny proposed microfossils—nanofossils—found in some Earth rocks. This photograph was part of the report that the McKay team published in *Science*. The largest of the fossil-like shapes are less than 1/100th the diameter of a human hair in size while most are ten times smaller, or about 1/1000th the volume of typical Earth bacteria—a size many scientists say is far too small to contain the basic machinery of life. (NASA)

In the bitter hostilities that followed the August 1996 press conference, J. William Schopf of UCLA, a celebrated authority on ancient microfossils, became a primary critic of the McKay group's interpretation of the evidence in the Mars rock. (Courtesy of UCLA)

Andrew Steele moved across an ocean to work on the Martian meteorite, and found evidence that others had missed. (Kjell Ove Storvik/AMASE)

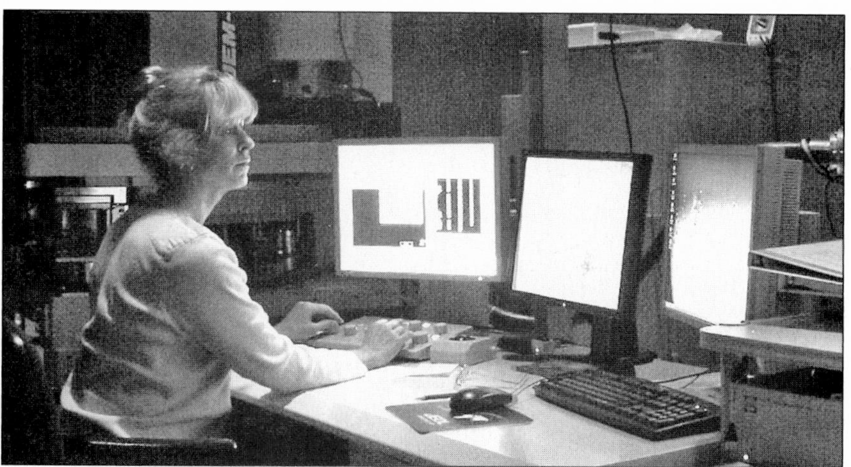

Kathie Thomas-Keprta is shown with a transmission electron microscope in Building 31. She and coworkers generated renewed headlines and rebuttals in early 2001, when they published new research on the mysterious magnetic crystals in the Mars rock, arguably the most resilient evidence put forward by the McKay group. (Courtesy Kathie Thomas-Keprta/NASA)

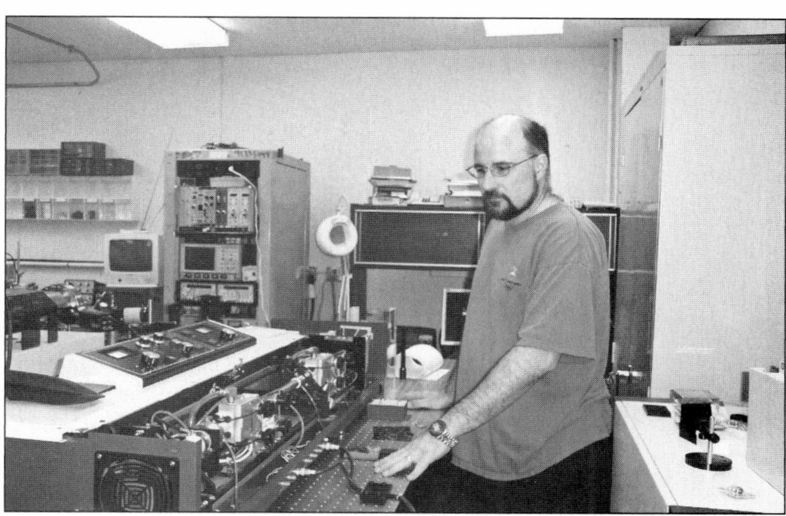

Simon Clemett, in the Building 31 laboratory. The former Zarelab researcher suffered sleepless nights over the vitriolic reaction to the McKay group's findings. He then joined others working with McKay to push the state of the art in further investigations of several Martian meteorites. (Kathy Sawyer)

Martin Brasier of Oxford speaks during the landmark confrontation with J. William Schopf of UCLA (right) at the NASA astrobiology conference in California, April 9, 2002. In the debate and in a published paper, Brasier challenged Schopf's 1993 descriptions of the oldest known fossil life on Earth. Like the debate over the Mars rock, this exchange highlighted major blanks in human understanding of the distinguishing characteristics of life. (Kathy Sawyer)

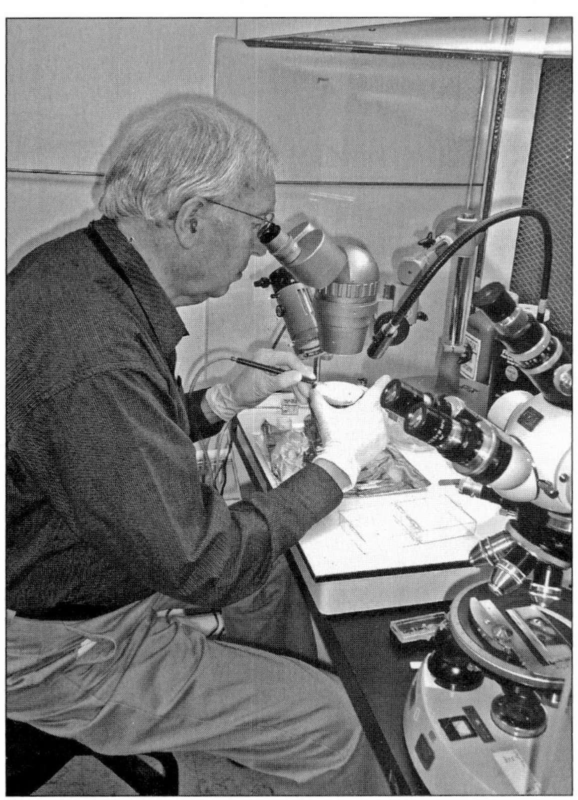

David McKay, in his microscope lab in April 2004, prepares a meteorite sample. More than one hundred metric tons of the world's most advanced equipment had been brought to bear on the 4.5-billion-year-old Mars rock, and the work continued to reverberate through several fields of study. (Kathy Sawyer)

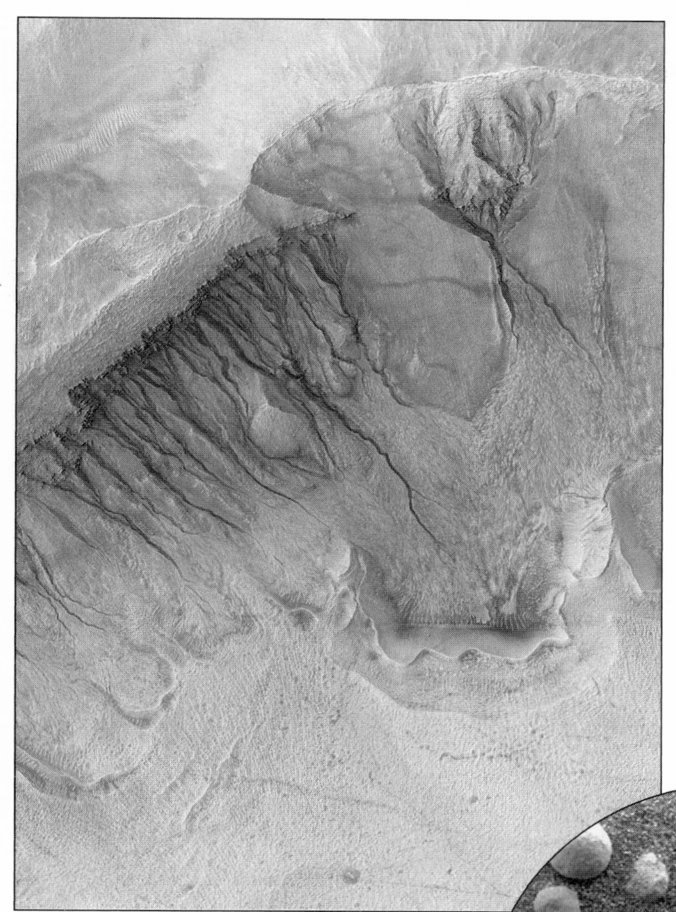

The Mars Global Surveyor craft, orbiting the red planet in late 2001, captured this image of gullies on a meteor impact crater in the Newton Basin, in Sirenum Terra. The discovery of this and similar features on numerous Martian crater walls startled scientists and triggered lively speculation about the gullies' origins, including the possibility that they may have been formed by the release of groundwater, or some kind of liquid, onto the Martian surface in geologically recent times. (NASA/JPL/MSSS)

The surface rover Opportunity, in 2004, sighted what looked like blueberries in a muffin, embedded on top of and within an outcropping of rock that was being eroded by windblown sand. The spherical grains contained a mineral that, on Earth, most often forms in water, and scientists concluded that this Martian rock formation had once been soaked with water in liquid form. (NASA/JPL/Cornell/USGS)

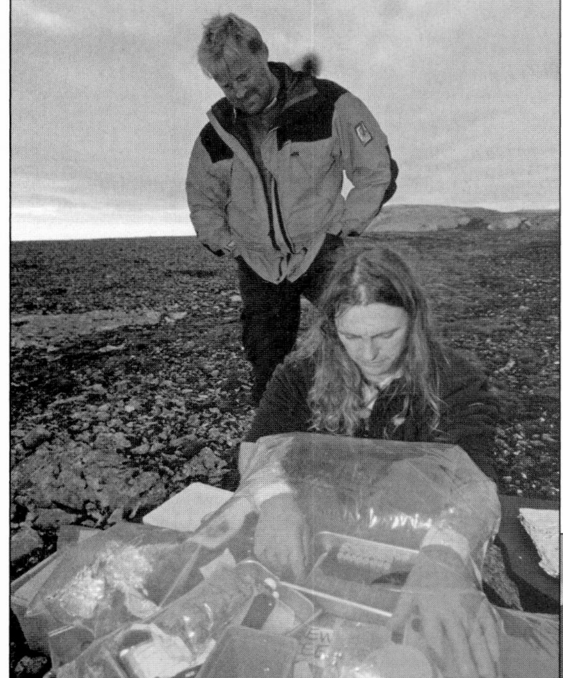

Andrew Steele, foreground, and colleague Hans E. F. Amundsen are among those developing techniques for robotic Mars exploration through fieldwork in Mars-like places on Earth, such as the archipelago of Svalbard, Norway, where hot springs climb up through permafrost. Found here are carbonates similar to those in the controversial Mars rock. (Kjell Ove Storvik/AMASE)

Allan Treiman, an outspoken critic of the McKay group's interpretations of the Mars rock evidence, at work at Svalbard. Antagonists from all sides of the dispute find common ground in the Mars exploration effort. (Kjell Ove Storvik/AMASE)

Steele moved to Houston alone, leaving Katja at the house they had bought in Portsmouth. He wouldn't be able to stand it for long. He would end up moving back home in December and staying put until February—feeling guilty but wanting badly to be with his blossoming family.

In December 1997, as he prepared to desert his post for a while and return to England to be with Katja, McKay gave him his second sample of the Allan Hills meteorite and asked him for a reaction. "What do you think that is?" McKay said, pointing to images from the rock samples and forms that to Steele's practiced eye looked exactly like filamentous bacteria.

This launched Steele, with Toporski, on what would be a pathway that no one before them had followed. Much of what Steele found would be both enlightening and painful for the McKay team.

The Brits would soon determine, unequivocally, that the Allan Hills rock was infested—"teeming like a pig's ass," in Steele's tongue-in-cheek phrase—not with little Martians but *with Earth life.* When he saw the emerging evidence, Steele took a mental stroll through all the published research papers that, by this time, had proclaimed there was "no life" in the rock. Organic contamination, yes, Earth contamination, yes, but "no life."

Everybody had missed it—until now. Once confirmed, this discovery would be the first documented account of terrestrial microbial activity inside any meteorite recovered from the blue ice fields of the Antarctic.

Steele delivered the news about the infestation to McKay in early 1998. As always, he thought, McKay took the news just the way a scientist should. He was disappointed—but the evidence was what it was.

Everybody on the team agreed that the discovery didn't eliminate the possibility of Martian biology. The monster challenge would be distinguishing among whatever "pools" of life did exist.

Steele and his coworkers looked at the surfaces of forty chips from the Allan Hills rock, as well as a variety of other Antarctic rock samples. They found that "their" Allan Hills bugs—highly branched, filamentous, with club-shaped endings to some of the branches—resembled the terrestrial actinomycetes bacteria, which had been found previously in Siberian permafrost and in the Antarctic.

The organisms had taken up residence in the outermost portions of the meteorite, near the fusion-crust surface. Steele concluded from this that they were "almost certainly terrestrial in origin" and most likely introduced in the Antarctic rather than in later handling. The bugs were feeding on organics in the rock. It was known that on Earth, the hydrocarbons (PAHs), which could be formed by bacterial decomposition, could also be an energy source for live bacteria. Also, another group had recently shown that the meteorite contained amino acids (organic compounds important to living things).

People had studied meteorites for a long time—for petrological, geochemical, and mineralogical purposes. Now, all of a sudden, there were organics from terrestrial life involved. As this project progressed, people in the meteorite game would show Steele images they had taken years earlier in which he clearly saw similar features. This phenomenon hadn't been spotted earlier because those investigators were not trained in the subject.

In the past, the mere whisper of the word *contamination* meant the end of a meteorite's career as a research subject for purposes of organic analysis. Once the rock was discarded, there would be no systematic follow-up research on *how* it had gotten contaminated. Other factors had to do with microscope viewing techniques and the random nature of microbial infestations.

Here was another instance in which artificial divisions in the academic and research worlds helped to create or prolong blind spots in human observations of nature. Such constraints could also explain why the painstaking efforts of Gibson and others in Building 31 to eliminate contamination from the rock had missed things. They had, for example, used only one kind of nutrient to try to culture the bugs. Steele knew you could give an organism the wrong thing to eat or too much to eat and that could yield a negative result even when there were bacteria present. What's more, the Houston team had relied on colleagues who were *medical* microbiologists, not *environmental* microbiologists like Steele—who came in with a different set of assumptions and knew different ways of looking.

Steele's approach to the rock covered different ground from that of the Zarelab team in their analysis of the organics (PAHs) in the rock. The Steele group was just tickling the surface, while Zarelab had been smacking much deeper holes in the material. Different techniques revealed different properties.

It was all about training the individual eye, Steele thought. In this case, David McKay had looked at the images and been unable to understand what he was seeing. But he understood that he didn't understand. So he handed the puzzle to Steele—an *environmental* microbiologist.

As Steele would write in one of the numerous papers he and the team published on the topic, with David McKay among the coauthors, "In the particular case of ALH84001, terrestrial organisms went undetected by all the techniques that claim to be able to detect life." There had been little impetus for research on this kind of contamination until the McKay group published its claims. It was then that "the issue of contamination from Antarctica came out of a backroom and was given serious thought, mostly by researchers seeking to discredit the work of McKay et al."

Hardly a huge surprise in *hindsight*, Steele thought. The discovery mirrored everything people were learning about life's robust nature.

The rock had been on Earth for 13,000 years, and on the surface of the ice (exposed to warming sunlight and snowmelt) for 500. Until about 15 years earlier, Antarctica had been regarded as a sterile wilderness. Among the factors that had allowed the microbes to escape detection, in Steele's view, was general "ignorance of the diversity of Antarctic microbes," which (scientists now knew) included bacteria, fungi, algae, cyanophyta, and lichens. As a survival strategy, Antarctic organisms were known to seek sanctuary inside rocks, which could contain liquid water from snowmelt, organics, and minerals they needed for growth.

The findings had implications for analyses done on other meteorites as well, since any contaminating organisms, eating away at the contents, digesting, and expelling material, could have muddied the picture for unwary researchers as to what was indigenous to the meteorite or its native home and what was terrestrial. These organisms needed stuff to eat; in other words, they weren't primary producers themselves.

And therein lay the crux of a big problem, Steele knew. Which was basically that "you are what you eat." So if you're eating Martian organics, you take on that carbon-isotope signature. How do you tell which is which?

Steele and his colleagues turned up a wealth of microbial species not only in the Allan Hills meteorite but in other types of meteorites from Antarctica. In one type, they detected "living cells all through the core of the meteorites." The shower of fragments known as the Murchison fall, which had landed in a farmyard and, by some accounts, in a manure-filled ditch, showed evidence of infestation by fungi and bacteria, Steele found. In some cases, he detected crystal growth that had occurred while the samples were in storage, growth that in certain cases might be mistaken for fossils.

And on a shelf in his lab, Steele was amused to see that, despite all precautions, old Chip, the speck of Allan Hills meteorite that had introduced him to his new career, was nourishing its own zoo of microbes. It was now carpeted with bacteria, right across its surface, and they were dining on and changing its minerals.

These revelations had profound implications for NASA's planning for future Mars missions. Steele's work, like other surprises flowing from the McKay group's 1996 claims, raised particularly serious concerns for the ambitious plan to bring back bits of Mars for study in Earth labs.

Steele and company argued that meteorites got contaminated immediately upon contact with the terrestrial biosphere, at which time the microbes reprocessed any organic material the rock contained. And yet none of the traditional approaches used to refute the McKay team's 1996 claims had detected anything alive inside the meteorite. Techniques that relied on electron microscopy and shape comparison, obviously, could take researchers only so far.

In pursuit of better tools, Steele got the idea of using biology's natural survival techniques to probe for life signs, terrestrial and otherwise. In a word: antibodies. These are the soldiers of the body's immune system—proteins that blood cells march out to repel invading bacteria or other foreign substances. The defense mechanism acts as a terrifically sensitive life-detection device. The little troops do their work by binding to a specific antigen (the invading molecule, fragment of protein, or whatever) like a handcuff around a wrist. Like so much in chemistry and therefore biology, this process depends on jigsaw shapes that fit together.

These techniques, called immunoassays, had become increasingly useful, as the technology advanced, in many areas of research—on everything from human health to ancient fossil biomarkers. For instance, antibodies could sometimes produce strong reactions in fossilized bone and bird-claw tissues. Antibodies had detected remnants of original proteins and other organic compounds in samples such as those of the dreaded *Tyrannosaurus rex*, dating back at least about 70 million years.

The important thing to Steele was this: research had demonstrated that the physical and chemical processes that follow the burial of an organism's remains on Earth, leading to their preservation as fossils in rock, leave products that can be detected, even from as far back as the time when Mars had a more favorable climate for life.

Now Steele and his colleagues started adapting the technique to detect past or present life on another world, and also to provide details on the organisms' metabolisms. There were a few little hurdles to be coped with, he noted, such as the fact that the technique was "Earthcentric" and might not work on weird extraterrestrials. In any case, the goal was an instrument called MASSE, a compact automated device that, they hoped, would fly on a Mars mission planned for 2009. His rallying cry, delivered at meetings where he described his project, would be: "Send your favorite antibody into space! Give us your blood!"

The tensions over the Mars rock were still on display at the next annual gathering of lunar and planetary specialists, in March 1998 in Houston. Naturally, the vitriol spilled into the social hours. One evening, the tourist attraction called Space Center Houston, next door to the real NASA space complex, was the site of a working cocktail party. Surrounded by Soviet space regalia, a partial mock-up of a space shuttle, and other exhibits in the giant hall, as many as a thousand people hoisted cold beers and dined on free cafeteria fare. Hundreds of posters—graphic displays of research results—were arrayed around the place, with the authors standing by to answer questions.

One of the schmoozers was John Kerridge, a cosmochemist and meteorite expert who was one of the most ardent McKay group critics. He remarked to a journalist that McKay and colleagues "have circled their wagons. They're not really trying to find out what's going on" with the evidence in the rock.

Not far away from Kerridge, a seething Kathie Thomas-Keprta walked up to David McKay with urgent news. According to an account in the book *Dark Life*, by Michael Ray Taylor, McKay's face flushed and he turned on his heel and "stormed off" toward another section of posters a few yards away, where he accosted a meteorite expert named Derek Sears, from the University of Arkansas. "How could you even use this image?"

Sears and his coworkers had compared the wormy shapes reported by McKay et al. with forms detected in meteorites from the moon and pronounced them similar. Since the moon is considered incapable of sustaining life, and therefore the forms could not be fossil microbes, the Sears group had concluded that the proposed nanofossils in the Mars rock must also be nonbiological.

McKay found the images Sears used to illustrate the point outrageously misleading. "It is absolutely nontypical for a lunar meteorite," McKay told

Sears. An intern who had worked with both McKay and Sears had "spent months, hundreds of hours on the microscope, finding this one image," McKay went on, his hand trembling as he pointed to the offending picture. "And it's the only one she could find. This is just—just unconscionable." And Sears had put next to it, for comparison, what McKay, among others, considered the least biological-looking of the images from the Mars rock.

"The pictures speak for themselves," Sears responded. "If you were impartial, you'd see that."

In the last years of the dying millennium, the divisions among scientists seemed to be hardening in place on the question of early Martians. People who looked favorably on the McKay group's claims were outnumbered and overshadowed by many more who decidedly did not. In the broader world, there was a creeping sense that the claims had effectively been buried. That was the buzz, the conventional wisdom.

More laboratories had received their allocations of chips off the Mars rock, as $2.3 million in funds from NASA and the National Science Foundation fueled a spate of new studies. Many of the researchers had come up with findings that cast doubts on significant chunks of the McKay team's hypothesis, and the McKay group gave ground on certain points:

- Some (though not all) of the wormlike microfossil shapes seemed to be bits of naturally formed clay or ridges of mineral.

- The rock was indeed loaded with Earth-born contaminants that apparently had seeped in with Antarctic meltwater.

The McKay regulars couldn't help but be distracted, they had to admit, by the continuing rain of personal, and, they felt, often unfair feedback impuning their motives and their competence. In private moments, McKay sometimes found himself focusing furiously on one of his critics, his stomach knotting, instead of on the work, and that made him even angrier.

At some point, McKay came to suspect—ironically, in light of speculation that NASA had staged the whole affair to boost its budget—the stigma of "political incorrectness" had attached itself to his hypothesis. Some people seemed to believe that their prospects for funding could be jeopardized if they worked on it, or allowed the McKay group to use their facilities.

McKay group antagonist Ralph Harvey, who himself was fonder of confrontation than most, quipped that the McKay group "felt that they were being mistreated when in fact they were being treated with the same contempt any scientist feels during that period. . . . It's contempt for anybody that would try and shake up the current paradigm."

Still, no single finding managed to deliver a definitive knockout blow for either side. Each answer led to more questions. Factions developed within the "pro-life" camp and the "anti-life" camp. The antis, especially, seemed to fight among themselves as much as they argued against McKay.

And the McKay team continued to argue, for example, that the organics could not be completely explained away as nothing but Earth contaminants, and that the organic matter in the carbonates—which Romanek showed had formed on Mars—would surely have been destroyed if the temperatures had soared as high as some proposed.

"You have to understand the whole system," Gibson would say. "They've been looking at it only from a geological point of view. Mother Nature had not only geology but atmosphere and water and other things in there, and that set the composition of the oxygen which allowed us to come up with a low temperature."

As the controversy rolled on, this would become a recurring theme of the conflict. Some people liked to consider all the various threads of evidence in the rock together as a single story, others piecemeal. Was one to look at the forest or examine each tree separately? Did the evidence represent a coherent portrait of a complex, interacting system—or a mélange of isolated elements that had happened to get trapped together in a well-traveled vessel? How a person answered such questions would affect his or her conclusions.

The variety of interpretations also seemed to flow from the fact that even this one simple lump of minerals had more than a single history. Just as parts of it were contaminated with terrestrial microbes while others might not be, different parts of it gave differing answers to other questions too.

Steele was among those who objected to people applying the result from one fragment of the rock to the whole thing. The rock varied tremendously from one part of itself to another. Simon Clemett, for another, argued that the meteorite contained at least five different forms of carbonate minerals, possibly deposited into the different locations at different times or in connection with different events, some at higher temperatures, some at lower temperatures. It was the cliché—one hell of a microcosm.

Painful as it could get, the dispute over the Mars rock was proving to be a valuable and forceful goad to scientists. Derek Sears, the pioneering cosmochemist who had so riled David McKay and Kathie Thomas-Keprta at the cocktail party, just months earlier had coauthored an editorial in a planetary science journal, saying: "The Antarctic meteorite Allan Hills 84001 may be at the center of a revolution in our thinking about the origin of life on Earth, Mars, and perhaps elsewhere . . . because it has forced a reexamination of the importance of microbes in the ecosystem, the nature of the smallest possible life forms, the nature of organic materials and structures that led to the origins of life and the temperature regime at which life originated."

What kept the process honest and fruitful, in the end, would be the collective interactions of the tribe of interested scientists. And the tribe was like a complicated organism.

To maintain balance and objectivity, McKay had formed two teams: the Red (for Mars) Team to elucidate the evidence supporting his hypothesis that life once swam on that planet; the Blue (for Earth) Team to gather evidence against the hypothesis.

Steele worked on the Blue Team. McKay had directed him to start out with the hypothesis that every bit of biological evidence in the meteorite was terrestrial contamination and see if he could prove it.

Steele considered the effort worthwhile because the McKay team's claims had affected such a wide range of research, including the "life on Earth" field. The Mars rock had opened so many gates. What Steele didn't know yet was that it would lead him back around to Bill Schopf. Steele would play a supporting role in a confrontation that would make some of the current spitball fights seem mild.

Steele would work in the Building 31 lab for fourteen months before eventually taking a job in Washington, D.C., on the leafy campus of the Carnegie Institution. McKay would continue to fund Steele's research. Among his other affiliations, Steele was linked to the NASA Astrobiology Institute through a McKay proposal. The two of them found ways to make it work.

Some of their conversations were tough, and the cramped internal politics of the space center could be annoying. But for now, in his exchanges with McKay in a quiet corner of Building 31—even when the evidence did not go the way McKay had hoped—Steele was finding the calm and rational discourse, the "sense of fair play" he had so ardently missed in that raucous introduction to his new life, and in the bitter public debate since.

CHAPTER THIRTEEN

BINGO

LATE ONE DAY in the last week of October 2000, Kathie Thomas-Keprta was on another expedition into the land of the Lilliputians, riding her electron beam into the deep interior of the Mars rock. She was heading further and further into frontier territory in her drive to solve the riddle that had captured her heart, brain, and soul.

In fact, Thomas-Keprta was about to experience one of those moments when the mind takes a leap that seems to change everything.

For David McKay, her breakthrough on this autumn night would stand out as one of the more satisfying moments in the marathon. With Kathie Thomas-Keprta's fretful insomniac insight would come, among other things, a sense of renewal.

Those seductive magnetic crystals—particles so small a billion would fit on the head of a pin—had become the most powerful witnesses in the rock. Thomas-Keprta and others on the McKay team regarded them as the closest thing they had to a smoking gun. For years the team had been pushing at the limits of technology as they probed the nuances of the structure. But now, thanks in part to a blunt lecture from a distinguished colleague who was critical of her group's work, Thomas-Keprta realized she was still not seeing the crystals "naked." She had *not* deciphered their true nature. She had to go deeper.

While the controversy over the rock churned on, the group had published nothing new since the original 1996 paper. Thomas-Keprta and her lab mates had been working hard, and occasionally described their progress on the magnetic crystals in presentations at meetings. But they had not published.

Now, finally, they had completed and submitted a long, detailed paper for one journal and were finishing up a second, more concise paper for a different publication. They were almost ready to ship that second one out.

But as she went through her paces in the lab, Kathie Thomas-Keprta felt ballooning discomfort about the whole thing. The crystals had her stymied, frustrated. Something about them wasn't adding up.

It was one thing to go public in the certain knowledge that your results would be attacked when you had results that you *believed* in. It was quite another to publish something that you yourself felt was on shaky ground. You had to be your own worst critic, she reminded herself. She imagined having to defend the work in front of thousands of people. She wouldn't be able to sleep that night.

Like virtually every aspect of the rock's deconstruction, the story of the iron-based magnetic crystals—the magnetites—had proved to be unexpectedly complicated. Thomas-Keprta and her coworkers learned, as Bill Schopf and others had, that in order to hunt successfully for the most primitive, ancient forms of life, they needed at least a passing understanding of multiple fields of study—and help from specialists in those fields.

The work of the McKay group and others on this labyrinthine topic would interweave theories of evolution (natural selection), emerging evidence about planetary magnetic fields, the complexities of crystallography, and the nuances of transmission electron microscope techniques, to name just a few.

So they had set about educating themselves on the microbiology and, particularly, on the esoterica of magnet-making bacteria and their environments on Earth. The magnetic crystals in the Mars rock were so mesmerizing precisely because they bore such a striking resemblance to the magnetic structures manufactured by Earth bacteria.

Prominent on their list of go-to specialists was Dennis Bazylinski, a geomicrobiologist at Iowa State University, who had spent years studying the organism known as MV-1, discovered in 1975. It was just one of the several strains of Earth bacteria that somehow grew these iron-based crystal structures inside themselves, giving them an evolutionary advantage over their uncompassed cousins in the survival game. Magnet-making bacteria were devilishly difficult to grow in the laboratory for study, but Bazylinski had developed ways to culture this particular strain.

Thomas-Keprta reviewed the available data. The Earth bacteria, by using these internal magnets to orient themselves relative to Earth's strong magnetic field, could navigate more efficiently up or down in the water column, to the zone of food and energy where they could best thrive.

The magnetic crystals these organisms grew had special properties, different from those made without benefit of biology. Ordinary nonbiological magnetic crystals tended to be a hodgepodge—formed by precipitating out of water or condensing from vapor. They came in a range of different sizes and, as often as not, interlaced with one another. They tended to be rounded—that is, of a shape that would fit inside a sphere, with corners touching the sphere's walls—or, as the experts would put it, cubo-octahedral.

By contrast, this unique group of microbes (called magnetotactic) worked from a genetic blueprint. Many of their magnetic crystals tended to be elongated and free of some of the chaotic defects found in the nonbiological ones. In addition, they consisted of highly pure magnetite without the inclusion of impurities common in nonbiologically formed magnetite crystals. The microbes used membrane compartments to wall off a space where they could control the concentrations and interactions of chemicals. They tended to align their magnetic crystal chains head-to-tail, causing the entire chain to behave as a single large magnet rather than a collection of individual small magnets. Microbes, it seemed, tended to grow their magnetic crystals within a crucial size range—just big enough to have a permanent magnetic moment, or directional tendency, but small enough for a number of them to fit into and orient the cell. The result of all this organization was efficiency, more magnetic bang for the buck.

But could counterparts on Mars have evolved in the same way? Had they?

When the McKay group first posed the question in 1996, one of the objections was the lack of any known magnetic field in Martian history. Throughout the space age, researchers had tried without success to detect one like Earth's, with strong north and south magnetic poles.

So: to what end would a microbe develop a biomagnet on a planet with no magnetic field? The answer arrived fortuitously in 1998, courtesy of the robotic spacecraft Mars Global Surveyor as it studied the red planet from orbit. Sensitive instruments aboard the Surveyor surprised planetary geologists with measurements indicating that Mars had indeed once sported a strong, Earthlike magnetic field—but it had disappeared at least 3.7 billion years ago.

The young Mars had a magnetic field, it seemed, at the same time that it still had vast quantities of liquid water and a carbon dioxide–rich atmosphere—a suitable environment for the rise of microbes. A magnetic field would have protected the Martian surface from the lethal "wind" of charged particles from the sun, in the same way that Earth's magnetic bubble has continued to shield it—and us.

The magnetic field might have persisted well after the geological cataclysm some 3.6 to 3.9 billion years ago—most likely an asteroid impact—that shocked and cracked the Martian rock formation whose fragment had landed on Earth, pieces of which Thomas-Keprta now studied. That was the impact that, in theory, allowed mineral-rich water to flush through the rock's fissures and evaporate, leaving behind the carbonate globules—with the enigmatic magnetic crystals strewn in and around them. Now researchers on all sides agreed that the magnetic crystals in the rock, at the very least, contained an important record of the vanished magnetic field of early Mars.

The timing was right: the magnetic field on Mars had apparently been present during the period when, in the McKay team's hypothesis, Martian microbes might have evolved with little compasses to take advantage of it.

It was a point of ongoing strain, vis-à-vis the critics, that Thomas-Keprta was focusing only on a unique subpopulation of magnetic crystals. All sides of the controversy had agreed that there were varied "populations" of these grains in the Mars rock and that as many as roughly three-quarters had been formed in geological, not biological, processes. Several people objected that Thomas-Keprta focused selectively on just the fraction that fit the biological model.

Wary that this subgroup might be some form of terrestrial contamination, both the McKay group and independent researchers carried out extensive examination and testing designed to assess that possibility. In the view of many scientists, they had fair evidence that this little magnetic clique was Martian.

One reason was that the oxygen of Earth should have changed the iron into a different form from what the researchers found. "If you take some of these magnetites and lay them out on a table," McKay would say, "they would probably oxidize [into another form] possibly in a matter of minutes but certainly in a matter of a few months. So there's no way these little magnetites could have been floating around . . . in Antarctica, or washing around in the meltwater, which also has air dissolved in it. . . . These little guys would not remain as tiny magnetites for very long."

So Thomas-Keprta had pressed on, confident that this particular one-quarter of the magnetic material she was seeing embedded in the carbonate globules, whatever it signified, had been formed on Mars. She was back to the primary point of contention. Were these Martian magnetites made in biological processes?

Opposition viewpoints, as painful as they could be, had continually driven Thomas-Keprta and the others back to do an even more careful, detailed study of the magnetites. They continually felt compelled to try to push their techniques past the state of the art.

"We take our cues from David," Thomas-Keprta told a visitor to the team's disheveled War Room, next to McKay's office. "He is our leader. He's totally honest, totally trustworthy. We follow David's cues on this. He tells us if it's right, if it's truthful, if it's honest, you go for it. That's what you do."

When she wanted to avoid distraction in order to work on a paper, Thomas-Keprta retreated to her favorite sanctuary, a Starbucks just up the road. She had been doing that a lot lately.

The paper the team had recently submitted was a long and technically detailed description of its work so far on the chemical composition and geometry of the magnetic crystals, based on a study of some six hundred crystals the researchers had separated out of the rock. With Thomas-Keprta as lead author, the paper would be published in the December 2000 issue of *Geochimica et Cosmochimica Acta.* It would generate no more than modest notice in the press but would be warmly welcomed by interested scientists—if only because it finally put the team's arguments on a factual footing, with firm data all laid out. It gave skeptics some targets to shoot at. And while it would not persuade the staunch opposition, the evidence it presented would be enough to intrigue some of the fence-sitters.

With the first magnetite paper on its way to publication, Thomas-Keprta and her coworkers in the lab were working toward the second, focusing with unprecedented intensity on the geometry of the crystals—possibly the key to resolving the enigma. What was the true shape? Technology that might answer the question had become available only a few years earlier. And the alien microscopic landscape remained devilishly difficult to interpret from a vantage in the macroworld using the fledgling techniques at hand.

Thomas-Keprta knew this. But she had never fully appreciated the nature and extent of the challenge until she received a pointed warning from a colleague on the other side of the argument, veteran meteorite specialist Peter Buseck, a mineralogist from Arizona State University.

Buseck was firmly in the skeptics' camp on the question of whether the magnetic crystals had been created by Martian microbes. Buseck was also the very same strolling scientist who had inspired Dick Zare to get involved with meteorites that sparkling night several years earlier as they'd walked across the Stanford campus. And that had led eventually to Zare's and Clemett's involvement with Mickey, Minnie, and Goofy.

This time, Buseck cautioned Thomas-Keprta that the limitations of the technology might fool her into making a mistake about the crystal shape. After their conversation, she realized that she needed to do considerably more high-resolution imaging of the crystals, and pay a lot more attention to their orientation in three dimensions.

So Thomas-Keprta went on a new hunt—a hunt for "the perfect crystal," the one that could best help her make a strong case about the three-dimensional crystal structure. She began a survey of the magnetic crystals as they sat inside the rock samples—still embedded in the carbonate moons. Attempting to look at them as they sat in their carbonaceous matrix, rather than dissolving them out of the carbonates for separate mounting, would make the detailed observations much more difficult and time-consuming. But she felt it was better to look at them this way because the nature of their surroundings—their context—made a difference in the arguments about whether they had formed biologically or nonbiologically.

She and her coworkers spent three months taking hundreds of thin samples from the rock in their hunt for this ideal crystal. Finally, the rock yielded up a beautiful specimen that fit the bill. It was in just the right orientation. It could be tilted through ninety degrees, so the observer could look down the top of the crystal and down the side. And yet it was still completely embedded in its carbonate surround.

Thomas-Keprta nicknamed the wonder particle Bingo.

With the power of her cathedral-tower microscope, its beam of electrons taking the place of light, an observer could actually manage to see the infinitesimal object—a bit of iron oxide, something like a speck of rust, about a millionth of an inch in diameter. But the mighty transmission electron microscope could show only two dimensions, revealing the object essentially as a dim silhouette. It was, Thomas-Keprta thought, like standing in bright sunlight looking into shadow. The task, again, was to translate the two-dimensional information into a three-dimensional shape—to distinguish a flat face from a tapered edge, for example, when the two looked the same in profile. She and her coworkers would aim the high-powered electron beam at Bingo and tilt the sample painstakingly by hand through a sequence of dozens of angles in a carefully chosen and executed pattern.

The situation reminded Thomas-Keprta of a scene in the movie *Toy Story,* which she and her son had watched too many times to count. The kids are coming up the walkway to a party, one holding what looks like a tiny gift. As the kid turns, the gift shape-shifts and, it becomes clear, is enormous. It was all in the viewing angle. That's what Thomas-Keprta was coping with in the crystals, only more so. From one angle, you saw a square; viewing it edge-on, you saw a rectangle; and if you were looking at an end point, you saw a hexagonal shape.

The team would figure out the orientation of a crystal—which axis they were looking down—by studying the planes of its atoms. Then they had to look at the same crystal from dozens of different perspectives. They would rotate the crystal through a series of angles with a known relationship to the plane of the image. The final product—the imagery—was something like stop-action photography: a series of partial views that collectively, ideally, added up to a more complete picture.

Over the years, Thomas-Keprta had watched the apparent geometries of these magnetic crystals flicker and change as if in a kaleidoscope. Were there bubbles? Bullet shapes? Parallelepipeds? When she viewed a specimen from one angle, she could see that the atoms lined up in planes. *Tilt:* a supposedly telltale "whisker with a screw dislocation" (that geometry known to be nonbiological) would reveal itself to be a flat plate. *Tilt:* a rectangular crystal became a hexagonal prism—like an elongated stop sign.

Now she was able to see more clearly than ever before. And she could see that something was off. In the early 1980s, a British chemist had established the shape of the magnetic crystals in the MV-1 bacteria to be sort of a hexagonal column. But she and her coworkers, huddled over their new images of those MV-1 crystals, thought that wasn't quite right. That wasn't the shape. When they compared the MV-1 images with the images of Bingo, their perfect crystal from the Mars rock, the two sets looked exactly like each other—but neither seemed to fit the widely accepted model for the shape of the crystals made by the terrestrial microbe MV-1.

Thomas-Keprta, with major assistance from Clemett (who had left Zarelab to work in Building 31), puzzled and fretted over the problem. Well, maybe at the edges the shapes were just fading away because the carbonate was covering them. Maybe, in other words, these crystals did come to a point at the edges, just as the British model suggested, but the observers weren't able to see it clearly. But—no, the shapes did seem to fade at the edges. There was something they were missing.

It was one of those irritating misfits, one of those confounding obstacles that sometimes pointed the thoughtful investigator toward new ground.

Thomas-Keprta and Clemett debated what to do. They were eager to submit their new paper to the National Academy of Sciences journal, but how were they going to handle this conflict over the shape? Just note it and move on?

Thomas-Keprta left work at the end of that day in a state of frustration. Clemett had drawn up an image of the crystal shape. She took that home with her. She sat down at her dining room table and started cutting out shapes like paper dolls, trying to make diagrams that would fit, trying to overlay the shapes. The hours passed. *Was* that 1980s model correct? Her MV-1 images didn't fit it, and neither did her Mars rock images, but the crystals in her MV-1 and Mars rock images looked like each other. Why isn't this right? she asked herself. What's going on? She went through the information, over and over and over.

Her husband found her still slumped there at the table the next morning. "What are you doing?" he asked, exasperated. She told him what she had decided. "This isn't right. This doesn't work. I can't put this paper out. This just isn't right."

She showered, changed, grabbed some breakfast, returned to Building 31, and told Clemett the same thing. "This doesn't work. There is another 'face' in here."

But now she had the key. *"We've got to shave off the corners!"*

Cutting corners would be a good thing in this case. If they cut the corners off their model of the crystal shape, leaving new geometric faces where before there had been points, she told him, "It will all make sense. It makes sense for MV-1. It makes sense for [the rock from] Allan Hills. Everything will fit perfectly. The old model is wrong."

The realization came equipped with another incantation, like the one that had floated through Duck Mittlefehldt's head during the steamy summer of 1993, leading him away from asteroids and on the trail to Mars. This time, instead of "Shergotty, Nakhla, and Chassigny," the phrase was "truncated hexa-octahedron." It was a shape with eight octahedral faces, six hexagonal faces, and six cubic faces. Kathie Thomas-Keprta figured hers would be the first report of any crystal having this particular shape.

But why would nature "select" such a configuration?

The answer her team proposed was that the addition of extra faces, in the case of MV-1, served to elongate the shape, and this elongation enhanced the magnetic pull of the little compass, making the bugs more efficient at locating food and energy sources. The truncated hexa-octahedron could be described as a microbe's evolved design for building the best magnet possible with the least amount of iron.

So, of the hundreds of crystal grains they'd studied from the rock, the team concluded that some 28 percent were identical to those found in the Earth bacteria—and they had to have come from Mars. In the bargain, the team was presenting evidence that the geometry of biomagnets manufactured by these Earth bacteria was different from what scientists had thought for some fifteen years.

Again, it was well known that magnetic crystals could be formed in nonbiological processes. But the ones formed in that manner seemed to be demonstrably different not only in shape but in other ways—size, purity, and quantity of defects—from those produced by bacteria. No one had ever detected any nonbiological crystals, natural or laboratory made, that displayed all the properties of the biologically formed crystals Thomas-Keprta's group had studied and was describing in the paper.

Taking into account those properties—the chemical purity and lack of defects, the distinctive size and shape—Thomas-Keprta and her coworkers concluded that there were no known nonbiological sources for this special population of magnetic crystals nested inside the carbonate moons inside the Mars rock.

Thomas-Keprta, for one, was completely convinced she was seeing signs of Martian biology. To her, the evidence was stunning. The crystals in the rock were Martian "magnetofossils," she wrote in the new draft of the paper, which she was now ready to release to the world—and defend.

What she felt after her night of inspiration, however, fell a bit short of the classic, theatrical *Eureka!* thrill. It was something more akin to sweaty-palmed relief. "You caught your paper just before it would have gone out wrong!" she told herself. "Oh, my gosh, I just avoided disaster!"

On February 27, 2001, the second, shorter paper was published in a special astrobiology issue of the *Proceedings of the National Academy of Sciences*, with Thomas-Keprta as the first of ten authors. "Unless there is an unknown and unexplained inorganic process on Mars that is conspicuously absent on the Earth and forms truncated hexa-octahedral magnetites, we suggest that these magnetite crystals in the Martian meteorite ALH84001 were likely produced by a biogenic process," the paper concluded. "As such, these crystals are interpreted as Martian magnetofossils and constitute evidence of the oldest life yet found."

There in the journal was Bingo in all its glory, albeit looking more like a smudged thumbprint than a history-making crystal form.

Joseph L. Kirschvink, a Caltech geobiologist who was one of the coauthors, said of the paper, "The process of evolution has driven these bacteria to make perfect little bar magnets, which differ strikingly from anything found outside of biology. In fact, an entire industry devoted to making small magnetic particles for magnetic tapes and computer disk drives has tried and failed for the past 50 years to find a way to make similar particles." He added, referring to the Martian crystals, "A good fossil is something that is difficult to make inorganically, and these magnetosomes are very good fossils."

MV-1 expert Dennis Bazylinski, another of the paper's coauthors, suggested that the magnetite crystals might turn out to be broadly useful to astrobiologists and geobiologists as markers for the presence of biology.

Their claims got an independent boost from a second finding described in the same issue of the *Proceedings*, by an international team led by biologist E. Imre Friedmann, of Florida State University, another veteran of Antarctic research campaigns. One of the first to use the scanning electron microscope for studies of bacteria, Friedmann was best known as the discoverer of rock-dwelling microorganisms. He had detected blue-green algae, or cyanobacteria, living inside desert rocks in Israel's Negev and the American Southwest. And in sixteen or so field expeditions to Antarctica, he had done extensive studies of life in rock.

Friedmann had been impressed and intrigued by the McKay group's work on the Martian meteorite. While he knew it was unlikely that one random rock could provide "the answer" to the riddle of Martian biology, he believed their 1996 paper had precipitated events that would lead, at the very least, to a clarification of the question of whether life had been present on Mars. Now the Friedmann group's studies seemed to fill in a key piece of the magnetic puzzle.

For years, skeptical scientists had challenged Thomas-Keprta's claims of biological origins for the magnetic crystals on grounds that she had never detected the telltale choo-choo train formation that characterized magnetic devices found inside Earth bacteria. Now Friedman was claiming they had it.

Friedmann's group studied the crystals under an electron microscope using a technique that enabled them to "see" the tiny chains in position inside the meteorite. What they saw were the crystals lined up "somewhat like a string of pearls," as Friedmann put it, or, as another scientist observed, sort of like "teeny backbones." This was how they appeared inside organisms on earth—like vertebrae running along some fraction of the creature's length.

The team could see fossilized outlines of both the chain and the membrane that had shaped it, Friedmann reported. "The chains we discovered are of biological origin. Such a chain of magnets outside an organism would immediately collapse into a clump due to magnetic forces," he said. The chains, in this scenario, were preserved in the meteorite long after the putative bacteria themselves decayed.

The group also concluded that the individual crystals were of similar size and shape and did not touch each other, and that the chains they formed were flexible, presumably in order to be able to move with the organism as it swam—all further signs of biological origin.

Friedmann said the discovery marked a potential turning point for understanding all life. "Until now, studying life has been like trying to draw a curve using only one data point—life on Earth. Now we have two data points to draw life's curve." He said the next step would be to find the remains of the bacteria themselves.

This time (in contrast to the quiet that greeted the McKay group's longer, technical paper published a month or so earlier), attention was paid. NASA, the National Academy of Sciences, and the universities of participating investigators issued press releases. The rock vaulted back into the headlines, and jump-started a new round of investigations in far-flung laboratories.

The new findings were widely professed to be intriguing. But, inevitably, many scientists resisted the pull of the Martian magnets, raising questions about the latest claims—particularly Friedmann's—and gearing up to put them to the test.

This was how the process worked. All these people labored in the borderlands of the unknown, on the cusp of the contentious assertion, where every answer led to more questions. Once they reached the territory of textbook certitude, they would be out of their element, and they would move on to another frontier.

Astrogeophysicist Chris McKay (no relation to David McKay) had long been in the "why waste time" camp when it came to the debate over the Mars rock. The evidence seemed conclusively inconclusive.

Now, he thought, people might move from *debating* David McKay's hypothesis to regarding it as a *working* hypothesis for further research. "I guess you can move me from the skeptical camp," he said, and into the one that says, "*maybe* this is starting to get serious."

One of the most outspoken critics, once again, was the leader of the Antarctic meteorite search, geologist Ralph Harvey. He told the Associated Press, space.com, and other outlets that the story of the rock was still far from conclusive. "It has all boiled down now to this magnetite." So what if these things have never been found in an inorganic setting, he asked. "The truth is that we haven't looked. It's hard to prove a negative hypothesis. It's hard to test it, but it is essential. It's the difference between science and faith."

Planetary scientist and biogeochemist Andrew Knoll remarked in an e-mail, "I wouldn't touch Mars magnetite with a ten-foot pole. There's simply a lot we need to know before making firm interpretations of that material."

Several studies soon challenged the biological scenario anew.

In an awkward turn for some of those involved, one of the most direct assaults on Thomas-Keprta's conclusions came from another group of scientists inside Building 31. They were hard at work trying to show that magnetites with the same "uniquely biological" geometry found in the Mars rock could be made in the laboratory—without biology. And if that wasn't close enough to home, the supervisor of the work was Gordon McKay, David's younger brother.

It was not exactly a coincidence, since David McKay had commissioned that group as part of his Blue Team—one of the two opposing groups he had set up to provide balance.

David McKay had envisioned the collaboration as a collegial, open one, in which information was shared and discussed, in marked contrast to the secrecy of the past. Now he felt angry and pained that some members of the team had taken such an aggressive stance against him—and that his own brother was securing their funding. It was also awkward for those with friends on both sides. For instance, a woman in David McKay's laboratory was in a romantic relationship with a man on the Blue Team.

As the principal investigator on this project, Gordon McKay had gone into it with the idea that his group could simply contribute new information to the debate. He soon saw that some of his coworkers were more combative in their pursuit of the case against his brother's claims. He realized they saw the setup as deeply adversarial and, for example, would decline to tell David McKay their results until they were ready to publish. This remained a sore subject in the building, where—in keeping with the tradition growing around the rock—players even within the same family could have differing interpretations of its meaning.

Gordon would later note that he had much less of an emotional investment in the answer than his brother, and therefore did not feel the pain over their differences that his brother clearly felt. Like many who had seen combat in the battle of the rock, Gordon was mainly tired of it.

The Blue Team contingent reported in 2001 that they had created something resembling the key subpopulation of magnetites in the Mars rock—and that no biological influence had been required. At the same time, they acknowledged that they had not—yet—shown that their homemade magnetites had the unique characteristics of biologically formed crystals. So the hypothesis of Martian life was still alive.

Once again, the rock had revved up lively animosities among its human hosts—and pointed the way to a harvest of new knowledge.

In the flush of renewed and often relatively upbeat attention that followed Kathie Thomas-Keprta's Euclidean leap and the public debut of Bingo, McKay's regulars never paused to celebrate. They were too focused on peeling the next layer of the Martian onion.

As they pushed the technologies, the team developed a new technique. Actually, it was an adaptation of an old technique, tomography, which was widely familiar from MRIs (magnetic resonance imaging) and other such nondestructive probes used in medical treatment. (*Tome* is Greek for "slice.") Instead of having a person lie in a tunnel while the machine moved around him, Simon Clemett would put samples in the transmission electron microscope and move *them*. The effect was the same. The group would use the tomographic reconstruction to interpret the three-dimensional shape of the magnetic crystals like doctors diagnosing a patient's innards.

Thomas-Keprta was delighted with the results. For those who said, "We don't like your truncated hexa-octahedron," she and Clemett felt they could hold up their tomography and let the pictures tell the story. But it still wasn't enough.

As the debate wore on, it revolved around ever smaller and more technical details. It was science at the fringes of what could be detected. "We're trying to solve one of the great questions by pickier and pickier details of technical analysis," meteorite specialist Allan Treiman would lament. "Shouldn't we be looking for much broader evidence?"

Dick Zare, advising and encouraging the McKay regulars from a distance, remarked that the whole mêlée was enough to make one's head hurt. "How in the world am I going to teach this stuff now?" he wondered. "It's a real frustration!"

McKay's regulars weren't hanging their hats on the magnetic crystals alone. McKay encouraged the group to go back to each of the original threads of evidence and see if they could push further.

For some time, McKay had been considering the possibility that the buggy shapes were organelles, or particular fragments of microorganisms, rather than whole bugs. He had a feeling that the next battlefield in the war of the rock would be the organic material. If the group was ever going to convince the wider tribe of scientists, it would have to come up with indisputable signatures of life, that is, compounds that (a) could have come only from Mars and (b) could have been created only by life. This would require more pushing of the state of the art and thinking outside the much-maligned box.

McKay was hoping to put together a narrative that tied the organics to the magnetic crystals and also to any underlying bug slime. As he pressed on through the rock with his microscope, he continued to detect fossil-like features and hollows called "molds"—places where cells might have resided but where only cavities remained.

There were cracks and veins all through the rock—cracks that had formed on Mars—and almost always when you broke the rock open along those cracks, you found some orange globs of the water-deposited carbonates, the sites where most of the interesting stuff was concentrated. But the team had also found the hydrocarbons (PAHs) in the rock (the pyroxene) itself—outside the orange carbonate moons. After intensive study of these surfaces under high magnification, McKay thought these organics might be the result of a biofilm—a kind of slime made up of microbial cells and the protective substance the cells produce. To buttress this interpretation, he would need detailed chemistry, such as the detection of well-accepted organic biomarkers called hopanes.

Then, of course, he would have to show that the biomarkers were not terrestrial, not from Antarctic melt or laboratory contamination or any other source on teeming Earth. The technology to do all this was just on the ragged edge of the possible, he knew, and it was advancing rapidly. The industrious Clemett was on the case, building new instruments.

McKay wanted to make the case. There was no denying that. But he constantly felt the need to obey the prime directive for anyone in his line of work: Make sure you are not fooling yourself.

One day in April 2002, Thomas-Keprta walked into McKay's office and found him poring over new data. He told her, "I wanted to look at some chips and understand what I was seeing. . . . I wanted to reassure myself that what I am seeing is true, and real."

At home, McKay's wife and daughters treated him with new regard. When he made some little domestic decision, they were not so quick to quibble. They looked at him differently. He had, after all, taken on a world-class controversy, changed the thrust of research, weathered quadruple-bypass heart surgery, and stuck to his guns through fire, smoke, and brickbats.

In the evenings, McKay would try to forget about all of it for a few hours, to get the kinks out of his stomach. In their glass-and-cedar house in the woods, he and Mary Fae would settle in to watch TV and relax. Every now and then, Mary Fae would turn to her husband and ask, "Hey, could you be wrong about all this?" And he would lay it out one more time, and end by saying, "No, we're not wrong."

CHAPTER FOURTEEN

SCHADENFREUDE

ANDREW STEELE HAD crossed an ocean and changed his life to answer the siren call of the Mars rock. He had applied himself with wit and zeal to his role as devil's advocate for David McKay and mobilized all his fresh-minted skills to try to ferret out the secrets of life at its smallest and most primitive.

Now six years later, in the spring of 2002, this work brought Steele full circle in a sense, from the soaking waters and hot cataclysms of early Mars back to Earth during the same epoch. He could see the arguments about the potential for life on Mars intermingling, like gully washes flowing into a river channel, with arguments about the beginnings of life on the young Earth. It was a confluence that raised the general depth and breadth of the discourse but also flooded the terrain with fresh doubt—the mother's milk of scientists.

A significant event in this flux, certainly one of the more entertaining, was about to take place in the drafty expanses of a vast dirigible hangar—large enough to contain three full-scale *Titanics*—in Silicon Valley.

It was Tuesday, April 9, 2002. Steele and David McKay were part of the crowd that assembled in a state of high anticipation in Hangar 1. Built for the navy in the 1930s, the structure was part of Moffett Federal Airfield and the site of Ames Research Center, NASA's nerve center for studies on the origin, history, and distribution of life in the cosmos. All this week, Ames was sponsoring the second annual international meeting in the emergent field of astrobiology—a field that had taken wing as a direct result of the McKay team's 1996 announcement.

In competition with the periodic roar and whine of jets on an outside runway, many of those in attendance were buzzing about the confrontation set for this afternoon between Bill Schopf, the god of the Precambrian, and Oxford don Martin Brasier. It was almost too delicious. The venerable Brasier was daring to assail Schopf on his key discovery all those years earlier: the zoo of ancient remnants that had brought him acclaim and had long held a place of honor in textbooks and museums and the Guinness record book as Earth's earliest known fossils.

The jousting had erupted into public view a month earlier, when Brasier published a paper in the March 7 issue of the journal *Nature*. In it, Brasier strongly suggested that Schopf had blundered when he'd claimed a biological origin for the fossil shapes he'd found in Western Australia; these "fossils" were actually imperfections in pieces of the rock that bore a faint resemblance to bacteria. *Nature* had published both the Brasier challenge and a reanalysis by Schopf, in the same issue. News of the battle had shaken the world of paleobiology.

The irony was not lost on the McKay regulars. The dispute was like a fun house–mirror image of the controversy over the Mars rock, and in particular of Schopf's challenge to the McKay group almost six years earlier: Schopf had misconstrued nonbiological mineral shapes as fossils of living things (Brasier suggested), and the features that *might* be fossils were not the type of bacteria Schopf said they were; these fossil-like structures had not formed at moderate temperatures on a wave-washed beach as Schopf had claimed, but in a high-temperature hydrothermal vent, or hot springs; the presence of organic molecules in the structures was not necessarily a sign that these were remnants of live organisms; and so on.

The Brasier findings went a shocking step beyond merely challenging Schopf's interpretation to allege that Schopf had failed—either deliberately or by mistake—to disclose fully in his published papers the complicated shapes of the supposed fossil structures. He had, in effect, ignored the parts of these forms that did not fit with his claims, Brasier suggested.

Brasier had set out not to attack Schopf but merely to update his textbook on microfossils. Eventually, however, he had come to the view that Schopf had failed to map the rocks in detail, study them comprehensively, or arrive at a model that fit all or most of the data. "It seems that he looked at huge numbers of samples from many sites whose story he and his colleagues simply did not understand fully, and scanned them for anything vaguely fossil-shaped," Brasier said, adding that Schopf's group "seem to have had an image in their minds and transposed it onto the rocks. It is a common problem in historical and material sciences, and none of us are immune from it."

To some, this turn of events represented condign comeuppance for the tenacious and driven "Bull" Schopf, who did not suffer criticism lightly, and who, some felt, had escaped close scrutiny because of his lustrous reputation and his perceived talent for writing successful grant proposals that kept money flowing.

In a moment of particular satisfaction for the McKay team, an article about the dispute that had appeared recently in *Newsweek International* alluded to Schopf's critique of McKay's work on the Martian meteorite. "While Schopf and his critics go back to the [terrestrial] rocks for more and more studies," the article said, "it's hard to imagine the supporters of the case for the Martian meteorite fossils being so noble as to avoid a faint glow of *schadenfreude* as their foremost critic gets criticized himself."

David McKay tried not to say anything that might sound like gloating. But Mary Fae McKay started signing her e-mail messages to friends, "With a faint glow of *schadenfreude.*"

At the very least, today's session promised to be great theater.

Bob Hazen, a Washington, D.C.–based astrobiologist, was among those salivating at the prospect. Over lunch that day, he described the Schopf-Brasier confrontation as "Shakespearean." The bard always made his characters kings and princes because that way they had farther to fall, Hazen mused. Now it was Schopf—god of the Precambrian, paleobiologist, geologist, microbiologist, organic geochemist, director of UCLA's Center for the Study of Evolution and the Origin of Life, prizewinning author—who stood at the precipice.

Hazen made a point of getting to the session early. He told a companion, "I want to sit up front. I want to see the spittle."

For Steele, the story of this extraordinary turn of events began when McKay asked him to work on his "anti-life" Blue Team, pinning down the characteristics of biomarkers—the telltale signatures that enable scientists to distinguish life from nonlife.

The two had coauthored papers together and, McKay thought, kept each other honest. Steele would prevent McKay from going too far over to unsupported claims favoring biological origins for certain features in the rock, and McKay would point out to Steele some of the finer nuances that distinguished possible Martian features from terrestrial contamination. Steele typically ran all his findings past McKay, and they looked unfailingly first-rate to the older man.

In the course of his research, Steele developed strong opinions that Schopf's descriptions of the earliest known life were probably wrong in at least one significant aspect. Oxygen-producing cyanobacteria—Schopf's "pond scum"—were too amazingly complex to have formed as early in Earth's evolution as Schopf suggested. As Steele would joke, it was like giving his three-year-old a set of Legos and expecting her to construct the Parliament building. After he waxed lyrical on the topic to a friend at a meeting, word got back to Brasier, who contacted Steele. Steele was soon working with the Brasier team on the geochemical analysis of Schopf's evidence.

The Ames meeting organizers had invited David McKay as well as Steele and others from Building 31 to give talks at other sessions during the week on new evidence or strategies related to their work on the Mars rock. In particular, they were interested in Kathie Thomas-Keprta's intriguing magnetic crystals.

The day before, McKay had given a talk on the vital importance of bringing Martian samples back to Earth and of identifying those elusive biomarkers—measurable properties that suggest life "is or was" present—in them. As the experience with the Mars rock showed, he told his audience, "Unless we develop [better] standards, we'll have the same trouble" when the robots come home with the new bits of Mars.

During the question-and-answer period, Baruch Blumberg stood up to speak. Blumberg had won the Nobel Prize for his discovery of the hepatitis B vaccine and was now director of the NASA Astrobiology Institute. Regarding McKay's quest for definitive biomarkers, he said, "You may be setting up a category you'll never find. . . . I think you're setting yourself an impossible task." He added, "Your data [on the Mars rock] are becoming more convincing as you go forward. This just takes time."

McKay had adopted the Nobelist as his latest mentor. Blumberg had been even more direct recently during a private dinner. "Stop looking for the smoking gun," Blumberg had advised McKay. "You don't need it. You will persuade by a gradual accretion of evidence, a weight of data."

In contrast to the months following the 1996 announcement, and despite the heart surgery, McKay looked slimmer, tan, and relaxed—whatever the bowlines and monkey fists coiled in his innards. Mary Fae admired her husband for the way he had handled the trying affair. Those closest to him thought the whole thing, having failed to kill him, had made him stronger, more confident.

McKay saw the tone of the combatants in the controversy changing. At least there was growing awareness, he thought, that the question of life signs in the rock was more complicated than anyone had dreamed. And he took pride in the fact that the flap had triggered a burst of learning on numerous fronts.

The McKay team shared a keen interest in the Brasier-Schopf shoot-out. The personal antagonism between them and Schopf had ballooned in the years since their 1996 encounter. Taking issue with the team's hypothesis about the Mars rock was one thing. But some thought Schopf had gone beyond disagreement all the way to ridicule. This perception was not limited to McKay and his advocates. Several scientists not in McKay's camp murmured privately that they thought Schopf had gone too far. The final chapter of his book had, as one said, "mocked the McKay group, comparing them to outright frauds."

On Tuesday, McKay joined the crowd in the "auditorium," actually a large white tent erected within the cavernous hangar, and took a seat. Interest was so high and the crowd so big that the organizers had relocated the confrontation from a smaller room.

Schopf, in gray turtleneck and brown pants, his name tag hanging around his neck, entered the big tent while it was in darkness, as one of the warm-up acts showed a slide presentation. Schopf hitched up his pants and waited a bit before taking a chair.

Finally, the lights came up and session moderator Jerry Lipps introduced Schopf and Brasier as "our primo speakers." Behind him, a backdrop as big as the side of a barn depicted the cratered, icy-blue surface of a planet.

As Schopf took the stage, he asked that the lights be turned down for his slide show, adding, "I don't need all that light on me." His first slide showed an ancient Chinese proverb, which he read aloud: " 'It's better to light a candle than to curse the darkness.' . . . My aim here is to cast a little light on this subject with a minimum of heat."

Schopf proceeded to demonstrate once again why he was a popular lecturer—confident, plain-spoken, emphatic, fluent. As his talk built to its finale, he would gather momentum until he was coming on like a force of nature.

Using a pointer beam, he began by showing ten of the putative fossil objects from the formation in Western Australia, which he had first reported in 1985 and formally described for publication in 1992 and 1993. The slide showed a series of dark, segmented snakelike shapes with their chemical analysis.

In his book, written before the controversy with Brasier erupted, Schopf had acknowledged that two of his graduate students, as they'd analyzed the material, had found nothing. And he himself had characterized the putative fossils as "scrappy"—charred, shredded, cooked, generally difficult to make out.

"So, the question before the house," Schopf told the audience, "is, are these old fossils truly fossils. Either they are or they are not. Either life then existed or it did not. The job at hand is to ferret out the evidence by looking at the facts."

Brasier stood nearby, listening intently.

It was because of the mighty impact of Schopf's work that Brasier had felt obliged to focus on it in the first place, beginning in 1999, as he went about updating his own textbook on microfossils. Schopf's specimens were, after all, in the *Guinness World Records* as the world's oldest fossils—almost 3.5 billion years old. Although there was other evidence of life's presence during the same period, Schopf's microfossils were by far the earliest life-forms that had been reported.

Most important, as Schopf seemed to interpret them, these were not simple, primitive organisms but amazingly complex ones—the cyanobacteria (formerly known as blue-green algae) that concerned Steele. Schopf's claims meant that life had evolved much more rapidly than previously suspected.

The claims shook theories of life's early evolution—and also the planet Earth's. The cyanobacteria would have pumped oxygen into the atmosphere as a by-product of their photosynthesis. And yet, geochemical evidence suggested that levels of oxygen in Earth's atmosphere remained quite low until a billion years or so later than this. Schopf's interpretation forced theorists to scramble to account for the incongruity.

Schopf had left his samples from the Australian site in the care of the Natural History Museum in London. They had been prepared for analysis in the usual way—cut thin enough so that a researcher could shine light through them for examination under the microscope.

As Brasier had studied these thin sections in the summer of 1999, it had struck him that there was something strange about the structures. The work was especially difficult, Brasier found, because the material containing the possible microfossils was "exceedingly complex, . . . highly broken and fragmented" and with curious fabrics not familiar to most paleontologists.

Brasier spent close to one thousand hours peering down the microscope at the record-setting Schopf specimens. Starting in July 1999, Brasier's group began searching for the discovery site in Australia and in 2001 examined the rock around the fossil beds in order to understand the context in which they occurred. Brasier found that the formation was not a continuous sedimentary formation but was split by lava flows.

In 2000, the Brasier group used a new computer program to build a digital database of the structures.

By late September 2000, Brasier had studied the shapes so intensively that he had started to see them in his *sleep.* It was then that he realized the structures were wrapped around spheres of glass, much as he had seen in other such formations in hydrothermal settings and hot springs. Brasier stared into the samples, not sure where to spot Schopf's "microfossils." The researchers began to draw and map every little structure in the rock, and saw that the supposed fossils were actually part of a wide spectrum of odd-looking shapes, most of them far too chaotic to be called fossils. And the appearance of cells as cited by Schopf, they concluded, was an illusion.

Brasier's team decided, based on its own tedious geological mapping and chemical analyses, that Schopf's specimens were actually nonbiological artifacts with no detectable organic matter, formed at a place where a primordial spring poured volcanically heated water from the seabed—a setting too hot for life and devoid of oxygen-producing photosynthesis.

The clincher, for Brasier, was that many of Schopf's specimens were much more complex than he had portrayed them in his own published papers years earlier. Sure enough, Brasier saw some shapes that resembled cyanobacteria. But he was also amazed to find that some of the putative fossils were oddly branched or otherwise complicated—unlike cyanobacteria—in ways that Schopf had failed to describe. When seen in this full new light, they no longer looked like the bacteria with which Schopf had initially compared them.

It made you wonder if there had been a "terrible mistake," Brasier would tell his audience in the hangar.

Now, in the first of two significant concessions, as he addressed the audience in the tent, Schopf acknowledged that he had initially mischaracterized the setting from which the fossils were taken. But this was based, he noted, on other people's work. At first, in 1983, geologists had classified the locale "as a shallow marine near-shore environment," but the site had been "recently reinterpreted . . . as being a hot springs fumarolic deposit. Everybody agrees on that."

He and Brasier also agreed, he said, that he had published microscopic images of some forty-four specimens of these fossil-like structures, that they were petrified like other Precambrian microscopic fossils, that they were cylindrical, curvy, and had cell-like segments, and "that they range in form from what I would call good fossils to clumps of organic matter. We even agree on their composition."

It was the exact nature of the organic matter about which the two sides differed, he said.

"There are some perceived problems here, of course," he said, but he could explain them away. For example, as to Brasier's claim that the filamentous shapes were not fossils, because they had formed in a hot springs fumarole, Schopf asked, "Why not?" Such specimens occur today in Yellowstone and at deep-sea fumaroles.

What about the apparent high temperatures at the site, which would be beyond the limit for life? That degree of heating had come later, he suggested, not at the time when the fossilized organisms were alive.

Another "perceived problem," he said, was Brasier's contention that the organic matter inside the fossils was chemically identical to organic matter not associated with supposed fossils but in the rock at large. This seemed to support the idea that they were all nonbiological artifacts. But Schopf countered that the particles scattered through the rock were bits and pieces of microbes, with the same geological history and coming from the same environment as the microbes that formed the fossils, "so of course they are chemically identical."

Brasier and company had submitted the first version of their grenade of a paper to the journal *Nature* in February 2001, and they'd kept working on the samples. Meanwhile, the *Nature* editors sent Schopf a copy of the manuscript, as a courtesy.

Schopf counterattacked. First, he persuaded the Natural History Museum in London to ship him the complete set of his original Australian specimens for reanalysis. (The Brasier group, by contrast, had been obliged to spend several days traveling back and forth between Oxford and London, because the museum would allow them to borrow only three thin sections at a time. The one-way trip by bus took an hour and a half.) Schopf had recently teamed up with some of the few specialists in a new laser technique that, according to Schopf and his collaborators, could distinguish one type of carbonaceous deposit from another.

Rebutting Brasier's claim that the fossil material had no detectable organics, Schopf reported in his own paper, also sent to *Nature*, that his new imagery showed that the fossil structures were rich in organic molecules and therefore must be microfossils.

Schopf sent Brasier his new findings. Brasier then revised his own paper to admit the presence of some kind of organic matter in the fossils. However, as the dispute over the Mars rock had so dramatically highlighted, not all organic molecules are produced by living things. Brasier argued in his revised paper that an extreme hot springs environment could have formed the organic molecules around the edges of crystals, from carbon monoxide and hydrogen, without the presence of life. His argument against Schopf's "microfossils" was similar to the ones marshaled against the McKay group's "nanofossils."

Schopf was arguing that if the features were shaped like bacteria and were made primarily of organic carbon, this meant they had once been alive. Brasier, by contrast, compared that notion to "the Canals on Mars, or the Face on Mars. A superficial examination might make [Schopf's argument] seem plausible." But when viewed in their proper context, from the perspective of multiple disciplines, they failed to stand up to scrutiny based on the criteria that Schopf himself had put forward as far back as 1983.

Jabbing at Brasier rhetorically under the darkened tent in the hangar, Schopf went over what he called "inaccuracies" and "errors" in the Brasier paper and took up one of Brasier's major points. First, he clarified some key definitions. "There's been a great deal of confusion as to the use of these words and we ought to straighten that out," Schopf said. "Kerogen is the fossilized insoluble organic matter in rocks. It is not a mineral. Graphite is a mineral. It is the crystalline form of carbon." The point, he said, was that "these things are *not* made out of graphite."

To prove it, he showed spectral data from twenty-five petrified fossils, ranging from as young as 400 million years to 3.5 billion years old, ranked from well preserved to poorly preserved, which he equated with "increasing graph-it-i-za-tion" (in professorial mode, he enunciated each syllable). The Apex chert—a type of flinty quartz—from which he had taken his fossils was in the lower midrange on this continuum, he indicated with the pointer, and was separated from the graphite objects. "Here's the Apex chert. Here's another kerogen, *another* one, *another* one, *another* one, *another* one, *another* one, and here, you start to get into graphite." His fossils, he said again, with force, "are not made of graphite. They are made of kerogen."

Now Schopf got to a particularly touchy issue—the true shapes of the supposed fossils. Brasier had argued that some of them rambled on and branched in ways that showed they were not blue-green algae, and in some cases not biological at all but actually a mineral extension of the quartz itself. Moreover, Schopf had failed to mention or depict these extended, filament-like shapes in his original papers a decade or so earlier.

On the question of branching, Schopf now asserted, "Dr. Brasier has used a new technique called Auto-Montage confocal microscopy. It is a really, really neat technique. The problem with it is it hasn't been used in this science before. It condenses the focal plane so that objects that are not on the same focal plane can be seen as though they are. It gives an optical illusion, unless you're rather careful."

As McKay and Kathie Thomas-Keprta had learned, staring into the realms of the Lilliputians could lead to the same sort of parable as the blind men describing the elephant.

The shapes Brasier had taken to be branched—like the letter *Y*—were actually filaments that had folded back on themselves, Schopf argued—something like the AIDS-awareness ribbons worn on celebrity lapels. This structure (he pointed to a slide of a putative fossil) was suggested to be "anomalous. Here is the upper part of the filament. It curls down beneath it. Here is the upper focus. Here is the lower focus. It is not branched at *all.* It, in fact, is a folded-over filament."

By now he was speaking with apparent irritation, as if trying to explain something to a frustratingly slow child, his voice rising occasionally.

"The third example is more difficult to show. . . . But there is no attachment here whatever. Yet if you look at their paper, it is, uh, suggested, it is *said* that the thing is branched and attached. Well, there is no attachment. There is no organic matter there.

"And the fourth example is shown here. This is said to be a branched filament. And the question is about this little bit here, and you can see that in fact it is a torn filament. This little bit fits back into the side of the filament *just like Madagascar fits into the southeastern corner of South Africa, for goodness sakes!*" He was fairly growling by the time he got to the word *Africa*. "This thing is not branched. It's made of cells, as you can see.

"Okay." He took a breath. "Well, the question before the house then becomes, how does one solve the matter of biogenicity?" Were these fossils from Earth's earliest life or were they formed without benefit of biology? "Now, this is a quote from the Brasier et al. paper, and what it says is you reject fossils if a possible nonbiologic source can be proposed. And I suggest to *you* that this represents flawed reasoning. I suggest that, because, the way this game is played, you propose what the fossils might be but you do *not* show in fact what they are."

Schopf next invoked his own youthful initiation rites.

"Now, we've been down that path before. And that war was won by Elso Barghoorn in the 1960s. He stood before the world *by himself* and he showed the Precambrian fossils are indeed fossils. And how did he do that? He said that the biologic origin of the fossils is shown by the traits they exhibit that are *unique to life*, traits shared by fossils and living organisms but *not by inanimate matter*. That is the way that this matter ought to be solved. That's the way to do this science.

"So the point of disagreement, the single important point of disagreement, comes down to simply this. Dr. Brasier and his colleagues say the fossils are pseudofossils, graphite formed nonbiologically by . . . an abiological synthesis.

"We say that the fossils are microbial fossils, kerogens formed by ancient organisms. So, let us look then at the plausibility of those two proposals.

"In the first place . . . no nonbiological particulate organic matter has ever, ever been recorded in the geologic record of the Earth. So this is an extraordinary claim.

"Secondly, there is no plausible environment, no natural plausible environment"—he gestured with both hands palms up, beseeching—"in which such things could produce such amounts of organic matter." He noted ways to do the type of synthesis in question in the laboratory, adding, "But, folks, those things don't occur normally, hardly at all, in nature." The sites that organic geochemists and micropaleontologists "have both worked the *most* on this planet are petroleum deposits. We'd *know*"—here he extended one arm to point at the slide on the screen—"if there were such 'sports' of nature there. They are not there.

"And finally the isotopes show that it's not *at all* abiological carbon" or it would have a different signature, he said. "Look, these things are made out of coaly kerogen"—he punched each of the next eleven words as if it were a separate sentence—"*just like every other petrified Precambrian fossil that's ever been found.*" Schopf took a quick, deep breath, then noted the few exceptions.

"The complexity rules out a nonbiologic source. There are many specimens, one hundred and eighty specimens . . . fifteen examples on average of each type. They *vary* like living microbes, they *grew* by cell division, they *inhabit* a livable environment, and their isotopes show they are biological."

He next outlined evidence from other deposits, within 100 million years of the same age, in which fossils were found, in order to show that his fossils were not an isolated phenomenon.

Then his voice subsided. "So," he summed up, "they're made out of [coal-like] organic matter, the complexity rules out a nonbiologic source, there are many specimens, they vary like living microbes, they grew by cell division, they are in a livable environment, their [isotopic signature] is unquestionably biologic. In addition, they are chemically the same as *undoubted* fossils in twenty-four other units [of rock] that fit the continuum of geochemical maturation. We *understand* the history of that organic matter. It fits all evidence of a similar age. And I come back to Professor Barghoorn. He solved this problem. He said, if it looks like life, if it's made of the molecules of life, if it has the isotopes of life, if it fits with all other evidence of life, well, folks, most likely it's life."

The audience applauded, and sat back for round two.

Jerry Lipps, the emcee, thanked Schopf "for a rousing good introduction of the two hypotheses we are dealing with this afternoon." He welcomed Martin Brasier, who with seven colleagues was "questioning the evidence for earth's oldest fossils."

Brasier took to the stage looking something like Sir Laurence Olivier as King Lear, a patriarch with a mane of white hair on a large head and a tuft of white on his chin. His delivery was as understated as Schopf's had been dynamic, but Brasier had the advantage of also sounding like Olivier. While Schopf had referred to him as "Dr. Brasier," Brasier called Schopf "Bill."

"Well, thank you, Bill," Brasier began softly, as he positioned his material on the projection machine and jabbed his rhetorical rapier in Schopf's direction. "A truly hydrothermal performance. . . . More heat than light, perhaps." Laughter rippled through the audience.

"What we want to do in this group, which is based in Oxford, is to question very deeply all those mantras, those understandings we've had about the nature of the early fossil record. And it's as much a shock to us as it might be to Bill about the conclusions that we've had to come to. It was no easy decision, but each time we inquired deeper, we found more that was unsettling about the existing paradigm. Of course there is always a lot of fighting when a paradigm has to shift. . . .

"We have until now all assumed that the reports of microfossils and stromatolites from rocks about 3.5 billion years old represent the beginning of the fossil record," Brasier said. This assumption also dated the beginning of the process—the metabolic activities of Schopf's microorganisms—that gave rise to Earth's oxygen atmosphere.

As Schopf had acknowledged in his book, this was a "puzzling scenario" that some scientists privately wished he had been wrong about. It meant that the basics of the world's ecosystem had evolved with dazzling rapidity.

"And of course," Brasier continued, "we have noticed over the last few years that, in some people's view, there's a major mismatch between the appearance of such complex fossils and the emergence of an oxygenic atmosphere" perhaps millions of years later, although the timing was uncertain. "And we've heard other reasons too . . . as to why cyanobacteria might not be expected so early. So we have to be very careful with those kinds of arguments."

Brasier had gone to Schopf's material "with great excitement," he said, "because I did my doctoral research in evaporitic lagoons around the Caribbean, and had worked on microbes and microfossils all my life. In fact, the reason I wanted to get images of these microfossils was to include them in the next edition of my textbook on microfossils, and I thought it would be great to have a look and get some good images using new techniques. And really we think that the change in view has come about by all the new techniques that we can now apply to these particular rocks: geochemical techniques, image analysis techniques and so on."

He showed a sample of the rock "fabric" that contained Schopf's microfossil-like features. "I was aghast, I must admit, to see just how complex and brecciated [rubbly, coarse, sharp-cornered] it was," he said. He described recent findings on the complex geography of the rock formation from which Schopf had taken the samples, where one kind of material transected another, and some layers had quite different histories of heating from others.

"We were intrigued to discover a range of minerals in this which turned out to be very suggestive of a hydrothermal setting," Brasier said. ". . . The more we did, the more consistent it seemed to be with a hydrothermal setting and the less consistent with any sort of stratiform wave-washed beach or river."

Now he raised a question about stromatolites—those geological formations that had fascinated Schopf from his early days, in China and elsewhere, with their artistic layering constructed by a zoo of microbial communities dominated by cyanobacteria.

"What is a stromatolite?" Brasier asked. "One of the things we are doing at Oxford at the moment is to explore the real meaning of the word *stromatolite.* Here is a sample we've been working on." He showed a slide of varying colors in curving layers. "It was originally in the collection of Sir George Taylor, curator of Kew Gardens, and is a beautifully digitate structure, branching, and so on."

Then he sprang his punch line.

"But we know this is actually made out of *paint.* It turns out to have been made probably in a car works. You can make out different colors here. There's pink and green and pale blue. Many of you will perhaps recognize cars of the late 1950s, early 1960s, in that color scheme."

Noting that his team had been conducting a series of experiments at Oxford on how the stromatolite structure develops, his voice rose with indignation as he added, "You can't just use stromatolite as a biogenic indicator."

As Brasier spoke, Schopf stood nearby, onstage, with his hands clasped behind his back, leaning so far forward to listen that he seemed on the verge of tipping over.

Brasier addressed the distribution of the microfossils within the material itself. He showed slides of the famous Gunflint Formation on the border between Minnesota and Ontario (the subject of the young Schopf's Oberlin honors thesis), with its "microbial mucilaginous dark layers here, and then you can see the various filamentous structures forming a distinct layer and then intertwining one around the other. Usually the structure's of about the same length and not showing complex organization."

Next, for contrast, Brasier showed a slide of one structure from the microfossil-containing fragment of Schopf's history-making Australian Apex chert. "This is a thing that is 'arrowed' as a microfossil [in the published paper], and these are the sorts of structures that float around it. It's quite varying in length, it's a rather ugly-looking little structure, and it contains various little chaotic, information-rich organic blobs and wisps throughout the matrix, and that is absolutely typical of the appearance of the Apex chert organic matter."

Then he showed images from the new Auto-Montage technique he had used to study the samples, revealing the new details in the alleged microfossils. A single threadlike structure in the Schopf paper viewed in this way became something else.

"We were surprised," he said, "to find that a great number of the structures, many more than just three or four, were more complex than were present in the descriptions, and in the illustrations" in Schopf's published papers. "In this particular one, it balloons out here . . . into a large bulbous mass which then turns this right angle. And this structure is continuous." Brasier took issue with the argument Schopf had just made—about the folding over, like an AIDS ribbon: "There's no illusion here, in which one is montaging one piece on top of another."

Brasier put up a topographic map that showed, in effect, the *altitude* of the features. Where Schopf had seen cyanobacteria, or something cyanobacteria-like, he said, "we found it was much more complex and branched than is 'normal' for cyanobacteria of that age"—i.e., the type alluded to by Schopf.

Brasier and Schopf were diametrically opposed in their arguments about the shapes of the fossil structures. Were they actually branched like a *Y*, as Brasier insisted, or just folded like the AIDS ribbon, as Schopf argued? For his part, Brasier was incredulous that Schopf, in his papers, had "never described folding," much less branching, in the fossil-like shapes. Schopf "ignored the folding," Brasier had concluded. But had Schopf missed it or simply failed to describe it?

Almost three months later, a former graduate student of Schopf's would voice her own reservations about Schopf's representations in his original papers. Bonnie Packer, who had first spotted signs of biology in Schopf's Apex chert samples, would tell the journal *Nature* that Schopf had indeed been highly selective with the evidence as he'd presented it from 1987 into the early 1990s. Schopf, she would say, had withheld from publication images of the supposed cyanobacteria that showed branching, and her attempts to challenge Schopf had met with stubborn resistance. "There wasn't a bloody thing I could do," she told the journal. To back up her account, she provided pages from her lab notes and described photos with Schopf's handwriting on them. Schopf would respond at first that Packer had never revealed the branched shapes to him, but he would later amend that to say that his memory on the topic was sketchy.

Brasier, as he looked out at the rapt crowd, discussed the process by which crystals form and showed what that process could do to organic matter to make it look like microfossils. "You get a spherical structure, with a kind of spherical symmetry to it, and as you move away you get a kind of radial symmetry or a bilateral symmetry, and then as the material gets scarce the matter self-organizes into information-rich filaments exactly in the way that complexity theory would predict, and these take up the appearance of microfossils." His group had found a lot of this, he said.

"When we looked at the whole spectrum—and of course we have imaged thousands of structures, and I have spent thousands of hours looking at this material," Brasier went on, "we went through the material imaging absolutely everything and not just hunting for things that looked like fossils"—an approach he pointedly contrasted with that of Schopf—"and we found structures which . . . clearly had branching that had never been discussed [by Schopf or anyone else in published papers], such as . . . the so-called AIDS ribbon–type structures. . . . And we found others which were so chaotic in their organization, so information rich"—he pointed to the images on the screen and revealed what his group had nicknamed the confused shapes—"like this one we call the 'little ballerina,' this one is the 'Loch Ness monster,' this one is the 'wrong trousers'—these are so complex, or so large . . . that you really have to scratch your head and ask, maybe there's some terrible mistake?"

He discussed the graphite and the particular way the quartz structure formed, which might explain other suggestive features—segmented shapes—without biology. "I don't think any of us would be confident in staking our reputations [on these features] as microfossils."

Brasier moved on to his last point: the light-carbon isotopes that seemed to imply living organisms. Noting that his group had pondered whether some kind of nonbiological synthesis could have produced the signature, he said,

"We don't believe the tools are yet at hand for distinguishing the one from the other." Brasier considered the current state of scientific knowledge so scanty that it seemed naive to think anyone could safely rely on isotopic signature as a sign of life. Carbonaceous matter and light-carbon isotopes were also abundant in comets and meteorites, after all, but in those cases they were certainly not believed to be signs of biology.

Brasier summed up: "So, the position we take, which is different from Bill's, is what I call the Oxford audit on astrobiology, and it's tentative." Brasier's default position was that no structures should be deemed to be biological in origin—whether from Earth, Mars, or elsewhere—until possibilities for their nonbiological origin had been exhausted.

Developing ways to rule out the nonbiological option "is an agenda for future work," he said. "It really says that we must explore all the [nonbiological] look-alikes in these various kinds of marker, before we can confirm the biogenicity of structures. Otherwise, the material you bring back from Mars, or the material we collect from early Earth, will be subject to long and futile debate."

After recapping his group's conclusions, he added, "We think our hypothesis fits better with a hydrothermal cradle for life . . . which Bill has now acknowledged, and it also opens the possibility that the geological record will contain much more of the evolutionary history of life than previously thought.

"So, astrobiologists, *go to it.* Thank you very much."

Lipps, the emcee, opened the floor for questions. Scientists and journalists were lined up at microphones placed in the aisles.

One asked whether certain parts of the sample material might represent precursors to life, if not life itself. Another raised a question about the stromatolites. There were exchanges about how much had been published on the topic of whether some of these features could be created without biology—and about how certain anybody could be either way.

Schopf, still standing away from the podium, off to Brasier's side, moved closer and leaned forward to hear better.

Brasier repeated his call for better criteria to discriminate between biological and nonbiological processes. He raised his voice: "When we're thinking about the early earth, we have to be thinking about how life originated."

Scientists had been accumulating evidence that the change had occurred along a chemical continuum. "We must therefore be thinking of a *spectrum* between abiogenic generation of organic matter and biogenic generation," he said. "It seems to me absurd to close the door on some of those processes before the discussion is begun."

Andrew Steele stepped to the microphone. His comments launched a fresh punch-counterpunch between Schopf and Brasier.

"Couple of points," Steele began. "On the carbon isotope signature work, we [in the Brasier group] are trying to say that it could be a biogenic or non-biogenic origin. It could be a hydrothermal vent community. One thing that's important for me as a microbiologist is I'm not hearing the word *cyanobacteria* at 3.5 billion years anymore. That's extremely important, and that's the point I wanted to make more than anything. . . . No cyanobacteria."

In other words, he was noting that in the weeks since the Brasier paper had appeared, Schopf had backed off his original claim as to the identity of the world's oldest known fossil life-form.

Schopf stepped up to the podium. "If I understood what you said, you are not hearing the word *cyanobacteria,* is that correct?"

"That's correct."

"Okay, let me explain," Schopf began, "because a whole lot of you are not professional paleontologists, paleobiologists, who have to deal with this problem. Put yourself back ten years ago and imagine that you have found a feathered dinosaur. Well, there *are* such things. Okay? It has the anatomy of a dinosaur, but it has feathers on it. But this is the first one that's been found. There are quite a bunch that now have been found. But ten years ago this would have been a great find. What would you do? You would say, well, it might be a reptile, might be a dinosaur, or it might be a bird." He used the term *incertae sedis* (of doubtful position), the Latin term applied when more data is needed in order to place a find in its rightful place in the taxonomic scheme of living things. "What that means is I don't know for sure whether it is a bird or a dinosaur. Well, that's exactly what I did. I compared these organisms with modern cyanobacteria and fossil bacteria. . . . I looked at some two thousand living species and strains of bacteria and cyanobacteria so I could make a good distinction. I came up with the conclusion that the things—I call some of them 'cyanobacterium-like,' and other ones I called 'bacterium-like.' . . . My best judgment is, what I could say without misleading anybody, is that these are prokaryotes [organisms in which the genetic material is not enclosed in a cell nucleus]; I believe they are members of the bacterial domain, . . . but I can't tell you for sure whether they are cyanobacteria, or bacteria, or some now-extinct group. And I did that even though one of my distinguished colleagues urged me to name them, like, proterazoic cyanobacteria. I resisted that. I did that because I thought it would be fairer to the scientific community."

Schopf had begun to enunciate emphatically again, his hands gesturing more widely. "I did not want to prejudice people. So I have been personally very careful about saying cyanobacterium-like, bacterium-like, but *by golly, folks have forGOT-ten . . . to . . . read . . . the . . . literature!*"

Brasier came to the podium as Schopf stepped back. Brasier said quietly, "What we do is more or less paraphrase exactly the sentences that were in the various [Schopf] papers. We were most careful to do that. And we say things like 'compared with,' not actually 'placed within.' So there's a slight difference of opinion there.

"What I would like to ask Bill, uh, then, is," Brasier said, and he reeled off a series of Latin names Schopf had given his organisms, "and these various other structures, when we imaged them, we found they had complex and perplexing structures about them which were not shown either in the figures [that Schopf published] or were not even referred to in the taxonomic descriptions. It would seem to me an essential part of taxonomic description to closely proscribe the nature of the structure that you're looking at and to help us scientists understand how much selection of information is going on and how—" Brasier interrupted himself abruptly and stepped back as Schopf moved quickly to the podium.

"Well, I appreciate the chance to answer that question," Schopf said, "because it is of some interest. It is of interest to me." He gripped the podium with his left hand, while the right swept air. "As you know very well, in the first instance, anytime one, uh, publishes a paper, you have a lim—" He cut himself off to say, "especially in *Science*, where this was published; this should have been in the *Journal of Paleontology*, Andy Knoll was right, I should have done that, but I was following Elso Barghoorn, and I thought maybe that he'd made the right decision. Anyway, I made the wrong decision, should have put it in a place where I had more space [to include more information]. But you're always constrained that way. The fact is, the rule is, that *you show in-for-ma-tion-containing* photographs and, in your taxonomic descriptions and all that. If, for example, Dr. Brasier had been correct—he's demonstrably wrong because I showed it to you on the slide—but if he had been correct that those . . ." Sounds of protest came from the audience. Schopf retorted, "No, I did. And if these things were in fact branched it would have been wrong if I had not put that in the paper. But the fact is they're not branched, they're curled back under themselves. I wasn't using Auto-Montage confocal microscopy. And thank goodness, because I could see the optical planes clearly and that's what I published.

"And let me say one other thing. Imagine that you are put in the position where you have found a new genus, or a new species of fossil wood. That's a good analogy because these fossils are preserved by petrifaction. Some folks get excited that the organic matter is, quote, wrapped around quartz [as Brasier had suggested in this case]. Well, of course it is. There's quartz inside the cells, quartz outside the cells, that's what petrifaction, permineralization, is all about.

"So imagine you have a petrified log. Now, it's not at all uncommon for a part of such logs to be poorly preserved. They've been eaten by fungi, they're decayed, and so forth. And you're going to name this new, petrified log. What you do is study the cells in it. You'd look at each tissue type, you'd show the variability in those tissue types, and so forth, but you would not waste a whole lot of space showing where it had been degraded by fungi, where it had fallen apart. And that's exactly what is done. It seems to me that those of us who've had a lot of experience looking at Precambrian microbial communities and looking at the taxonomy of modern microbial communities know very well that you have some living forms, you have some that are recently dead, you have some that are degraded, partially, you have some that are essentially bits and pieces of stuff," and other material, he said, that contains no useful "information."

A woman from Arizona State University asked, "In the big picture, if Dr. Schopf's fossils are disproven, what implications does that have for the atmosphere of Earth and the evolution of Earth's atmosphere through geologic time?"

Brasier replied that before this conference he had consulted a well-known expert on the topic, paleobiologist Hans Hoffman. "He looked at the material and he said that he would not have described many of these structures as biogenic, that he felt that those that were more suggestive might be 'dubio-fossils.' But what I was really interested in was how far back would we have to go to find an assemblage of comparable complexity, or I should say diversity. [The expert] said more or less what my thought was, that the foundation on which we feel we can all build, without controversy, is going to be the Gunflint chert," dated at 1.9 billion years old.

"Now, if that is true," Brasier continued, "that will be an astonishing stirring of the paleontological nest—from 3.5 to 1.9 [billion years]. I honestly believe that is *not* going to be the case. There are various indications—biomarkers, always questionable, stromatolites, always a little suspect, possible sheaths, always a bit vague—that things might start to come out of the mists about 3 to 2.7 [billion years ago]. Now people will debate here when the atmosphere actually became oxygenic. And I don't want to get involved in that. But there are others here . . . who feel that oxygenation took place relatively late, 2.3 [billion years ago] or afterwards. And that would be consistent with a later appearance of the cyanobacteria. . . .

"So whatever you do, it causes us to have tremendous pause, and thought, about that interval, between 3.5 and 2.5 in particular. I think we've got to look very carefully and intensely during the next two years to see if we can't pull out what the real scenario is."

Schopf, who had been standing nearby, listening with his head bowed, took the podium.

"I think that's a very interesting question," Schopf began slowly. "I think it's also important, probably, to make a distinction between facts and models. And maybe between what we see and what we wish to see, what we'd like the answer to be.

"Dr. Brasier is a good deal bolder than am I," he noted, triggering a round of laughter in the audience; Schopf paused, hiked his eyebrows, and grinned mischievously. ". . . I mean, I did show you, or at least mention the work of"—here he reeled off the names of prominent scientists in the field—"and that's the tip of the iceberg. And all of these people, it seems to me, are really quite experienced, and it would seem to me it's unlikely that they are wrong.

"But with regards to your specific question regarding, let's say, the Apex fossils do not exist. I don't think it makes much difference at *all.* The fact is—unless one wants to throw out about five hundred isotopic analyses, and unless one wants to say, even though they're consistent with everything else, for some reason we don't like them, and unless a person wants to throw out roughly forty Archaean stromatolitic occurrences with hundreds of stromatolites, including one at 3.49 [billion years ago] in the Dresser Formation . . . that goes over tens of square kilometers, unless you want to throw all that out, what the fossil record is telling us is that there was life on this planet.

"And we still have to fight the thing out, ferret out the evidence, to tell us whether it was oxygen-producing or not oxygen-producing. That's still a very important question. And what sorts of organisms they were. . . . We need more data. I must say in closing here that it seems to me that there are two things: Having Dr. Brasier and his group become interested in this problem is a great boon to us all. We need folks asking questions about the early history of the Earth, and we'll figure out what the answer is. That's the way science operates. If the data are good, they will stand up."

"And I'll say one other thing. I mean this seriously and I mean it respectfully." Here, Schopf cataloged, again, what he saw as the Brasier group's concessions. "I personally am very thankful that Dr. Brasier and his colleagues did the work that they did, because you see they started out saying there was no detectable organic matter in the rock, and now have come back and said yes, there is. . . .

"All these things indicate biogenicity, so the evidence is far *stronger* thanks to their questioning. And since they set out to *debunk* this, it makes it even more strong. 'Cause if someone's out to show you that you're wrong, and they come around to your point of view, it means, gosh, you might have a chance of being right."

The crowd was laughing again. Brasier had taken a position just behind Schopf and had been shaking his head emphatically during the last few comments. Brasier moved up to the podium.

"I'll just say that we didn't set out to debunk Bill Schopf's hypothesis at all," Brasier said, smiling. "It was a hypothesis which developed as we looked further, and the whole thing about science is that you keep looking and you gradually modify your approximation to the truth. We will never, probably, know the truth about all these things. All we can do is get approximations. . . .

"Finally I would say that I would applaud Bill for having produced a hypothesis which . . . is eminently testable and falsifiable. It was a bold hypothesis. We've attempted to falsify it. The last thing I would say is that Bill

Schopf—and he doesn't know this—was the man whose lecture I went to back in 1973 or 1974 that determined me that I wanted to be a Precambrian/Cambrian paleobiologist. So—compliments to you." He nodded and bowed slightly in Schopf's direction.

A *U.S. News & World Report* writer took the microphone to call the whole conversation a delight for reporters, saying "It's nice to see people with their heels well dug in." He asked Brasier if he had conceded there was "a plausible biotic scenario to explain these structures?"

Brasier answered no, drawing a distinction between the organics and the fossil-like structures. "Not to explain the structures. I don't think anybody who has spent more than a few hours looking at the range of materials that we have, and you can see this when we rotate these structures in three dimensions, would feel that you could have any confidence in these as actual morphological microfossils. I don't dispute for a moment that a biogenic interpretation for the organic matter is open. . . ."

Turning to Schopf, the journalist asked, "You're now an agnostic on whether these things produced oxygen?"

Schopf answered, "Agnostic as used in a theological sense. . . . Let me say that the data that are available at the present time are consistent either with oxygen-producing photosynthesis" or with other kinds of microorganisms.

As the crowd started to thin, a *Washington Post* reporter asked Schopf to compare his current position with that of David McKay in 1996, when Schopf was more or less "on the other side of the debate in terms of the evidence and its strength."

"Heavens to Betsy!" Schopf answered. "Heavens to Betsy, no! I made the statement at the news conference—it's very kind of you to remember I was even there—when the Martian objects and data were first revealed. I said at the time, I quoted Carl Sagan, and said, look, this is an extraordinary claim. . . . After all, we have no evidence of life on Mars, we have no evidence of fossils on Mars, we have no evidence of organic matter—*kerogen*—on Mars, we have no evidence of the isotopic composition of Martian material. This was a one-of-a-kind deal.

"Now I ask you to contrast that with what I'm talking about [in the terrestrial evidence]. I have just listed off the names of fifteen or sixteen people who have contributed to a biogenic interpretation between 3.3 [billion years ago] and 3.5 [billion years ago]. We have fifteen hundred carbon isotopic analyses on the Precambrian of this planet. We have seven hundred petrified microbial communities just like the one we're discussing. We have thousands of stromatolites. We know what sort of organisms ought to be there, and these are not otherworldly. They're the right size, the right shape, the right form, and the whole business. So this isn't extraordinary at all. It simply is an extension of knowledge at one time. Now it fits into that knowledge and fits in with what we knew before."

The face-off had lived up to expectations. Two intellectual virtuosos, quick, articulate, and worthy adversaries, had gone for each other's eyes. There was the tectonic sense of pillars swaying, edifices teetering, inky pigments dissolving in the pages of textbooks. The two men had touched on fundamental gears and levers of the scientific process, and on great burning questions of the day. They had left devastation and disarray in their wake, and this was much appreciated by their audience.

A few weeks later the journal *Nature* would report in its June issue, "Most judges gave a clear points victory to Brasier," while Schopf "had won few converts to his cause." A report in *Science* was kinder to Schopf.

The questions about the nature of Earth's oldest fossil life, it seemed, would remain in flux for the foreseeable future, in a kind of paradoxical unity with the even more vexed question of whether biology had flowered on the young Mars.

Andrew Steele greatly enjoyed Schopf's acknowledgment that his specimens were not oxygen-producing cyanobacteria. Others agreed, noting that if, as Schopf conceded, the source of the fossils was a hydrothermal vent deep beneath an ancient sea, there would not have been enough light for the organisms to use photosynthesis.

For many scientists, the notion of oxygen-producing bacteria at such an early time had been the crucial element of Schopf's claims, and the most difficult to explain. Without that, they would lose interest in the issue of whether those features were fossils or not.

Just as many scientists had lost interest in McKay's claims.

Andrew Steele would come to decide that all three men—Schopf, Brasier, and McKay—had pushed their interpretation of the evidence beyond the point where they could be definitively persuasive. But he believed, like many others, that pushing the envelope pricked and stung and stimulated the broader population of interested scientists to new learning. In Steele's view, however, David McKay and Martin Brasier had pressed their case for the "right reasons," and Bill Schopf apparently less so.

David McKay could have played it safe and simply claimed to have found evidence that Mars was "habitable" in its youth, but he and his group had pressed on to suggest that there could be signs of Martian biology. Martin Brasier could have stopped with his challenge to the nature of Schopf's fossil shapes, but he'd pressed further to suggest that the Australian formations held *no* signs of biology at all. The absence of biology in that case was as hard to prove as its presence.

In a funny way, Steele thought, Schopf had sealed his own fate when he'd set the bar so high for David McKay's group. He had not met his own standard.

Steele believed that with the weight of the "Schopf doctrine" removed, that area of research was refreshed as if a window had been opened to let the breeze blow through. Young, energetic investigators could take up the hunt without being tromped on by the establishment regime. It might turn out that there had been oxygen-producing pond scum at an amazingly early point in Earth's infancy, but acceptance of that notion would henceforth require more than a voice of one.

Even as researchers around the world—notably Japanese and Australian groups—took a fresh interest in the topic, Schopf would, with characteristic vigor, continue to defend his 1993 findings, and his way of presenting them. Meanwhile, he could take solace in the fact that he had accumulated a solid body of other work that he considered equally significant and that was widely admired by his colleagues. And, in an unwitting echo of Steele, he observed that he had always hoped his 1993 paper would "spur a lot of further work," but for a long time it had not. "Now," he noted somewhat dryly, "it has!"

On this spring day, as David McKay made his way out of the white tent after the debate, he still stood as Schopf's rival in claiming the oldest known signs of life anywhere. But neither man could boast textbook certitude. And both were passionate about defending their claims.

Not surprisingly, McKay personally found Martin Brasier's argument that Schopf's little shapes were not true microfossils convincing. In his view, Schopf actually had less specific evidence of biology than the McKay team had put forth in 1996.

McKay remarked to a walking companion that his research team was more cautious in its published claims than Schopf had been. "He put down genus and species for eleven of these little forms [in the Australian rock]. We wouldn't do that [for the fossil-like shapes in the Mars rock]. We were just more *conservative* than he was."

McKay was reluctant to say more about the shocking assault on his nemesis.

But for many observers, the debate that afternoon had usefully outlined in neon two obstacles frustrating their collective enterprise: that nobody had yet found a reliable way to recognize the signature of life in the geochemical record on this planet, much less on any other; and, more fundamentally, that there was no consensus about the very definition of life.

CHAPTER FIFTEEN

DOWN THE RABBIT HOLE

One afternoon in early April 2004, McKay—pushing seventy now—was in his lab on the southern corridor of Building 31, staring at the monitor attached to his microscope. He was at 15,000 volts, in the dark and down the rabbit hole.

McKay happened to be lending Kathie Thomas-Keprta a hand with the preparation of selected bits from the Allan Hills meteorite. A facility in Austin had a new technique she wanted to try out on her magnetic crystals.

But much of the time, McKay was keeping an eye on events well over 100 million miles away, where the red planet was suddenly fairly *aswarm* with Earth robots that had managed to slide past the Martian devils.

In January, the twin U.S. surface rovers Spirit and Opportunity had bounced onto equatorial flatlands on opposite sides of Mars and quickly become the most prodigious excursionists on another world since the Apollo astronauts. They covered more alien acreage than their human predecessors. And they had better equipment. As they prowled their dusty way across crater and plain, the rovers deployed grinders and microscopes and other analytical tools, collected data, and sent back tens of thousands of images.

In detailed three-dimensional color imagery, the robot eyes translated the alien landscape into visual poetry and reminded Earth-bound citizens once again of the dreamy, savage beauty of interplanetary exploration.

At the Jet Propulsion Laboratory in Pasadena, the site of Mission Control, you could put on the special liquid crystal shutter glasses, stand in front of a freshly transmitted, wall-sized panorama of the rovers' new habitat, scan the sci-fi pink sky and the play of light on gradations of browny-red in rock and dune—apricot, saffron, auburn, terra-cotta, russet, blood—and feel for a moment as if the ball of your foot had pressed into the cold sand, and you were poised to push off over that dune there toward the hills rising gently on the horizon. It was just a matter of stepping forward.

More important for the watching astrobiologists, Spirit and Opportunity were a species of water witch.

Decades of Mars observations had showed ancient riverbeds, valley networks, islands streamlined to the shape of teardrops, and other signs of flowing water from the period that ended some 3.5 billion years ago. There were even some hints of possible groundwater seepage onto the surface of modern-day Mars. Now the twin explorers were busy scraping up tantalizing signs of long-vanished water at both landing zones—the most definitive accumulation of water evidence ever pulled directly from Martian rocks themselves.

The rovers were cementing scientists' recognition that the surface of early Mars might have been habitable.

In keeping with NASA's "follow the water" mantra, scientists had recently shown that the Martian polar caps could contain a volume of frozen water equal to about 85 percent of the Greenland ice cap. In the northern hemisphere of Mars, the Odyssey orbiter had revealed concentrations of water ice filling up to 90 percent of the volume in the top meter of ground. In other words, the place was not a modern-day desert but frozen tundra—and a resource that could be used by future explorers.

(In early 2005, European scientists would report that their Mars Express orbiter had detected what could be vast bergs of frozen water only a few million years old buried beneath the surface at an unexpected location—quite close to the relatively balmy Martian equator. They would deem this "a place that might preserve evidence of primitive life.")

At the same time, a particularly intriguing explanation for ancient (and to some extent continuing) radical global climate change on Mars was gaining currency with influential players such as Michael Meyer, the astrobiologist who had become head of NASA's Mars program. The planet, it seems, has a habit of dipping like a dancer—though in very slow motion—on its sweeps around the sun. Though Earth and Mars are currently at about the same gentle tilt relative to the sun, advanced computer models (based on the accumulating data) indicate that the poles of Mars have bobbed to and fro dramatically over hundreds of thousands, or even millions, of years—sometimes making curtsies of as much as 60 degrees. (At a time of maximum tilt, for example, the summer polar cap would heat more as it pointed more directly at the sun. It could burn off the polar ice and contribute to a greenhouse effect.) As planetary investigators pondered the origins of the ancient branching channels, and the sources of the stunning, seemingly recent seepages from hidden aquifers, and wondered where all the Martian water had gone, they could point to growing evidence that the planet's rocking provides a mechanism that might drastically alter the climate and move water on a global scale.

But were the signs of ancient water flows the result of flash floods from catastrophic events—comet impacts, volcano melts—which would have refrozen too fast for life to evolve? Or did Mars once have rain, oceans, lakes—waters that lingered long enough for chemistry to become biology? That question had vexed Mars investigators for years.

Key pieces of the water puzzle eluded the investigators. But near the Martian equator, they had detected a potential water beacon—a concentration of a reddish iron ore (hematite) known to form in the presence of water (although it could also form in other ways).

The mission team landed Opportunity at the site of the water beacon in an equatorial plain called Meridiani. The robot had the stunning good luck to hit a geological jackpot. It fetched up near the first rock outcrop ever found on Mars—"essentially a road cut, a piece of time-history" recorded in chronological layers, as Mars mission scientist Maria Zuber, of MIT, described it.

In its landing zone at Eagle Crater, a small impact hole in the vast, wind-rippled equatorial plain, Opportunity photographed legions of peppercorn-sized mineral orbs dubbed "blueberries" scattered across the surface. They contained the iron ore that had summoned the rover to this spot. On Earth, these mineral BBs form in standing water.

Opportunity also measured the highest concentration of sulfur ever seen on Mars. On Earth, rocks with this much sulfur in the form of sulfate salts would either have formed in water or been significantly altered by water after they formed. Scientists were most excited by the discovery of a hydrated iron sulfate salt (jarosite), which hinted that the rock might have rested in an acidic lake or hot spring.

Such highly acidic waters could solve another riddle: Why hadn't the Martian explorers found the expected large deposits of carbonates, which (in addition to salts) should be left by any primordial seas or lakes? Because the acid could have dissolved them. The significance of the small deposits of carbonate globules in the Allan Hills meteorite, in terms of this bigger picture, remain unknown.

Once again, Mars was springing a surprise. Scientists had expected that the key to the Martian riddle would be carbon. Zuber said, "It turns out the real key is sulfur."

The bad news for prospects of life: the water apparently came and went repeatedly in this locale. That unreliability plus the highly acid, salty conditions would have made living there difficult—though not impossible.

McKay watched with keen interest and no little frustration as the golf-cart-sized rovers went about their business. He was once again, as he had been that day in Apollo Mission Control thirty-five years earlier, dependent on the hands and eyes of a remote emissary to do the fieldwork on another world. And this time, he was even farther removed. He managed a few consultations with Mars team members by phone, to discuss details that interested him, but mainly, like much of the population, he had to watch on TV and the Internet.

Alternative approaches to the Martian mysteries also attracted McKay's involvement. The acidic history exposed by the robot on the plain of Meridiani was mirrored in some ways at the highly acidic Rio Tinto (Red River) in Spain, where dissolved iron gives the waters the color of a Burgundy wine. McKay was detaching a young biotech engineer from his staff to work there with an international team of biologists and engineers as they tested deep drilling and other robotic techniques for Mars. They also hoped to find mineral-eating bacteria in a watery habitat deep beneath the surface—in conditions similar to those considered most likely for any microbial denizens of the other planet.

McKay was particularly interested in the Martian blueberries. He had joined a team led by a University of Utah scientist who was studying similar concretions in the desert Southwest, particularly Navajo sandstone in Utah. The researcher had found remarkably similar spheres—terrestrial blueberries—where groundwater had flowed through the sandstone and altered and dissolved the iron. Earlier published work convinced McKay that these things had precipitated from water at relatively low temperatures—mild enough for life.

Then there was an unusual patch of ground dubbed the "magic carpet," where Spirit's landing had disturbed the topsoil and left exposed crust that appeared to be "cakey" or matted, unusually cohesive and resilient. Rover team members speculated that there was some kind of electrostatic process at work or even possibly that sticky salt had been left when water had evaporated through the upper layer. But McKay thought the material looked like the mats formed by microbes on Earth. Had it been processed by Martian microbes? Of course, he knew, this was wild speculation, possibly even nutty. But he couldn't get it out of his mind.

As always, the instruments on Mars were limited, some of the findings were subject to dispute, and the whole story remained unclear.

The twin explorers on Mars, in exceeding the fondest hopes of their human sponsors, incidentally provided the U.S. space agency the balm of good news after a year dominated by another *Challenger*-like calamity: the fiery February 1, 2003, disintegration of the shuttle *Columbia* over the southern United States, which killed seven more astronauts and crippled the human spaceflight program for years. (In January 2004, NASA had named Spirit's landing site Columbia Memorial Station, in honor of the fallen crew.) In February 2004, President George W. Bush proposed to return people—working cooperatively rather than in competition with robots—to the moon and eventually on to Mars. (As one result, McKay found himself in demand once again because of his old specialty—moon dust.) This would not be a Kennedyesque all-out push but was billed as a long-term national commitment to proceed in manageable, affordable increments. The initiative quickly fell into the political gears, its fate uncertain.

But the Mars twins were a hit. The new plan called for NASA to keep sending new missions to the planet at every opportunity, and to continue the search for signs of life. The still-young field of astrobiology had grown fivefold in a decade. But the planners had learned the need for patience, and were devising small steps to be carried out before any direct attempt at life detection. Viking had proven the wisdom of this approach.

Besides, the Allan Hills rock had taught everybody about the tricky complexities of the enterprise. As a NASA committee reiterated ominously in a recent Mars exploration report, "completion of all the investigations will require decades of studying Mars. Many investigations may never be truly complete (even if they have a high priority)."

On August 12, 2005, the $500 million Mars Reconnaissance Orbiter lifted off from Cape Canaveral on course for arrival in March 2006. Spacecraft instruments were to tackle the question of whether—and in what places—Martian waters had persisted long enough for life to arise. They would search for minerals that form in such conditions and for shorelines of ancient seas or lakes, and they would study the character and depth of the recently detected subsurface ice deposits. The most powerful orbital "spy" camera ever sent to another planet would zoom in for close-ups. All of this would be plowed into landing-site selections for robots in the pipeline.

The labs in Building 31 were being remodeled, and McKay and his group were anticipating the arrival of two big new pieces of equipment: a $2 million ion microprobe and a new transmission electron microscope more advanced, and with more bells and whistles, than the one Thomas-Keprta had been using all these years.

Next door to McKay, Simon Clemett was trying to get his new laser lab up and running. But there had been setbacks, such as a power outage that had ruined an expensive pump. The lab door, with its combination lock, was almost completely covered up with warnings: DANGER: CLASS IV LASER SYSTEMS; DANGER: INVISIBLE LASER RADIATION; DANGER: CARBON DIOXIDE LASER; DANGER: LASER RADIATION, AVOID DIRECT EYE EXPOSURE; DANGER—HIGH VOLTAGE, 25,000 DC.

Farther down the hall, Everett Gibson was in his office day and night this week, pushing hard to finish a proposal to study salts from evaporated waters in the dry valleys of Antarctica, to see if they were analogous to the salts the U.S. rover team was currently seeing on Mars. This was Gibson's Plan B. He had hoped to be in London now, with his friend Colin Pillinger, working on data flowing from Pillinger's Beagle 2 on the surface of Mars.

Just months earlier, on December 19, 2003, Gibson had returned to the Royal Geographic Society hall in London, to the podium where he had felt supremely honored to stand not long after the 1996 announcement. This time, he was part of a group that included Prince Andrew. The happy occasion was the arrival of the signal that Beagle 2 had separated from its mother ship, the European Space Agency's Mars Express orbiter, and was on course for a Christmas Day landing near the Martian equator.

Gibson had been selected as an "interdisciplinary" scientist on the project. Beagle 2, named after Darwin's ship, was to be the first such life-detection mission since the Vikings in the 1970s.

The key experiment, in Gibson's view, was one designed to look for methane in the Martian atmosphere. Since this gas was destroyed by the sun's ultraviolet radiation, its presence would suggest a source of recent renewal, such as outgassing by bacteria surviving beneath the lethal Martian surface. (On Earth, as fans of humorist Dave Barry well knew, methane is a major product of cow flatulence and other processes associated with single-celled organisms—but it can also come from nonbiological sources.)

In characteristic fashion, as he had done for the Mars Meteorite Group, Gibson ordered polo shirts for fellow team members, at his expense, with the legend "Beagle 2 Mars, the search for life." It was important, he believed, for the people in the trenches, the unsung technicians, to feel connected to the mission, and the shirts were his way of doing that. Gibson also outfitted his Lexus SUV in Houston with a special license plate frame that read: "Beagle 2 Mars" on top and "Search 4 Life" on the bottom.

Unfortunately, on Christmas Day 2003, at Beagle's landing time, engineers listened in vain for the Beagle call sign—a nine-note composition written for the mission by the rock group Blur. The Martian devils had snared another Earth craft. Despite repeated attempts and the assistance of other spacecraft operating at Mars, they never heard from it again.

Gibson felt a special agony when, weeks later, in March 2004, instruments on Earth and aboard the European Express orbiter at Mars indeed detected signs of methane in the Martian atmosphere. One explanation was Martians. But there were good reasons to suspect there are other ways of producing methane, such as hidden modern-day volcanic activity, for example, or mineral chemistry. Oh, he thought, what the Beagle instruments might have done with that methane data. He held out hope that NASA might be persuaded to incorporate Pillinger's backup instruments (still sitting in the London workshop) into a future mission, to help answer the big questions about life on Mars.

Now, as Gibson worked on his Antarctica project, the nail on the middle finger of his right hand still bore a small, fading black mark where he had accidentally struck his hand on a water spigot in the London apartment he'd rented for what he'd hoped would be months of Beagle 2 operations. It was a very minor battle scar, but a sad reminder of what might have been.

Andrew Steele moved his wife and daughter, now six, from England and into a home in the suburbs of Washington, D.C. (Soon they would be expecting a second child.) As the high-pitched white noise of the emerging seventeen-year cicada horde rose in the old trees outside his lab at the Carnegie Institution, he was busy juggling work there with the occasional field trip and with "big-picture" jobs he had taken on. As chairman of one of the committees appointed to advise NASA on the next round of Mars exploration missions, he was rushing to finish a report.

The previous year, in August 2003, the riddles in the rock had prompted Steele to join an annual international field expedition to the archipelago of Svalbard, in the Arctic Ocean between Norway and the North Pole. It was another frigid desert, where glaciers lay like sheet cakes and stretched like fingers toward the Bock fjord. Here, in hot springs, the heated water climbed up through some twelve hundred feet (about four hundred meters) of permafrost.

Several types of Mars-like environments conveniently combined in this one place. There was rocky rubble—breccia—similar to what Spirit was seeing in Mars's Gusev Crater. There were sites with evaporates and concretions like the "blueberries" Opportunity was finding. And there were evocative "red beds"—elevations where reddish floodplain sediments were crosscut by gullies. They seemed so eerily similar to the startling gulley-wash formations detected on Mars by the orbiting Global Surveyor that Steele found they raised the hairs on his neck.

Here, on the island of Spitsbergen, sat a volcano named Sverrefjell, after an old Norwegian king. This volcanic system, raked over by glaciers, was the only known spot on Earth that manufactured carbonate globules very similar to those found in the Allan Hills rock—same composition, same concentric layering, same Oreo-cookie rims.

As Steele's Svalbard team leader enthused, "That poor little Mars rock traveled millions of miles," and it was just one rock, with no context. "Here, we have the 'crime scene' intact. We can go to these volcanic centers and look at mineral deposits identical to those you know exist on Mars and you can look at the whole scenario and try to figure out what on Earth was going on—and was, or was not, biology involved."

Carbonates are not usually found in volcanoes, but in this case the lava pipes that come from deep in the Earth apparently carried them up through the permafrost. The lava had spewed out, the pipes were left empty, and the empty tubes gradually filled with the hardened rubble. At some point, they apparently also filled with water.

When the lava cooled some million or so years ago, it released gases, which left vesicles, or pores, in the rock. The carbonate globules had been captured in these pores.

The carbonates appeared to have been deposited by water reacting with the basalt at low temperatures—in a moderate hot spring like the setting in which the Martian carbonates (in the McKay team's analysis) had most likely been deposited.

The large international team that included Steele was trying to decipher in detail these processes by which the complex carbonate rosettes had been created, at least for the terrestrial case. At the same time, they were field-testing strategies and tools for life detection, some of which would win berths on upcoming Mars missions.

When Steele and company examined the carbonate globules (taken from the cavities in the interior of the rock to avoid "analyzing bird shit"), it turned out they were full of organic material. The researchers were shocked. You were not supposed to find organic material inside volcanic flows. The researchers were not even sure yet what was the right question to ask in order to figure out if this betokened biology—or not.

Most terrestrial rocks contain microscopic life; the samples from the volcanoes were no exception. The striking thing was that life seemed to be closely and intimately associated with carbonate globules taken from inside the rock. When the researchers stained a sample of carbonate with a substance that binds to DNA, they saw microbes crawling all over it. In some samples, there was also biofilm, a snotty residue left by microorganisms, coating the walls of the cavities and of the carbonate globules themselves, and there were hydrocarbons "of unknown origins."

Here you had a rock very much like the rock from Mars, and you could see organic processes going on in that rock.

Back in civilization, one of the hard truths that Steele and many more-experienced Mars-exploration planners had been confronting, through all the surveys and meetings and teleconferences, was the degree to which they still lacked the tools for distinguishing life at its most ancient and primitive from mere chemistry, and Martian signs from terrestrial. There was still no consensus even about the basic definition of life.

"It's assumed we'll look," Steele said. "We have no idea what to look for, how to look for it, where to go to really look for it, or what to look with."

The important thing was that people were finally taking steps to address the abundant deficiencies. Steele, like many others, traced this promising turn directly to the rock from Allan Hills and the work of the McKay group.

Life on Earth represented a sample of only one. There should be organic chemistry happening on Mars, Steele thought. Life had formed or it had not. As Sagan had said, to learn of either outcome would be remarkable, and unimaginably fascinating.

McKay volunteered as a spear catcher in yet another controversy—one that returned him to nanobacteria issues, this time with a medical angle. The research focused on mysterious "autonomously self-replicating particles," spheres armored with a hard calcium phosphate coat, first detected in the early 1990s by Finnish researchers in fluids such as cow serum that were routinely used in labs to grow cells. In 1998, another published paper linked the spherical nanothings with diseases of calcification, such as kidney stones and hardening of the arteries—prompting interest from the Mayo Clinic and the National Institutes of Health, as well as profit-seeking commercial entrepreneurs.

In fact, the lead researcher on the Finnish studies credited the controversy connected with the McKay group's claims with his ability to get his work published. Previously, reviewers had simply rejected it.

Even though he had to scrounge to find the money to pay her, McKay brought one of the lead authors of the 1998 paper, Neva Ciftcioglu, to work in his lab. There was heated disagreement about whether these nanoentities, or nanobes, were actually living organisms—or something for which no proper nomenclature yet existed. You could grow them in the lab. And at 200 nanometers or smaller, they could pass through the filters intended to sterilize medicines used for vaccinations, getting into places where they were decidedly unwanted. But nobody had found genetic material in them. Though some people were convinced they were living bacteria, they could also be something else—some stage—between bacteria and viruses.

Ciftcioglu was hunting for evidence of even the smallest scrap of DNA in the mystery particles. Some people, McKay included, thought the research might lead to beneficial medical treatment in any number of diseases, including some ailments that afflicted astronauts in weightlessness.

But McKay, not surprisingly, had another reason for bringing Ciftcioglu to Houston. Her putative nanobes were roughly the same size as the fossil-like shapes he and his team had seen in the Martian meteorite—the ones deemed too small to contain the machinery of life.

In 1998, in response to a NASA request and in an effort to deal with the issues raised in the nanofossil skirmishes, the National Academy of Sciences had gathered a blue-ribbon panel of experts to try to define the minimum size for life. After months of discussion, they had agreed on a bottom line: the volume in a sphere 200 nanometers across.

At about that time, McKay was giving ground, conceding that entities smaller than 100-nanometer spheres were not "indicative of bacteria" after all. However, he said, the original fossil-like structures might be bits and pieces of defunct Martian microbes.

Some experts saw both sides yielding toward a new middle ground on the question of strangely small biological entities, thanks in part to work like that of Ciftcioglu's Finnish colleagues, and also to a new report of 50-nanometer nanobes in Australian rocks.

If McKay's revisionist argument sounded "like top-of-the-head improvising by one stuck in a tight corner," wrote science historian Steven Dick and biologist James Strick in *The Living Universe: NASA and the Development of Astrobiology*, "we must also note that, by the time the [academy] report on nanobacteria appeared at the end of 1999, their own 200 nanometer published figure was also being finessed to leave some 'wiggle room.' " The National Academy of Sciences experts adjusted their position to allow for the possibility of primitive unknown microbes as small as 50 nanometers. In a statement that must have pleased McKay, one panel member, John Baross, told *The New York Times*, "We have to think about them [nanobes] in a different way, and one is that they are components that function as a living organism only in totality, the whole being greater than the sum of the parts."

Down the rabbit hole, at unprecedented magnification, life took on all sorts of unexpected possibilities.

The mother of planetary rocks, somewhat diminished, sat serenely in its nitrogen vault on the second floor of the Building 31 annex. A sizable chunk of it was on display in the Smithsonian in Washington. Chips and shards of it resided in more than sixty laboratories around the world. More than one hundred metric tons of equipment, the most advanced technology the planet could muster, had been brought to bear on it.

People had published at least eighteen peer-reviewed papers in support of the McKay group's claims; some twenty-nine papers opposed those claims. Numerous others were hard to place cleanly on one side of the line or another, and still others studied the rock but were not particularly relevant to the debate about ancient Martian life. In any case, as Allan Treiman, one of the more prominent gadflies and critics of the McKay claims, noted, "A body count does not prove who won the war." The debate would continue and expand—on Mars.

The rock was a catalyst. It fell into a volatile brew of human endeavors and yearnings. It changed the recipe, and became part of the new brew. It changed people's lives and it changed their thinking.

Thanks in part to the rock, with its clear evidence of organics on early Mars, the American space program was revamping its Mars campaign and mainstreaming the search for extraterrestrial life there. Spurred by the rock and Dan Goldin, NASA had turned itself into a "biology-centric" agency, organized around the cosmic search for life's origins.

The controversy over the rock had driven forward key areas of astrobiology. And with the rock's riddles in mind, biologists of various stripes were tackling the history of life on Earth with fresh eyes and techniques.

The hostilities unleashed on Terrible Tuesday 1996 had:

- Helped expand the fields of geology, biology, and planetary sciences into the nanometer realm.

- Alerted Mars mission planners to how shockingly difficult it would be to assess the meaning of any bits of rock or dust brought back from Mars, triggering efforts to find solutions and to use Mars-like settings on Earth as rehearsal stages.

- Helped narrow the selection of future landing sites on Mars.

- Increased awareness of the potential for interplanetary contamination, and the danger that Earth explorers could upset unsuspected ecosystems on other worlds as well as the reverse possibility of a biological threat to Earthlings.

- Dramatized complexities involved in the struggle to understand how chemistry turns to biology (on any planet).

- Accelerated the useful cross-bedding of scientific disciplines (along with a proliferation of multisyllabic skills descriptions: cosmogeophysicist, geoastrophysicist, paleomicrobiocosmogeochemist).

- Enhanced understanding of Martian meteorites, provoking research that would aid in sorting out organic Earth contaminants from authentic alien material.

- Demonstrated lively public interest and funding support for the search for extraterrestrial life and related work.

In the persistence of researchers—on all sides—to press their own interpretations of the evidence, even through storms of vitriol, the drama exposed what one science historian called "the sociology of science at work" and the ways that "money, ambition and politics" can come into play when the stakes are so high. It had demonstrated vividly how one person's obsession can be another's high calling.

Like a living thing, the rock had altered its adopted habitat in mischievous and interesting ways. Like a Siren, it lured its discoverers irresistibly toward its treacherous and baffling source. Like a teacher, it instructed us in our ignorance and in the wondrous possibility forged by human audacity.

Still, and always, for McKay there were the meteorites. On some far-off day when NASA's sample-retrieval mission finally got under way, if it ever did, he knew the robots could bring back at best only a small fraction of the mass of Martian material that had already fallen to Earth as a gift. Despite the unsorted natural chaos the meteorites embodied, he found it remarkable that more people weren't studying this precious rubble.

The new explorations on Mars prompted McKay to take another look at the SNC called Nakhla, the chunk of Mars that had rained pieces around a village in Egypt one June morning in 1911. Because Nakhla had apparently formed as lava, crystallized within a few feet of the Martian surface, he thought, it should show some effects (such as the formation of sulfates) similar to those being seen right now by Opportunity on Mars.

But he would get back to that tomorrow. This afternoon, McKay was still exploring the oldest chunk of planet in captivity. As he stared into the monitor screen in the darkened lab, McKay used his left hand to tweak knobs, or sometimes to move the mouse, to navigate his peculiar brand of flying machine over the alien terrain.

McKay was soaring above the flat, hatch-marked plain that was actually the highly magnified brass mount that held the sample, hunting for one of the irregularly shaped fragments of the meteorite. A shadow appeared on the monitor—the sign he was looking for. It was the dark puddle of epoxy he had used to fix the sample in place. The Martian terrain rose suddenly into view from the brass flatland like a range of mountains, and he was cruising above pitted badlands, with crevasses and cliffs, twisted gullies and billowing pillow formations.

At "altitude," this newest sample of Robbie Score's rock from Allan Hills looked for all the world like the state of Texas, with a couple of other smaller states—possibly Kansas and Oklahoma—lying above it like pieces escaped from a jigsaw puzzle. Actually, it was the major portion of one of the carbonate pancakes that Kathie Thomas-Keprta had tapped from the meteorite. As McKay zoomed in on "Texas," Thomas-Keprta walked into the lab and whooped with delight. "Look at that texture. That is a thing of beauty!" Referring to the tedious, painstaking process of knocking out these chosen bits, she noted, "I worked a long time on that."

She and McKay traced on the screen what appeared to be the shadow of a ridge, which they judged to be the black-and-white Oreo rim of the carbonate in cross section. It seemed to intersect the rim of another one, as if two of the carbonate blebs had been welded together.

McKay, staring into the screen, said, "I need to figure out what that is . . . but it's cool."

He lined up one patch of sample after another, snapping photographs and accumulating those prairie-grass graphs of the chemical composition to be sent to the lab in Austin. He moved over a segment of Oreo rim. *Click.* He went to the compo setting to see where the iron-rich areas were. *Click.* He was still awed by the fact that these were supposedly patterns of nature's work on another planet almost four billion years ago—right there in front of him like ancient runes.

"Did you know that a face is the thing most people readily see in cloud shapes?" he mused. Then he added with a slight smile, "Usually when I start seeing faces, it's time to go home."

He left Building 31, climbed into his old Chevy van, and headed down the highway in the direction of the setting sun, toward Mary Fae and the sanctuary of a glass house on a floodplain in the piney thicket.

EPILOGUE

On an austral summer day in late 2001, Robbie Score stepped out of a helicopter that had touched down on the bottom of the Earth near a ridge composed of ferrar dolerite, in the Meteorite Hills region of the Darwin Mountains. Ferrar dolerite is a crystalline volcanic rock that researchers were studying as a stand-in for Martian surface geology, based on what they had learned from the growing collection of Martian meteorites. The official purpose of the chopper's stop here was to deliver a fresh crew of meteorite hunters.

Almost twenty years had passed since Score fell in love with the ice on that first visit. In 1996, after eighteen years as a meteorite curator in Building 31 at Johnson Space Center—and just as the public furor erupted around the igneous lump she had bagged that first season—she had left Houston and NASA to devote herself to Antarctica.

As far as that rock was concerned, she had tried to keep an open mind, a professional detachment. She had to admit she was thrilled that it had caused such a stir. But she had other preoccupations these days. She managed a research program that included labs at McMurdo Station, at Palmer Station, and at remote field camps. She was responsible for the planning and support of more than 120 research groups working in the world's most forbidding—some would say its most otherworldly—landscape.

But Robbie Score had not traveled to this desolate location just to swap out the team. She had a more personal reason to take in the scenery. In honor of her association with the most famous Martian meteorite and her contributions to Antarctic research, the authorities had renamed the spot: Score Ridge.

She paused there, looking from her own booted feet out toward the horizon. The ice was moving somewhere below, she knew, carrying dark nuggets toward the brutal light.

ACKNOWLEDGMENTS

As A STAFF writer for *The Washington Post*, I was part of the press pack covering the August 1996 story of a homely rock from Mars as it briefly emerged as a worldwide celebrity and set scientists quarreling over its meaning. Not until four years later, with the dispute still stewing, did I decide to write this book. My goal was to understand how so many dozens of very smart people, armed with the world's most advanced tools, could argue so fiercely for so long about the meaning of one fist-sized lump—albeit the oldest rock known from any planet including Earth. Why couldn't somebody win the battle over this tiny chunk of Martian turf?

The answer to that question is the subject of this book. The argument at its heart concerns not only the frustrations of the search for life on Mars but also those bedeviling the effort to understand how life began on Earth and what are the distinguishing characteristics of life wherever it may be found.

I write as a journalist; I am not a scientist or a scholar. For the sake of readability I have left out, or reduced to notes, many people who produced significant work on the rock or were otherwise touched by it. Still, I've tried to convey all sides of the case fairly while providing a taste of what cutting-edge research looks like from inside as it unfolds and a sense of the sea of uncertainties in which its practitioners often must swim. I've tried to impart an appreciation of the secret drama, the individual yearning, confusion, fear, ambition, obsession, courage, and extraordinary diligence—in fact, the sheer messiness—that may lie behind any of the "discoveries" so glibly summarized (sometimes by me) in the iceberg tips of news reports and textbooks.

My mother, Ruth, has a forty-pound amethyst, a spiked and crenulated object full of violet sparkles, sitting on a velvet pillow the size of a small dog bed inside her comfortable double-wide mobile home near Santa Cruz, California. Early on, she taught my younger sister and me to appreciate rocks. While we were growing up in Tennessee, and on family trips to the national parks, we admired the layering, the granite outcrops exposed in road cuts, the pretty pebbles at our feet when we hiked. We haunted mineral shops around the American West, desert and mountain. The granite beauty of the High Sierra captured my sister, Sharon, and she has lived most of her life there in a log cabin by a lake. All of us have odd rocks strewn around our homes.

So maybe it should not surprise me to find such talismanic powers in a seemingly ordinary object like the Mars rock, when it is inserted somewhere in the mosaic of emerging knowledge these audacious investigators are assembling.

My deepest gratitude goes to *The Washington Post* and the Alfred P. Sloan Foundation (particularly Doron Weber), whose support made this book possible. Among my many former editors and coworkers at the *Post* who helped in various ways are Jackson Diehl, Leonard Downie, Steve Coll, Liz Spayd, Joel Garreau, Rick Weiss, Rob Stein, David Brown, Guy Gugliotta, Eric Pianin, Shankar Vedantam, Nils Bruzelius, Susan Okie, Ceci Connolly, Paul Richard (and his wife, Deborah), Linton Weeks, Tom Wilkinson, Tom Shroder, Bob Kaiser, Michael Abramowitz, Pat O'Shea, Chris White, and always on the alert at Cape Canaveral, Bill Harwood. And I thank former *Post* editor Peter Osnos, now at Public Affairs in New York, for his wise and classy advice.

I am grateful to the dozens of participants on all sides of the dispute, as well as those observing from a distance, who generously shared their time and insights, and in some cases great chunks of their lives, with me. Some participated in multiple lengthy interviews over a period of years. Special thanks to David and Mary Fae McKay, Everett Gibson, Kathie Thomas-Keprta, Richard Zare, Simon Clemett, Allan Treiman, Andrew Steele, Robbie Score, Ralph Harvey, Bill Schopf, David "Duck" Mittlefehldt, John Rummel, Martin Brasier, Sue Wentworth, Tim McCoy, Wendell Mendell, David Salisbury, Steven Dick, and Carleton Allen; and for the assistance of those who were in the thick of the action at NASA and the White House as the story unfolded, including Wes Huntress, Dan Goldin, Ed Weiler, Michael Meyer, Jim Garvin, Don Savage, Laurie Boeder, Alan Ladwig, and Rick Borchelt; and also to Lynn Rothschild, Kathleen Burton, Bill Jeffs, James Hartsfield, and Cathy Watson at NASA affiliates around the country. My thanks to the NASA history office, including Roger Launius, Jane Odom, and John Hargenrader; and Richard Faust in the NASA library; and to Steven Maran of the American Astronomical Society; John Logsdon of the Space Policy Institute at George Washington University; Richard Obermann, senior staff member of the House Science Subcommittee on Space; and Louis Friedman, executive director of the Planetary Society.

At the National Science Foundation, my thanks in particular to Scott Borg and my friend and former colleague Curt Suplee, who helped open my eyes to the Antarctic and other wonders. I also want to thank the Carnegie Institution of Washington, especially Maxine Singer, Richard Meserve, Robert Hazen, Susanne Garvey, and Claire Hardy. I owe a special debt to my friend and former editor Boyce Rensberger, director of MIT's Knight Science Journalism Fellowship program, for his insights and advice.

I thank my husband, John Atkisson, for his generous assistance and affectionate forbearance during the hair tearing, and I thank a constant muse named Beacon. Loving thanks to the awesome Ruth Weiland, beyond words, and to Sharon Sawyer, whose sharp wit extends even to minerals. For their warm encouragement throughout, I thank Jackie White, Marilyn Elkins, John and Barbara Holum, Bob and Binnie Holum, John and Vicki Thorne, Dick and Dorothy D'Amato, Ruth Pontius, Erica Atkisson, Paul Thompson, Jim and Helen Ashford, the Cones and the Weilands, Greg Giles, Tracy Barrett, Laura Beth, and Patrick, Marguerite, and Teresa. To deserving people I have inadvertently omitted, my apologies.

I am grateful to all those who read some or all of the manuscript and offered expert comment. Any errors remaining are entirely my own.

Finally, the highest praise to Jonathan Karp, my outrageously talented editor at Random House, and his able assistant, Jonathan Jao, who deserve enormous credit for the shaping and polishing of this narrative, and to my energetic agent, Gail Ross, who made it happen.

NOTES

CHAPTER ONE: SCORE

3 **In this stretch of** • The description of the meteorite hunt and discovery of the rock on this day is based primarily on the author's multiple communications with participants, particularly telephone interviews and e-mail exchanges with Robbie Score, John Schutt, Ralph Harvey, and William A. Cassidy. Cassidy's *Meteorites, Ice, and Antarctica* (Cambridge: Cambridge University Press, 2003), and Harvey's online Antarctic Search for Meteorites (ANSMET) site provided much of the history and context. The ANSMET site is at: http://geology.cwru.edu/~ansmet/.

3 **They were working** • According to the National Science Foundation, the Allan Hills were named for Professor R. S. Allan of the University of Canterbury, New Zealand.

3 **In places, the** • Cassidy, *Meteorites*, p. 60.

6 **She arrived on the** • "Highest" in the sense that the continent averages 1.5 miles in height above sea level, making it almost a mile higher than the global average land height, according to "Glacier," a Web site developed by the National Science Foundation, the Education Development Center, and Rice University, at www.glacier.rice.edu/.

6 **The society of** • The discovery of the first known Antarctic meteorite, in Adélie Land in 1912, became a minor reference buried in one of the most harrowing tales of heroic survival ever to unfold on the Antarctic frontier. It is recorded in Douglas Mawson's *Home of the Blizzard* (London: W. Heinemann, 1915), vol. 2, p. 11: "To avoid crevasses, we steered first of all to the southwest on the morning of the 5th, which was clear and bright. After six miles, the sastrugi [dunes of stiff snow] became hard and compact so the course was changed to due west. Shortly afterwards, a piece of rock which we took to be a meteorite was found on the surface of the snow. It was approximately 5 by 3 by 3½ inches and was covered in black scale, which in places had blistered. Most of the surface was rounded. . . . There was nothing to indicate there had been a violent impact." The finder was a

member of the grim Antarctic expedition led by the author, Australian geologist Douglas Mawson.

7 **And then there is** • John S. Lewis, *Rain of Iron and Ice* (Reading, Mass.: Addison-Wesley, 1997), p. 175.

7 **Because people had** • Cassidy, *Meteorites,* p. 2.

7 **Then a few alert** • Ibid., pp. 1–2 and chapters 2 and 3.

8 **In December 1969** • Ibid., pp 16–21. Japanese glaciologist Renji Naruse picked up and recognized the first meteorite in the area. The Japanese team found eight more the same day, in an ice field five by ten kilometers.

8 **These facts led** • Ibid., p. 17. The Japanese (M. Yoshida et al.), in a 1971 paper, were the first to write about the possible concentrating mechanism.

8 **In 1973, at a meeting** • Ibid., p. 16. The 1973 event attended by Cassidy was the thirty-sixth annual meeting of the Meteoritical Society, in late August, in Davos. There, Japanese chemists who had analyzed the stones described the discovery.

8 **After bouts of rejection** • Ibid., pp. 17–21, 29, and 55.

8 **Sending teams to hunt** • The inflation-adjusted figure for the total cost of the Apollo program was provided to the author by Richard Obermann, professional staff member of the House Science Committee's subcommittee on space and aeronautics, based on an analysis by the Congressional Research Service which says: "The $25 billion is from a 1975 NASA fact sheet on the costs of 'manned' space programs." The figure was determined using the gross domestic product "chained price index" from the latest Office of Management and Budget historical tables: the analyst divided 25 by .2563 × 1.1045.

8 **The astronauts brought home** • Don E. Wilhelms, *To a Rocky Moon* (Tucson: University of Arizona Press, 1993), p. 355: "Almost 382 kg of rock, soil, and core samples—38% of a metric tonne and 42% of an English short ton—were returned from the moon, given 2,196 sample numbers, and cut up (so far) into 80,000 pieces."

8 **Once those flights** • Cassidy, *Meteorites,* p. 228.

8 **Like dashboard clocks** • Ibid., p. 228.

8 **One of Cassidy's first** • Ibid., p. 35. In the early days, the meteorite hunters were regarded as "intruders" in the small society of Antarctic research. And resources were severely limited. Because there were no snowmobiles, the searches were conducted on foot.

9 **Some of the rocks** • Ralph Harvey, meteorite search leader, on Web site: http://www.cwru.edu/1785680/affil/ansmet/faqs.html.

9 **For the preceding six** • Author interview with Score.

9 **Having passed** • Cassidy, *Meteorites*, p. 15.

10 **The sun was up** • Ibid., p. 8.

10 **However, it had also** • Ibid., p. 14. For more on temperatures at select Antarctic sites, see: http://www.coolantarctica.com/Antarctica%20fact%20file/antarctica%20environment/climate_graph/vostok_south_pole_mcmurdo.htm#McMurdo.

11 **Knowledge of the whole** • "Of the world's 61,000 nonfiction papers and books published about the Antarctic since the earliest papers dating from the 1600s, 91 percent have been published since 1951," the National Science Foundation has noted. (See *The United States in Antarctica: Report of the U.S. Antarctic Program External Panel* [Washington, D.C.: National Science Foundation, 1997], p. 17.) A commercial ship made the first discovery south of the Antarctic Convergence (where polar waters meet temperate ones) in the 1670s. Sailors first spotted the Antarctic landmass itself sometime in the early 1820s, but historians disagree as to which nationality—a British, Russian, or U.S. ship—could claim the honor. Norwegians were the first to document the sighting. As the twentieth century arrived, a wave of exploratory expeditions tackled the forbidding ice desert, the most famous of them led by Roald Amundsen of Norway and Robert F. Scott of England. Amundsen beat Scott by a few weeks to become the first to reach the geographic South Pole. The United States got into the fray with the privately financed expeditions of Richard E. Byrd in the 1920s and '30s.

11 **Later, Cold War military** • The U.S. Operation Highjump in 1946–47 was "the largest single expedition ever to explore Antarctica, involving 13 ships, numerous airplanes, and more than 4,700 men," according to the National Science Foundation (see "U.S. Antarctic Program Participant Guide, 1998–2000 Edition, NSF 98–117, Arlington, Va., 1998 p. 2). Its mapping operations paved the way for the International Geophysical Year. Then came the navy's Operation Windmill, which used ship-based helicopters. A six-nation race was touched off briefly by reports of uranium deposits.

12 **His reputation had grown** • Cassidy, *Meteorites*, p. 85.

12 **The sledge broke** • The 1912 expedition led by Mawson encountered unexpectedly vile conditions even for the Antarctic, with hurricane-force winds and fierce blizzards. One party, trying to cross a treacherous, snow-covered crevasse, lost a man, a sledge, and six dogs. Peering into the gaping rift, the survivors could see only a ridge some 150 feet below where one

of the dogs lay whining, its back apparently broken. Beyond that they glimpsed the abyss. For more than three hours, the two survivors yelled into the hidden depths. Gone along with the sledge was their tent, most of their food, their spare clothing, and most of the other supplies. They were 315 miles from the main base. In order to survive, they ultimately began to kill and eat their remaining dogs, stewing even the paws. These and other ordeals are recounted in Mawson's *Home of the Blizzard.*

14 **It would advance** • Cassidy, *Meteorites*, p. 317.

14 **Under certain circumstances** • Ibid., p. 44. One night after dinner, Cassidy writes, two field team members disposed of some hot water by pouring it into the snow at the edge of the canvas floor inside the dark, opaque-walled utility tent. "Suddenly a beam of lovely blue light dimly illuminated the interior of the tent! . . . The light was streaming upward from the melt-hole in the snow." The team realized that sunlight was penetrating the exposed ice surface. Because the ice was so clear and many meters thick, the visible light, composed of all the colors of the rainbow, had "a long path length before being reflected back toward the surface. When it emerges it is predominantly blue because the ice has preferentially scattered and absorbed the other colors."

14 **This progression gradually** • The meteorite hunters dubbed these shifting, changing regions "stranding surfaces." An area no bigger than a tennis court might yield as many as 250 meteorite specimens, according to the National Science Foundation.

15 **In places along the** • E-mail to author from meteorologist Matthew Lazzara of the University of Wisconsin's Antarctic Meteorological Research Center (September 2, 2005). The record wind speed measured by the program's Automatic Weather Stations (funded by the National Science Foundation) was a maximum of almost 130 miles per hour (58 meters per second), at Cape Denison on August 24, 1995. The monthly mean wind speed for that location is 55 miles per hour. The McMurdo area recorded 143 mile-per-hour winds during a storm in May 2004. As for the historical record, Lazzara and coworker Linda Keller noted, the intrepid Australian explorer Mawson in 1912 reported days of sustained 80 and 90 mile-per-hour blasts at his Cape Denison hut (not far from what is today the location of the Research Center's weather instrument cited above). Then he wrote: "Having failed to demolish us by dogged persistence, the hurricane tried new tactics on the evening of May 24, in the form of a series of Herculean gusts. As we learned afterwards from the puffometer, an in-

strument for determining the velocity of gusts, the momentary velocity of these was of the order of two hundred miles per hour."

For a report on the May, 2004, storm at McMurdo, see http://www.southpolestation.com/mcm/storm.html. For current surface weather conditions in Antarctica, see http://uwamrc.ssec.wisc.edu/realsfc.html.

15 **Early Antarctic explorers** • Mawson, *Home of the Blizzard,* vol. 1, pp. 116ff. Cited in Cassidy, p. 46.

15 **The ones big enough** • Cassidy, *Meteorites,* pp. 325–26. In addition to the meteorites, searchers have found smaller stuff: tiny extraterrestrial grains in fine sediments in water and surface debris. One group separated thousands of micrometeorites and other cosmic microdetritus from Antarctic ice by melting tons of it and filtering the meltwater.

Meteorite collection in such places as the northern Sahara is not as scientifically useful, Harvey told the author, in part because profiteers pick the most commercially valuable specimens in order to sell them on the black market and leave behind the "ordinary" meteorites. Thus, in those settings, scientists are deprived of the Antarctic-style systematic approach.

16 **One consequence was that** • *The Old Farmer's Almanac* estimates that a typical shower uses 15 to 30 gallons (see http://www.almanac.com/edpicks.0698/waterused.html).

16 **In the worst** • Although some people assumed they would find the solitude great for thinking "big thoughts," this was not necessarily the case. Some of the men, in particular, would pass the time in endless speculation about what aspect of being in Antarctica caused the fart rate to shoot up. They toyed with such theories as chaotic effects associated with the polar magnetic field, according to geochemist E. Julius Dasch, "The View from the Crevasses," *Ad Astra,* (Nov.–Dec. 1994): p. 49.

16 **Score shared** • Author interview with Schutt.

16 **Over the years** • Author interview with Ralph Harvey, and Cassidy, *Meteorites,* pp. 89–90.

16 **And older hunters** • Cassidy, *Meteorites,* p. 83.

16 **Cassidy would sit** • Ibid., p. 48.

17 **Therefore, when venturing** • Author interviews with Harvey, Scott Borg, and Curt Suplee of the National Science Foundation. Visitors were sometimes prohibited from leaving anything behind in the ice that could jeopardize the ecology. As a consequence, the hunting teams were supposed to carry with them significant detritus from their bodily functions. The

transport planes might fly a season's worth of poop, also known as "used food," back to civilization in cold storage, along with ice cores and other research trophies.

17 **At "night,"** • Author interview with Suplee.

18 **The field notes** • C. Meyer, "NASA Mars Meteorite Compendium," NASA/JSC, online at http://www-curator.jsc.nasa.gov/curator/antmet/mmc/84001.pdf. As for the "Yowza," Score recalls in an e-mail to the author that "Scott Sandford and I were trying to spice up the field notes. I think Scott [was] the instigator of the wording for that entry."

19 **The team had already bagged** • Author interviews with Score and Schutt.

20 **After the season's haul** • Author interviews with Carlton Allen, Cecilia Satterwhite, and others at NASA's Johnson Space Center meteorite laboratory.

CHAPTER TWO: MOON DUST

22 **It reminded him** • Kathy Sawyer, "Armstrong's Code," *Washington Post Magazine*, July 11, 1999, p. W10; Armstrong's accounts, at University of Cincinnati and NASA Apollo archives; and Andrew Chaikin, *A Man on the Moon* (New York: Viking, 1994) pp. 208–11.

22 **The events unfolding** • Chaikin, *Man on the Moon*, p. 190; Gene Kranz, *Failure Is Not an Option* (New York: Simon & Schuster, 2000), p. 288. The events of the *Apollo 11* mission described in this chapter are also reported in NASA's extensive Apollo mission records, and in the *Apollo 11 Lunar Surface Journal*, available online at http://www.hq.nasa.gov/office/pao/History/alsj/a11/a11.launch.html.

22 **Sitting in the back** • Author interview with McKay; see also Kranz, *Failure Is Not an Option*, p. 278.

23 **The stomach-clenching** • Chaikin, *Man on the Moon*, pp. 200–1; Kranz, *Failure Is Not an Option*, p. 292.

23 **From the moment** • Author interview with McKay; see also Chaikin, *Man on the Moon*, pp. 212, 215–16. The landing, lunar surface excursion, and delivery of specimens to JSC are captured in NASA videos, including JSC Production 554 and PMU: 11/49473-HQ 194.

24 **Apollo in the early 1960s** • Charles Murray and Catherine Bly Cox, *Apollo* (New York: Simon & Schuster, 1989), p. 79 (citing an account by *Life* magazine writer Hugh Sidey).

See also Walter A. McDougall, . . . *the Heavens and the Earth* (New York:

Basic Books, 1985), pp. 141–95. In 1961, President Kennedy committed the United States to putting "a man" on the moon before the decade was out, and returning him safely. The move had been spurred by a spasm of public awe and fright over the "space gap" first revealed by the Soviet launch of Sputnik in 1957. Sputnik demonstrated that the Soviets had the capability to launch nuclear warheads into our cities (the threat alone was a valuable commodity). And the ongoing "gap"—manifest in a series of Soviet firsts in spaceflight—suggested the enemy might also win in the Cold War theater of public relations. Capitalism and democracy were facing off against communism and tyranny in a battle for world dominion. The stakes couldn't be higher, and the "high frontier" was, for a time, the primary arena for a symbolic victory.

24 **It proved to be** • Tom Wolfe, "The Tinkerings of Robert Noyce: How the Sun Rose on the Silicon Valley" *Esquire* (Dec. 1983): pp. 346–74.

24 **Author Norman Mailer** • Walter A. McDougall, . . . *the Heavens and the Earth*, p. 412.

25 **David McKay was** • Donald Goldsmith, *The Hunt for Life on Mars* (New York: Dutton / Penguin, 1997), p. 67.

25 **Still, two of the** • Author interviews with David McKay and Mary Fae McKay.

25 **In the early sixties** • Author interviews with Wendell Mendell, of Johnson Space Center; see also Murray and Cox, *Apollo*, p. 131.

26 **"For eight months** • Tom Wolfe, *The Right Stuff* (New York: Bantam, 1980), pp. 295–96.

26 **But the resistance** • Murray and Cox, *Apollo*, pp. 130–33.

27 **(It was being invented** • Don E. Wilhelms, *To a Rocky Moon* (Tucson: University of Arizona Press, 1993), pp. xii, 58; see also Chaikin, *Man on the Moon*, p. 385.

28 **First, there were** • Author interviews with McKay, Wendell Mendell, and others. See also Wilhelms, *To a Rocky Moon*, pp. 76–78 and 80–84.

28 **Then there was another** • Author interview with McKay.

28 **At the birth of Apollo** • Wilhelms, *To a Rocky Moon*, p. 58, 192–93; see also Chaikin, *Man on the Moon*, pp. 383–410.

28 **President Kennedy had** • McDougall, . . . *the Heavens and the Earth*, p. 309, 315–24; see also John M. Logsdon, "An Apollo Perspective," *Astronautics & Aeronautics* (Dec. 1979): pp. 112–17, and Logsdon's *The Decision to Go to the Moon: Project Apollo and the National Interest* (Cambridge, Mass.: MIT Press, 1970), pp. 94–100, 111. This subject has been covered in a number of space

histories and is preserved in copious congressional records and other documents. See also online material for Apollo at the NASA history site: www.hq.nasa.gov/office/pao/History.

29 **The space agency** • McDougall, . . . *the Heavens and the Earth,* p. 374, regarding pork-barrel politics.

29 **But once it dawned** • Wilhelms, *To a Rocky Moon,* pp. 55–58.

29 **In order to force** • McDougall, . . . *the Heavens and the Earth,* pp. 381–82, citing Webb's words in *Space Age Management: The Large-Scale Approach* (New York: McGraw-Hill, 1969), pp. 113, 117; Webb's address at the Milwaukee Press Club Gridiron Dinner, April 17, 1963, as reported in NASA, *The American Space Program: Its Meaning and Purpose,* HHN-47 (Dec. 1964), p. 25; and an interview with Webb by Harvey et al., March 16, 1968, pp. 16–17, at NASA headquarters.

29 **Some of the most** • Wilhelms, *To a Rocky Moon,* p. 84; Chaikin, *Man on the Moon,* pp. 383–97.

29 **Many in the astronaut** • Wolfe, *The Right Stuff,* pp. 317–18.

29 **As they taught** • Author interview with McKay.

30 **But many astronauts** • Wilhelms, *To a Rocky Moon,* p. 122, and Michael Collins, *Carrying the Fire* (New York: Farrar, Straus and Giroux, 1974), pp. 72–75.

30 **But there were** • Author interview with McKay; see also Wilhelms, *To a Rocky Moon,* p. 122.

30 **And the broader tensions** • Chaikin, *Man on the Moon,* p. 389.

31 **McKay was admitted** • Author interview with McKay.

31 **To a degree that is** • Wilhelms, *To a Rocky Moon,* pp. 67, 97, and 146. Eventually, it would be shown that as early as 1964, based on findings from the robotic Ranger craft, Harold Urey and Gene Shoemaker had solved much of the riddle, introducing the term *gardening* to indicate how the pulverized surface layer was formed (in churning by constant impacts), and correctly predicting its properties, such as a thickness that varied up to a few tens of meters and a bearing strength that increased with depth (meaning that the astronauts and vehicle would not sink in). Later, after studying the findings from Surveyor spacecraft, Shoemaker named the lunar surface layer *regolith,* a term previously applied to loose material on Earth's surface. Shoemaker would maintain that the Surveyor missions provided better information about the lunar surface debris than did Apollo.

32 **For example, was** • Wilhelms, *To a Rocky Moon,* pp. 67, 97, and 146; Chaikin, *Man on the Moon,* p. 180. The reference is to astronomer Thomas

Gold. Despite images from the Surveyor series, which had landed safely on the moon, he argued that the moon was covered by a deep layer of powder so fine it amounted to airy fluff and advised that the astronauts drop colored weights during landing to see whether the objects sank in, aborting the touchdown if necessary.

32 **As one had observed** • Chaikin, *Man on the Moon*, p. 180; reference is to NASA's Elbert King.

32 **During his two-hour** • Wilhelms, *To a Rocky Moon*, pp. 202, 206. Wilhelms notes that in the book *Moon Rocks* (New York: Dial, 1970) *New Yorker* writer Henry S. F. Cooper recorded some of the "skirmishes in the tug-of-war between scientists' emotions and the facts" during the lunar surface explorations and the arrival of the first specimens.

32 **He reported that** • NASA archives, *Apollo 11 Lunar Surface Journal*, p. 74; Back on Earth, Armstrong would praise his geology teachers and add modestly that, although he enjoyed geology, "Had I been a better geologist, I might have seen some things [on the lunar surface] that were important, that I missed. If that's true, I regret it. But in the time we had available, I think everyone did a pretty credible job."

32 **By the time** • Chaikin, *Man on the Moon*, p. 218; see also NASA *Apollo 11* archives.

33 **Less than a week** • Chaikin, *Man on the Moon*, p. 233. Also, Wilhelms, *To a Rocky Moon*, p. 208, expresses disbelief that the only scientific debriefing of the *Apollo 11* crew was one he participated in while the astronauts were still in quarantine, on August 6, 1969. If that's true, he said, "you have a fine illustration of NASA's attitude toward science." He also remarks on the "stingy" nature of the camera NASA provided for the mission, which resulted in those "fuzzy, ghostly images."

33 **Armstrong had decided** • Author interview with McKay.

34 **Two nights later** • Chaikin, *Man on the Moon*, pp. 233, 355–66; Wilhelms, *To a Rocky Moon*, p. 344. Until the lunar samples arrived, Nobel laureate chemist Harold Urey and others known as "cold mooners" had championed the theory that the moon was a primordial lump essentially unchanged since the formation of the solar system (except for craters and plains formed by violent impacts). The opposition "hot mooners" held that volcanic activity had shaped the face of the moon. The majority of geologists had straddled the two camps, allowing for a combination—"cosmic impact catastrophes" alternating with gentle volcanic extrusions and an occasional fire fountain from deep in the interior, according to Wilhelms, p. 344.

34 **The art of handling** • Author interviews with Mendell, Gibson, and others.

34 **The rock custodians** • Gibson, in interviews with the author, said the scientists working to eliminate potential contaminants ran into resistance from engineers, who argued that they needed certain of the offending materials—"until they were told and understood" the reason for the changes. The cleanup group decreed that only stainless steel, aluminum, and Teflon could come into contact with the samples.

34 **After three landing missions** • A legend would persist within NASA for decades about a common Earth bacterium (*Streptococcus mitis*, typically found in the human mouth, throat, and nose) that was found nestled inside a camera from a 1966 unmanned Surveyor mission that astronauts Pete Conrad and Alan Bean retrieved from the lunar surface in 1969, during the *Apollo 12* mission, and brought back to Earth. Analysts at the NASA lab in Houston concluded that the microbe had flown from Earth to the moon and survived all those years in the vacuum. However, scientists at the federal Centers for Disease Control and Prevention soon determined that the microbe had never left Earth, but that someone had most likely contaminated the camera during postflight examination in Houston. NASA concurred. (See Kathy Sawyer, "Hardy Microbes Appear Able to Survive in Space," *Washington Post*, Oct. 4, 1999, p. A11, and F. J. Mitchell and W. L. Ellis, *Analysis of Surveyor 3 Material and Photographs Returned by Apollo 12*, esp. "XI. Microbe Survival Analyses, Part A, 239–251"; note in particular "Results," on p. 250 [Washington, D.C.: NASA; Scientific and Technical Information Office, 1972].)

34 **McKay was in** • Author interviews with McKay.

34 **Because McKay knew that most** • Wilhelms, *To a Rocky Moon*, pp. 97, 146, 342; regolith included not only soil but fragmented bedrock, bits of glass from impacts, and other loose material.

35 **McKay teamed up** • Author interview with McKay. McKay's coauthors were two other young geologists, Don Morrison and Bill Greenwood.

37 **As McKay and his** • Author interviews with McKay and dozens of published papers, including: D. S. McKay and A. Basu, "The Production Curve for Agglutinates in Planetary Regoliths," *Journal of Geophysical Research* 88 (1983): pp. B-193–99; and L. A. Taylor and D. S. McKay, "Benefication of Lunar Rocks and Regolith: Concepts and Difficulties," in *Engineering, Construction and Operations in Space III*, vol. I (New York: ASCE, 1992), pp. 1058–69. The studies would show that the micrometeorite bombardment melted and vaporized the grains of lunar topsoil; the vapor was redeposited, and the

melted material, before it resolidified, scavenged dusty material from nearby and incorporated it. The result of all this business was new and different particles (the agglutinates), aggregates of minerals, rocks, and glasses welded, or cemented, together. The glass in the fragile material further broke down, producing ever-increasing amounts of glass in the soils as they aged. The higher the proportion, the longer the soil had been exposed to the impacts—and therefore the more "mature" it was.

See also Wilhelms, *To a Rocky Moon,* p. 342, about the "strange optical properties of the lunar surface and the full moon." The brightness "depends mainly on the quantity and composition of glass-bound agglutinates that are created by incessant small impacts in regoliths. The agglutinates get their darkness and color from the iron and titanium in the rock from which they formed."

In 2004, President George W. Bush's proposal that NASA start working toward a return of robots and humans to the moon triggered a new round of seminars and studies about the lunar surface, particularly how best to produce up to one thousand tons of simulated lunar soil for training and the like, given its unique properties. See L. A. Taylor, "The Need for Lunar Soil Simulants for ISRU Studies," University of Tennessee, for the Space Resources Roundtable, Board of Directors, 2004. Taylor mentions McKay, along with himself and others, as " 'lunatics' of the first order, experts in the physical, chemical and geotechnical properties of lunar regolith, and the best qualified for this endeavor."

37 **Inside his tribe** • Goldsmith, *Hunt for Life on Mars,* p. 69. See also Taylor, "Need for Lunar Soil Simulants."

38 **In 1971, when Coulter** • Mary Fae Coulter, in interviews with the author, said she was not particularly religious—David was more of a believer—but she wanted a church ceremony, not a civil one. They decided to hold the service in the Rice Chapel. The university chaplain was a Methodist, and he would officiate only with the Methodist text. But Coulter, her 1960s feminist consciousness freshly raised, wasn't comfortable with the patriarchal tone of it, with its "honor and obey," and its "man and wife." She was pleasantly surprised, then, when her husband-to-be, after reading the Methodist wording, told her, "I don't like it. We need to find a Presbyterian."

38 **But many people** • Wilhelms, *To a Rocky Moon,* pp. 232–43; Murray and Cox, *Apollo,* pp. 447–49; Chaikin, *Man on the Moon,* pp. 335–36, 349, 505–06.

39 **Their haul would** • Wilhelms, *To a Rocky Moon*, pp. 329, 333; Chaikin, *Man on the Moon*, pp. 521–22. Meteorite and lunar specialists Larry Nyquist and Wendell Mendell, at NASA's Johnson Space Center, told the author that, while the oldest lunar specimen reported in the published literature was found to be 4.57 billion years old, it was "an outlier." The most ancient lunar rocks tend to be in the age range of 4.42 to 4.43 billion, they said. See Marc Norman, "The Oldest Moon Rocks," *Planetary Science Research Discoveries*, April 21, 2004, at http://www.psrd.hawaii.edu/April04/lunarAnorthosites.html. The article discusses Norman's research (with Nyquist and Lars Borg), and Mendell noted that, though Norman "never cites an age older than about 4.40 billion, deep in the article he mentions 'an unexpectedly large range of ages (4.29–4.54 billion years)' from other measurements." The article suggests that "the unexpected large range of ages' on these rocks may be due to disturbances in the isotopic compositions due to the effect of impacts on the rocks over time," Mendell said. "As you can see, this is fuzzy business."

39 **Toward the end** • Wilhelms, *To a Rocky Moon*, p. 330; Chaikin, *Man on the Moon*, pp. 527–53.

40 **A nearby McDonald's** • See Stephen Harrigan, "Heaven and Earth," *Texas Monthly* (Apr. 2003): p. 128, for the height and makeup of the faux astronaut.

40 **He had become** • Author interviews with geologist Abhijit Basu (now a professor of geological sciences at Indiana University, in Bloomington), Chris Romanek, and others familiar with McKay's microscope work. Basu recalled the story of McKay fetching the engineering drawings.

40 **McKay's natural bent** • Author interview with Basu.

41 **He had accepted the** • Mimi Swartz, "It Came from Outer Space," *Texas Monthly* (Nov. 1996): p. 122.

41 **By now, McKay had** • McKay shared not only his expertise but his samples with fledgling scientists who came to work in his lab. Over the years he nurtured many, including Abhijit Basu from India. In 1975, just a few years after the lunar samples arrived on Earth, McKay gave Basu some of his—no strings attached, no stipulations about who would get credit for the work—even though McKay knew the younger man would not be coming to the Houston lab as a postdoctoral fellow. (Basu opted for Harvard instead.) Basu said in an interview with the author that he considered this an act of remarkable generosity, given how rare it was for investigators to swap the precious samples so freely.

41 **Wendell Mendell, a** • Author interviews with Mendell.

43 **Civil servants, for instance** • Author interview with Tim McCoy, Smithsonian meteorite curator.

43 **As one occasional participant** • Author interview with Basu.

44 **But McKay preferred** • Author interview with McKay.

44 **As he would remark** • Goldsmith, *Hunt for Life on Mars*, p. 69.

CHAPTER THREE: ODD DUCKS

46 **In one of these** • This chapter's account of Mittlefehldt's key research on the Mars meteorite comes primarily from author interviews with him and an account he wrote in *Planetary Report*, vol. 17, no. 1 (Jan.–Feb. 1997): pp. 8–11.

48 **And in the years** • The author thought the admirable phrase "keys that can unlock the vaults of cosmic memory" (found in her notes) came from Ralph Harvey or the website he operates for the meteorite search program, but he has denied parentage and she has been unable to determine its source.

48 **After Robbie Score's hunting** • Author interview with Cecilia Satterwhite of NASA contractor Lockheed Martin, a veteran "hands-on" meteorite curator at Johnson Space Center. She breaks space rocks for a living. Regarding the level of cleanliness in the meteorite processing lab, curator Kevin Righter told the author in an e-mail that it was a Class 10,000 facility. In 1997–1998, the air handler system would be replaced and upgraded, he said, "and it has effectively operated as Class 1,000 or better since then (even though the official rating is 10,000)." The class refers to the number of dust particles and other impurities per cubic foot that the air filters admit at or above a certain size. Class 1,000 meets a higher standard than Class 10,000, admitting fewer particles larger than 0.5 microns.

49 **Here, the curators weighed** • Author interviews with Carlton Allen and Satterwhite, of the JSC meteorite facility. See also *Antarctic Meteorite Newsletter* vol. 8, no. 2 (Aug. 1985): p. 5, and Charles Meyer, *Mars Meteorite Compendium* (Washington, D.C.: NASA/JSC, 2001), p. 107; on Web at: http://www-curator.jsc.nasa.gov/curator/antmet/mmc/84001.pdf.

49 **There, the task** • Author interviews with Glenn MacPherson, Chris Romanek, and Tim McCoy. (As of this writing, MacPherson was chair of the department of mineral sciences at the Smithsonian's National Museum of Natural History.)

50 **His finding appeared** • Interested research labs routinely received these bare-bones descriptions of the season's new meteorite arrivals by means of the newsletter circulated by the Houston archives and based on the Smithsonian analysis.

50 **But this was in part** • Author interview with MacPherson.

51 **Otherwise, his initial bulk** • The rock was mostly (90 to 95 percent) made of a silicate material (orthopyroxene) similar to the material found in other offspring of Vesta—except it was a bit richer in iron. Mittlefehldt noted that it had been banged up and there were cross-cutting veins of granular material, but there were also still patches of the original igneous rock—the material into which the primordial magma had cooled and solidified. In that sense, it was quite different from most igneous meteorites—those formed from molten rock. Most had been so banged up by impacts on the parent body that there was nothing left but fragments from the original material, a mixture of grain sizes from microns up to perhaps centimeters. (From author interviews and Mittlefehldt's reports.)

51 **But it took a** • Mittlefehldt and his coworkers would later write a paper about the lessons learned from the misclassification of meteorites. See M. M. Lindstrom, A. H. Treiman, and D. W. Mittlefehldt, "Pigeonholing Planetary Meteorites: The Lessons of Misclassification of EET87521 and ALH84001," *Lunar and Planetary Science*, vol. 25 (1994): pp. 797–98. It suggested that they and their fellow rock detectives had been too quick to "pigeonhole" rocks in rigid little categories. "If we had opened our minds, . . . might we have discovered ALH84001 sooner?" They noted that certain types of Martian meteorites, even if they fell in Antarctica, might go uncollected because the items would look so much like Earth rocks—so ordinary—to pigeonhole thinkers.

51 **He was still** • Mittlefehldt started out with a fellowship at the space center, then was hired as staff geologist for a NASA support contractor—what is now Lockheed Martin Engineering and Sciences Co. In 2001, he was hired as a NASA scientist.

53 **In the early 1980s** • William A. Cassidy, *Meteorites* (Cambridge: Cambridge University Press, 2003), pp. 159, 229. One achievement of the Antarctic meteorite hunt was the 1982 discovery of the first stone identified as having been catapulted from the moon to Earth. It showed little shock effect, suggesting that debris could, after all, be violently dislodged from a world without melting. This helped to change mainstream thinking on the matter. See also William K. Hartmann, *A Traveler's Guide to Mars* (New York: Workman, 2003), pp. 258–59, and Steven J. Dick and James E. Strick, *The Living Universe* (New Brunswick, N. J.: Rutgers University Press, 2004), pp. 180–81, regarding Bogard's work with Pratt Johnson to identify the first Martian meteorite.

53 **Bogard, working at the** • Donald Bogard and Pratt Johnson, "Martian Gases in an Antarctic Meteorite," *Science* (Aug. 12, 1983): pp. 651–54. Author interview with Bogard. See also Harry Y. McSween, Jr., *Stardust to Planets* (New York: St. Martin's, 1993), pp. 99–100. The oddball rocks, the SNCs, all resembled volcanic rock found on Earth, indicating they had experienced a similar pattern of melting and crystallization. And yet they had apparently crystallized a mere 1.3 billion years ago. This had raised suspicions that these stones could not have come from Earth's moon or from asteroids, which had cooled billions of years earlier to the point of ceasing their volcanic activity.

54 **Through "guilt by** • Michael Meyer, chief NASA astrobiologist, in interview with Steven Dick, Feb. 4, 1997, NASA archives.

54 **Humanity had finally** • Decades earlier, Gene Shoemaker and other planetary scientists had established the importance of impact cratering throughout history, while Stephen Jay Gould and other biologists published evidence of a stepwise history of evolutionary change called punctuated equilibrium. (See Gould, "The Evolution of Life on the Earth," *Scientific American*, vol. 271 [Oct. 1994]: pp. 85–91.) In 1979, Luis and Walter Alvarez and colleagues made the crucial identification of extraterrestrial material at the boundary layer separating the Cretaceous and Tertiary periods of geological history—evidence that the impact of a comet or asteroid at least six miles (ten kilometers) across had triggered the mass extinction that ended the age of the dinosaurs. Soon afterward, mainstream science accepted the notion of short-term catastrophic changes in geological and biological history. Scientists had linked the course of biological evolution on Earth to the planet's cosmic environment.

In the dinosaur extinction event, at least half of all Earth's species swiftly vanished. That extinction event cleared the decks for the advance of the weaselly mammals that were the early ancestors of humans.

The largest space rock encounter in the last century was the 1908 explosion of a huge fireball about four miles above Siberia, releasing the force of one thousand atomic bombs like the one dropped on Hiroshima and flattening evergreen forests in the Tunguska region. About once a month, according to declassified data from military satellites, some extraterrestrial object detonates at high altitude with the force of a kiloton or more of TNT.

54 **Human awareness of** • Several groups of "killer rock" specialists around the world began to mount modest efforts to detect and possibly fend off a pending catastrophic collision. They estimated at least a couple of thou-

sand objects larger than six-tenths of a mile (one kilometer) in diameter were on trajectories that could someday intersect with Earth's. So far only a fraction of those had been detected, none of them known to pose a threat, according to NASA.

The experts said it would take an object at least a mile in diameter to cause global damage and disrupt civilization. They estimated the odds that such a killer rock would smash into Earth in the next century at slightly less than one in one thousand—cause for study and alertness, but not alarm.

54 **In this context** • The idea that meteorites might deliver life to Earth had been around since 1834. William Thompson (Lord Kelvin) in his "Presidential Address to the British Association for the Advancement of Science," *Nature* (Aug. 3, 1871): pp. 262–70, made passing reference to the probability "that there are countless seed-bearing meteoric stones moving about through space"; see also Steven J. Dick, *The Biological Universe* (Cambridge: Cambridge University Press, 1996), p. 326.

The Swedish chemist Svante Arrhenius proposed a variation called panspermia in 1901, suggesting that isolated biological spores drifted among the stars. If these kernels of life fell on a suitable world, they might evolve and thrive there. Late in the twentieth century, scientists focused instead on fresh evidence that space was full of traveling bits of prebiotic chemicals—building blocks of life rather than life itself. See Dick, *Biological Universe,* pp. 324–30, 341, 348, 351, 366–77; see also Donald Goldsmith and Tobias Owen, *The Search for Life in the Universe,* 3rd ed. (Sausalito, Calif.: University Science, 2001), pp. 196–97.

A series of spaceflight experiments indicated that certain microbes were capable of surviving prolonged exposure to the vacuum of space. In the late 1980s, Gerda Horneck and colleagues at the German space organization DLR flew spores (a dormant state of *Bacillus subtilis,* a common, harmless organism found in soil and fresh water on Earth) aboard NASA's Long Duration Exposure Facility (LDEF) spacecraft, which stayed in space for six years. They found an 80 percent survival rate in the vacuum—if the spores were shielded from radiation. Beginning in 1994, microbiologist Rocco Mancinelli, of the SETI Institute in California, collaborated with Horneck using flights of the European Space Agency's Biopan experiment, launched on Russian Foton rockets. Their small experiment would suggest that hardy salt-loving microbes found in Baja California might eke out a small survival rate—just a few percent of the specimens—during a two-week exposure to the vacuum of space and ultraviolet radiation. Horneck found that the most damage, especially during the short term, came from ultraviolet radiation, "but heavy ionizing radiation had a greater probability of being lethal." See Kathy Sawyer, "Hardy Microbes Appear Able to Survive in Space," *Washington Post*, Oct. 4, 1999, p. A11; see also R. L. Mancinelli et al., "Biopan-Survival I: Exposure of the Osmophiles *Synechococcus sp.* (nageli) and *Haloarcula sp.* to the Space Environment," *Advanced Space Research*, vol. 22, no. 3 (1998): pp. 327–34; G. Horneck, G. H. Bücker and G. Reitz, "Long-Term Survival of Bacterial Spores in Space." *Advanced Space Research*, vol. 14, no. 10 (1994): pp. 41–45; and G. Horneck, "Responses of *Bacillus subtilis* Spores to Space Environment: Results from Experiments in Space," *Origins of Life*, vol. 23 (1993): pp. 37–52.

In a later study of patterns of magnetization within the Allan Hills rock as it was heated and cooled during its journey, Benjamin Weiss and Joseph Kirschvink of the California Institute of Technology used Vanderbilt University's SQUID (superconducting quantum interference device) to determine that the core of the meteorite never reached the killing temperatures felt by its exterior. The surface would have reached thousands of degrees at certain stages, but much of the melting shell would have blown away, never allowing the intense heat to reach more than a few millimeters inside. See B. Weiss, et al., "A Low Temperature Transfer of ALH84001 from Mars to Earth," *Science* [Oct. 27, 2000]: pp. 791–795.) They reported that the rock's heart got no hotter than 104 degrees Fahrenheit (40 degrees Celsius)—not much worse than a summer day in Houston, and without the humidity.

54 **As Mittlefehldt fretted** • The total of known Martian meteorites has risen to thirty-four as of this writing, and scientists are evaluating a possible thirty-fifth.

57 **A landmark report** • "Summer Study" report by the Space Science Board of the National Academy of Sciences at Iowa State University, 1962. Cited by science historian Steven Dick at George Washington University Space Policy Institute symposium, November 1996, in connection with events related in this book.

57 **To the young scientist** • Carl Sagan, *Cosmos* (New York: Ballantine, 1985), pp. 99–101, 103, 106, 107.

58 **With the two Viking** • Bruce Murray, *Journey into Space* (New York: Norton, 1989), pp. 68–73. See also Goldsmith and Owen, *Search for Life*, pp. 315, 337–47, and Sagan, *Cosmos*, pp. 101, 103–106. In order to prevent the Viking instruments from detecting Earth microbes that had hitchhiked aboard the lander, mission planners carefully sterilized the spacecraft

with heat. They assumed that Martian life, like Earth life, would be based on carbon chemistry in water and that these microbes must take in food and give off waste gases, or they must take in and convert atmospheric gases into something they could use.

Each of the two landers carried an instrument that analyzed the atmosphere and found it to be 2 to 3 percent nitrogen (nitrogen exchange between living things and atmosphere on Earth is fundamental) but detected no telltale signs of life, such as methane. The soil-analysis instruments—gas chromatographs and mass spectrometers—found zero organic compounds at both sites (and they were capable of detecting even a few parts in a billion if they were there). The mere presence of such organics would not have proven the presence of life, because they could be made in nonbiological processes.

In addition to these indirect indicators, each of the two landers carried three biology instruments specifically designed to detect certain signs of life, if present. These were:

1. The gas exchange, also known as the "chicken soup" experiment, which treated a Martian soil sample to a broth of selected nutrients known to appeal to terrestrial organisms. An instrument would then look for gas emissions that might result.
2. The labeled release, which dripped a mixture of what Murray called a "more basic, if less tasty menu" of organic compounds "labeled" with radioactive carbon atoms onto the soil sample to see whether any life-forms processed the material and released it back into the atmosphere.
3. The pyrolitic release, which moved Martian soil into a test chamber similar to the external Martian environment, but with atmospheric gases inside that had been labeled with radioactive carbon. The instrument then baked the soil and analyzed the released gases. As Goldsmith and Owen summed it up (p. 343), the third experiment "aimed at roasting the remains of Martian microbes to release carbon atoms that the microbes had incorporated through biological activity."

Each of the three biology experiments initially seemed to yield positive results—as if microbes were metabolizing the "soup," or as if respiring, photosynthesizing microbes were turning atmospheric gases into organic matter. But the signals were too strong, and the scientific team was skeptical. Further study led them to believe that the results reflected exotic inorganic chemical reactions that only mimicked the effects of biological activity. For instance, it seemed, stable water was so completely alien to Martian surface materials that the addition of even a few drops produced reactions.

58 **They'd found no organic** • In 1996, planetary geologist Bruce Murray, a veteran of Viking and other planetary missions and a former director of the Jet Propulsion Laboratory, would tell a symposium that Viking was "a complete success because it unexpectedly showed the surface of Mars is self-sterilizing. . . . Viking produced a powerful result."

58 **A few scientists** • In the mid-1990s, there was a resurgence of interest in reanalyzing some of the Viking data, but NASA officials preferred to focus on new instruments and conducting related research to "be better placed so that when data does come back from Mars, you have a better system to work with" (NASA chief exobiologist Michael Meyer interview with Steven Dick, Feb. 4, 1997, NASA archives).

58 **In October 1993** • D. W. Mittlefehldt, "ALH84001, a Cumulate Orthopyroxenite Member of the SNC Meteorite Group," *Meteoritics*, vol. 29 (1994a): pp. 214–21. When Mittlefehldt told the meteorite curator about his discovery, she urged him to get one more confirmation, from Bob Clayton, at the University of Chicago. Clayton was recognized as one of the leading experts on isotopic compositions of meteorites, and had developed key approaches for distinguishing Martian meteorites from all others. His laboratory confirmed that the rock had the unique Martian oxygen isotope fingerprint.

Mittlefehldt's paper got processed with unusual speed: it was received by the journal on December 3, 1993, and accepted after revisions on December 21.

59 **Score, working in** • Author interview with Score.

59 **This was the first** • Ron Cowen, "The Case of the Misclassified Meteorite," *Science News*, vol. 145 (1994): p. 206. See also Cassidy, *Meteorites*, p. 121. Mittlefehldt told *Science News* that the meteorite apparently incorporated its carbon dioxide at higher temperatures than the other SNCs, suggesting that it could have acquired the gas "from magma fluids. The higher pressure below the surface probably prevented more carbon dioxide from bubbling out of the rocky body; a significant fraction remained trapped" as carbonates.

59 **Geologists in Germany** • Meyer, *Mars Meteorite Compendium*, p. 116. See also Donald Goldsmith, *The Hunt for Life on Mars* (New York: Dutton/Penguin, 1997), p. 53. Emil Jagoutz and colleagues at the Max Planck Institute for Chemistry in Mainz, Germany, determined the rock's age. See E. Jagoutz et al., "ALH84001: Alien or Progenitor of the SNC Family?" (abstract), *Meteoritics*, vol. 29 (1994): pp. 478-79.

Isotopes, stable and unstable, would be important in the drama of the rock. The Germans' "clock" was a variation on the familiar carbon-dating technique, a way of figuring out the age of certain materials based on the natural tendency of atoms, like people, to seek the most comfortable balance of forces within them—that is, to seek stability. If some traumatic event renders them unstable, atoms try to regain their balance by giving birth to offspring. Each type of unstable atom produces offspring at such a predictable rate that you can set a clock by them. Carbon dating is used to calculate the ages of archaeological finds, hair, cloth, bone, plant fibers, and other materials—but it only works as far back as 50,000 years or so. Other elements are required to measure cosmic time scales.

Each chemical element is defined by the number of protons in the nucleus of its atom. This number determines how it interacts chemically with other elements. The number of neutrons, on the other hand, determines atomic weight and which isotope of an element it is.

All the isotopes of a given element behave the same *chemically*. But they behave differently at the level of *nuclear* interactions, which occur under certain violent and usually intensely hot conditions. In those events, naked atomic nuclei actually collide and the type of isotope does make a difference.

Mittlefehldt's Martian meteorite apparently hardened out of a hot flow of volcanic magma, whose intense heat would have sent inert gas bubbling out into the atmosphere while unstable (radioactive) isotopes formed in the cooling, hardening material. In the decay process, typically, the unstable "parent" isotope would give up part of its atomic nucleus in the form of energy (an alpha- or beta-ray emission, for instance) and form another isotope—the "daughter." The rate of decay would be expressed in terms of the isotope's half-life, a known unit of time. This is the number of years it takes for half the original radioactive isotope supply to turn into the daughter substance.

The clock in this case started when the rock crystallized on Mars, marking the last moment it harbored parents with no offspring. As the unstable parents decayed and produced stable daughters, the resulting gas would be trapped in the rock, unable to escape (until the rock was again heated to the melting point).

To pin the time the rock crystallized at 4.5 billion years ago, the German group compared the amount of rubidium 87, the parent isotope, with the amount of strontium 87, the daughter isotope. (This parent was known to decay into the daughter with the unimaginable half-life of 49 billion years. The age of the known universe is currently put at no more than about 14 billion years.) In this case, other isotopic data also indicated that the rock had been partly remelted some half a billion years after it first crystallized.

In contrast to unstable isotopes, stable isotopes stay the same over billions of years. They are prevalent in nature. People study stable isotopes not to find out the age of something but to pinpoint its birthplace. Various places, on Earth and in the universe, have different ratios of isotopes of a given element—an isotopic signature. This isotopic diversity occurs because minerals, water, and gases favor one isotope over another, or because living things can metabolize one type of isotope more efficiently. This is what tied all the Martian meteorites together—a common isotopic signature found in the Martian atmosphere and nowhere else. Meteorites from the moon have been identified using the same technique, based on Apollo lunar samples.

60 **They estimated how** • See Meyer, *Meteorite Compendium*, p. 116. See also A. J. T. Jull et al., "^{14}C Terrestrial Ages of Achondrites from Victoria Land, Antarctica," (abstract), *Lunar and Planetary Science*, 25 (1994a): pp. 647–48, and Jull et al., "Isotopic Composition of Carbonates in the SNC Meteorites Allan Hills 84001 and Nakhla," *Meteoritics*, vol. 30 (1995): pp. 311–18.

60 **Some researchers studied** • Cassidy, *Meteorites*, p. 127, citing the work of Nadine Barlow, University of Central Florida, "Identification of Possible Source Craters for Martian Meteorite ALH84001," *Proceedings of the SPIE*, vol. 3111 (1997): pp. 26–35.

Analyses of how material got ejected off Mars had shown that near-vertical impacts producing craters larger than sixty-two miles (one hundred kilometers) in diameter and low-angle impacts (less than fifteen degrees relative to the horizon) that created elongated craters greater than about ten kilometers were the only ones that met the criteria. One such site was in the Sinus Sabaeus region of Mars, south of the Schiaparelli Basin, some fourteen degrees south of the equator, with the possible river channel on its northern side, according to Barlow. Another possibility was a sharp-rimmed crater east of the Hesperia Planitia region, twelve degrees south.

Specialists at the Jet Propulsion Lab in Pasadena estimated that the rock had to leave Mars at more than 11,000 miles per hour (about five kilometers per second) to escape Martian gravity.

61 **And in time** • Cassidy, *Meteorites,* p. 126.

61 **Mittlefehldt and others** • The carbonates in the rock were of several "types": *calcium* carbonate, *iron* carbonate, *magnesium* carbonate, siderite, magnetite, and calcide. The researchers recognized the importance of the carbonates, at this point, as the first abundant substance anyone on Earth had ever seen that resulted from the interactions of rock with the Martian environment.

Various evidence indicated the age of the carbonates at between 3.6 billion and 1.3 billion years, but most investigators leaned toward the older age because Mars began to lose its water and its atmosphere around 3 billion years ago.

61 **Like several others** • Author interview with Mittlefehldt.

61 **Late one evening** • Author interviews with Mittlefehldt and Romanek.

62 **(The intriguing carbonate** • *Mars Meteorite Compendium,* NASA/JSC (2001) on the Web at http://curator.jsc.nasa.gov/curator/antmet/mmc/mmc.htm, p. 123.

62 **Slender, athletic** • Author interviews with Romanek.

64 **It was only** • As noted earlier, Bogard and colleagues, in 1982, had been the first to identify a meteorite as Martian. See McSween, *Stardust to Planets,* pp. 99–100.

64 **The Murchison meteorite** • Regarding the discovery of amino acids in the meteorite and its significance, see Christopher Wills and Jeffrey Bada, *The Spark of Life: Darwin and the Primeval Soup,* pp. 88–92 and 120–21; also Dick and Strick, *Living Universe,* pp. 75–79; and Goldsmith, *Hunt for Life on Mars,* pp. 148–49.

Amino acids come in left-handed and right-handed types, each a mirror image of the other. All life on Earth selects the left-handed variety exclusively—a state of affairs that allows chemical reactions to work more efficiently. This property of handedness is thought to provide a key signature for distinguishing biologically generated stuff from nonbiological. In theory, a living system would pick either right- or left-handed molecules to run on, but *nonliving* systems are just as happy with equal numbers of each. That's what was found in the case of the Murchison meteorite—some of each. The discovery in Murchison showed, among other things, that these building blocks of life had been manufactured by purely chemical processes somewhere besides Earth.

65 **Romanek, in this case** • Goldsmith, *Hunt for Life on Mars*, pp. 49–57. As noted earlier, people study stable isotopes not to find out the age of something but to pinpoint its birthplace. Various places, on Earth, Mars, and elsewhere in the universe, have varying ratios of isotopes of a given element—an isotopic signature.

65 **In the course of the** • Author interviews with Romanek.

CHAPTER FOUR: E.T.'S HANDSHAKE

67 **Everett Gibson, the** • Author interviews with Gibson.

67 **For sport, Gibson** • Every fall, Gibson (although he had not served in the military) coordinated the military portion of the fourth-largest air show in the country, Wings over Houston, at nearby Ellington Field.

67 **These attributes** • E-mailed list from office of NASA/JSC meteorite curator Carlton Allen (compiled by database analyst Terrie Bevill) at author's request (April 22, 2002); also, Romanek provided to author a list of sample numbers, recipients, and issue dates sent by Robbie Score to Romanek (Aug. 8, 1994).

69 **Gibson considered** • More on the story told by the isotopes in the rock: the carbonates held a signature of the gases (the carbon dioxide atmosphere of Mars, presumably) that had interacted with the fluids that flowed through the rock. Over time, as the chemical interactions reached equilibrium, they should have set up a particular ratio of the carbon isotopes and a particular ratio of the oxygen isotopes—like fingerprints of sorts. The carbon isotopes told one part of the story, and the oxygen isotopes another.

THE OXYGEN ISOTOPES

From the oxygen ratio, researchers could get a handle on the temperature at which the carbonates formed. (If you cool water down it becomes enriched in the light isotope. If you boil water, the light isotope escapes, leaving the water enriched in the heavy isotope.) Romanek and Gibson, and their British colleagues, found that the sample was enriched in the lighter oxygen isotopes. Using a computer model to reveal the conditions that likely would have produced this particular isotopic signature, they concluded that the rock was probably formed between freezing and boiling temperatures—a range within which life could exist. Later, others would interpret the same evidence differently, sometimes concluding the temperatures were much too high for life. Eventually, the consensus would settle toward the milder end of the temperature scale and the Gibson-Romanek position.

Romanek and Gibson initially assumed there was a state of equilibrium. (If you "cooked" the material at a certain temperature and pressure over time, the stuff should assume a certain stable mineral composition because there were no longer chemical reactions going on.) It later became clear that the minerals had formed under conditions decidedly *out of* equilibrium. One way to achieve such a state is to more or less freeze the ingredients very quickly. You "catch them in the act," as McKay would put it later, of changing from one set of minerals to another set of minerals. Another way to achieve it is biologically. Biological systems are typically out of equilibrium. They can manipulate their environment. The McKay team, investigating this possibility, looked at experiments done by Henry Chafetz, of the University of Houston, who had been growing carbonates out of equilibrium. He proposed that the carbonates grew only when bacteria were present. The McKay group found the Chafetz evidence remarkably similar to what they were seeing in the Mars rock.

THE CARBON STORY

When the researchers measured the carbon 12–to–carbon 13 ratio, they found out it was enriched in the carbon 13 fraction—heavy carbon. Until then, researchers had not known very well the carbon isotopic composition for Mars's atmosphere (which had reacted with the fluids in the rock).

The ratio was measured in parts per thousand ("mils," in the jargon). As the number deviated from the terrestrial standard of zero, it showed that a given mix was either enriched in heavy carbon (a "plus" value) or depleted in heavy carbon (a "minus" value). The range for Earth carbonates went only as high as about plus 20 parts per thousand—and that was only in very unusual situations. What Romanek and Gibson (and the group in England) saw in the Mars specimen was a value of "plus 40" parts per thousand—well beyond the range of anything resulting from natural processes on Earth. The way the carbonates were enriched in heavy carbon suggested that the light carbon—carbon 12—had perhaps been stripped out of the Martian atmosphere early in the planet's history.

But the carbon also told another story. From the carbon, it was possible to find out whether a substance was associated with life or not by the way carbon atoms were known to be used by living systems, whether plants, critters, or something else. The numbers told the investigators that this particular fraction of the carbonate was *not* a type of carbon associated with any known living matter, because living systems on Earth produced ratios that were typically down around minus 25 to 30 parts per thousand.

So the carbon in these measurements did not reflect biological origins, but did signal that it was Martian.

69 **He knew that a** • Author interviews with Gibson: In 1984–85, when Gibson wanted to learn more about stable isotopes, he was awarded a Leverhulme Fellowship for study in England. He sought out Colin Pillinger, of Britain's Open University, whose strong suit was isotopic analysis. Gibson studied with Pillinger for more than a year.

69 **The journal *Nature*** • In July 1994, after the transatlantic phone conversation, Gibson and Pillinger and their coworkers decided to work together to publish the isotope findings.

71 **Not many people** • J. H. Alton, J. R. Bagby Jr., and P. D. Stabekis, "Lessons Learned During Apollo Lunar Sample Quarantine and Sample Curation," *Advanced Space Research* 22, no. 3 (1995): pp. 373–82.

71 **In the mid-1980s** • Author interviews with Gibson. In the season of 1979–80, Gibson joined the meteorite hunt in Antarctica. But he was yanked out of the field in midseason when word came that his infant son had been diagnosed with spinal meningitis. Gibson and his wife, a biologist, turned much of their focus to their son's health.

72 **Gibson and the others** • See H. R. Karlsson, R. N. Clayton, E. K. Gibson, Jr., and T. K. Mayeda, "Water in SNC Meteorites: Evidence for a Martian Hydrosphere," *Science*, vol. 255 (1992): pp. 1409–11. Haraldur Karlsson had worked with Gibson as a postdoctoral fellow. And it was the expertise of one of the paper's coauthors, Robert Clayton of the University of Chicago, that also confirmed Duck Mittlefehldt's discovery of the strange new Martian meteorite the following year.

73 **Romanek had in mind** • Author interviews with Romanek; see also Michael Ray Taylor, *Dark Life* (New York: Scribner, 1999), p. 104.

73 **When Folk had** • Taylor, *Dark Life*, pp. 20, 37–45. (Folk preferred the term *nannobacteria*, but that spelling didn't catch on.)

75 **Led by a single** • Steven J. Dick, *Biological Universe* (Cambridge: Cambridge University Press, 1996), pp. 462–70. After years of struggle for money, NASA's SETI began operations in October 1992, and in September 1993, Richard Bryan of Nevada successfully led a move to terminate what he called (misleadingly) "the Great Martian Chase." SETI's supporters soon

revived remnants of the project by enlisting private-sector money. In the early 1990s, NASA had sponsored a series of workshops where a group of some twenty historians, scientists, behavioral scientists, and government policy experts discussed the implications of actual contact. But the report appeared only in preprint form and was not widely seen. (Steven Dick, in a talk delivered at George Washington University symposium, Nov. 1996, referencing "Social Implications of Detecting an Extraterrestrial Civilization: A Report of the Workshop on the Cultural Aspects of SETI" [Mountain View, Calif.: SETI Institute preprint, 1994]. The report was published in 1999 as *Social Implications of the Detection of an Extraterrestrial Civilization*, available for purchase at: http://www.seti.org/site/pp.asp?c=ktJ2J9MMIsE&b=180343.

75 **Legitimate research** • Dick, *Biological Universe*, pp. 141, 354–55, 370, 377–78. U.S. life sciences research relating to the possibility of extraterrestrial life (as opposed to astronaut health and survival issues) since 1960 had been centered at NASA's Ames Research Center in Mountain View, California, and the effort attracted the support of prominent scientists. (The Ames program included studies of the origins of life on Earth.) But the research was assailed repeatedly over the years, and even more so after the Viking probes for biology on Mars showed no persuasive signs. Evolutionary biologist George Gaylord Simpson, in 1964, had said exobiology was a "science" that "has yet to demonstrate its subject matter exists." Physicist Frank Tipler in 1987 compared bioastronomy to parapsychology, calling it a "pseudoscience" that should not be given institutional respectability.

75 **In 1977, three** • Victoria A. Kaharl, *Waterbaby: The Story of Alvin* (New York: Oxford University Press, 1990), p. 173, cited in William J. Broad, *The Universe Below* (New York: Simon & Schuster, 1997), pp. 105–06.

76 **People learned from these** • In May 2002, on the twenty-fifth anniversary of the discovery of hydrothermal vents with associated life-forms, the Woods Hole Oceanographic Institute, the National Oceanographic and Atmospheric Administration, and the National Science Foundation summarized the significance on a CD-ROM, *The Discovery of Hydrothermal Vents.* The author derived some descriptions from her own descent aboard *Alvin*, in the summer of 1998 with researchers on an expedition to the Juan de Fuca Ridge in the Pacific. Under the auspices of the National Science Foundation's Life in Extreme Environments program, the researchers were studying extreme life-forms around the vents and potentially below the seafloor.

76 **The microorganisms at** • Instead of photosynthesis, they used chemosynthesis.

76 **In the last quarter of the** • Woods Hole Oceanographic Institute et al., May 2002 summary.

77 **(More than a decade later** • See Glennda Chui, "Study Surveys Human Intestines," San Jose *Mercury News* wire service, posted April 15, 2005, at: http://www.miami.com/mld/mercurynews/news/local/states/california/peninsula/11401088.htm.

77 **Some of these organisms** • The traditional system of naming and classifying species (kingdom, phylum, class, order, etc.) has been evolving to take advantage of advances in evolutionary biology. The new, improved system is based on the comparison of the sequences of information-bearing molecules, such as DNA or proteins, from different organisms in which the molecules carry out the same function. See Christian de Duve, *Life Evolving: Molecules, Mind, and Meaning* (Oxford: Oxford University Press, 2002), pp. 100–04. The three domains of life are Archaea (discovered late in the twentieth century), Bacteria, and Eucarya. The first two are made up of prokaryotes, whose cells lack nuclei. Of all the domains, only a small fraction of the Eucarya branch is made up of "the subject matter of a conventional biology course, organisms large enough to be visible without a microscope," as noted by Donald Goldsmith and Tobias Owen, *The Search for Life* (Sausalito, Calif.: University Science, 2002), p. 214, Figure 9.1 (illustration of the three domains).

77 **One was the emerging** • J. William Schopf, *Cradle of Life* (Princeton, N.J.: Princeton University Press, 1999), pp. 166–67.

77 **Just months before Romanek** • Schopf published articles in 1992 and 1993 describing the discovery of the oldest known fossils. See J. W. Schopf, "Times of Origin and Earliest Evidence of Major Biologic Groups," in J. W. Schopf and C. Klein, eds., *The Proterozoic Biosphere: A Multidisciplinary Study* (New York: Cambridge University Press, 1992), pp. 587–93; and J. W. Schopf, "Microfossils of the Early Archean Apex Chert: New Evidence of the Antiquity of Life," *Science*, vol. 260 (April 30, 1993), pp. 640–46.

78 **Many people had** • Dick, *Biological Universe*, p. 331.

79 **What Mittlefehldt had** • Mittlefehldt told the author that he believes this initial allotment contained two different types of samples: four chips (which would be destroyed by such methods as Romanek's acid etching) and a thin section that could be used over and over, like a library book, for nondestructive microscope viewing. Records provided to Romanek by Robbie Score from the curation lab (faxed Aug. 8, 1994) indicate that Mit-

tlefehldt had received twenty samples of the rock by this time, the first one in 1987 and all the rest between November 1993 and March 1994.

79 **Mittlefehldt and meteorite curator** • Author interviews with Mittlefehldt and McKay.

79 **In most cases** • Author interview with Mittlefehldt.

79 **This afternoon in the late** • Author interviews with Gibson, Romanek, and McKay.

81 **When the staff tried** • Charles Meyer, *Mars Meteorite Compendium* (Washington, D.C.: NASA/JSC, 2001), p. 126; illustrations, pp. 120–22. (See also online site: http://www-curator.jsc.nasa.gov/curator/antmet/mmc/84001.pdf.) With photo and graphics, the curators have depicted the sequence in which samples were chipped or sawed out of the main body of the meteorite.

81 **But by the end of** • Curators records, as of August 8, 1994 (list faxed from Score to Romanek).

82 **It was only a few** • Author interviews with McKay and Gibson. Also McKay interview with Dick, NASA archives. Both McKay and Gibson had received NASA exobiology funding before this, and as they set up their collaboration they applied for a new grant to look for signs of biology in the Allan Hills rock and other meteorites from Mars. NASA rejected the proposal the first time but in the summer of 1996 approved the group's second request.

CHAPTER FIVE: THE CONVERT

83 **She was at work** • Accounts in this chapter rely primarily on the author's interviews with McKay, Gibson, Romanek, and Thomas, and an e-mail exchange with Dennis Bazylinski, a specialist in magnet-making bacteria.

86 **Her analysis held up** • See K. L. Thomas, B. E. Blanford, C. S. Clemett, G. J. Flynn, L. P. Keller, W. Klock, C. R. Maechling, D. S. McKay, S. Messenger, A. O. Nier, D. J. Schlutter, S. R. Sutton, J. L. Warren, and R. N. Zare, "An Asteroidal Breccia: The Anatomy of a Cluster IDP," *Geochimica et Cosmochimica Acta,* vol. 59 (1995), pp. 2797–2815.

86 **She would run into** • E-mail from Thomas-Keprta to author, April 21, 2005.

89 **Microbiologists had known** • The magnetic crystals Thomas detected in the rock had about the right size (40 to 50 nanometers), the right shapes (roughly cubical or teardrop), and the right chemistry, in the McKay group's view, to have been made by living organisms.

On Earth, *magnetotactic bacteria* churn out a line of nearly identical little magnets—virtually flawless compass needles. The microbes use them to navigate up and down through the water column to find food. These biomagnets derive their power from their iron atoms. (Such minerals typically are compounds of iron and something else, usually oxygen or sulfur.) These bacteria manufacture either magnetite (iron and oxygen) or greigite (iron and sulfur). What Thomas and the others found in the Mars rock were both kinds—iron oxides and iron sulfides.

For early descriptions of magnetite-producing bacteria, see S. Mann et al., "Structure, Morphology and Crystal Growth of Bacterial Magnetite," *Nature*, vol. 310 (1984), pp. 405–7. For a review on the magnetotactic bacteria see D. A. Bazylinski and R. B. Frankel, "Magnetosome Formation in Prokaryotes," *Nature Reviews in Microbiology*, vol. 2 (2004), pp. 217–30.

89 **Keenly aware that** • Author interviews with McKay and colleagues. See also Michael Ray Taylor, *Dark Life* (New York: Scribner, 1999), pp. 94–95. Among the experts whose published writings the team consulted were Chris McKay, Jack Farmer, and Michael Carr, on exobiology (a field soon to be expanded under the term *astrobiology*); Imre Friedmann on microbial communities in Antarctic rocks and the Sinai Desert; Henry Chafetz on bacteria in carbonates; Carl Woese and Norm Pace on genetic studies of ancient microbes; Thomas Gold's *The Deep Hot Biosphere* (New York: Copernicus, 1998); James McKinley and Todd Stevens, on microorganisms extracted from deep drill holes in Washington State's Columbia River basalt; and E. Olavi Kajander, on (claimed but controversial) nanobacteria in blood.

89 **McKay approached Hojatollah Vali** • Author interview with Vali. The National Research Council funded his fellowship.

89 **One night, feeling guilty** • The first type was magnetite—iron combined with oxygen. This kind appeared to be iron with sulfur—greigite.

90 **The array surprised** • Aside from the temperature questions, the presence of *carbonates* implied that the native solution from which the magnetic crystals formed was one of low acidity (alkaline). The presence of the minerals found *inside* the carbonates, by contrast, implied an acidic solution. Ordinarily, the iron oxide crystals (magnetite) would require an oxidizing environment, while the iron sulfide (greigite) would need the opposite, a reducing environment. Vali told the author that seeing these two types of crystal together raised the suspicion that biology might have been involved as a mediator.

90 **McKay and his team** • Author interviews with the team. See also Donald Goldsmith, *The Hunt for Life on Mars* (New York: Dutton/Penguin, 1997), p. 81.

90 **At the same time** • The McKay group studied papers published by others who argued that the key features in the rock had formed at temperatures too high for life. The McKay group reinvestigated and concluded the high-temperature arguments could not be right. The numbers didn't fit the profile, so to speak. They compared the data to high-temperature "phase diagrams" worked out earlier by Eric Essene at the University of Minnesota. The McKay group stuck by its low-temperature scenario.

91 **Gibson called and invited** • Author interviews with Gibson.

91 **(Schopf would later** • J. William Schopf, *Cradle of Life* (Princeton, N.J.: Princeton University Press, 1999), p. 304. In an e-mail to the author, May 6, 2005, Schopf clarified, saying that NASA's chief exobiologist Michael Meyer had called him first. "Only after he asked me to go to JSC did folks from there call me," Schopf said.

91 **When Schopf arrived** • Ibid., p. 304.

91 **The group had sworn** • Author interview with Gibson; e-mail to author from Schopf.

91 **Romanek would later remark** • Author interview with Romanek.

91 **Schopf told them** • E-mail from Schopf to author. Author interview with Gibson.

92 **Schopf would write** • Schopf, *Cradle of Life*, p. 304.

92 **Schopf's most disheartening** • Author interview with Romanek.

CHAPTER SIX: MICKEY, MINNIE, AND GOOFY

94 **In late 1994 and** • Author interviews with Zare were a primary source of material for this chapter (except for assurances of his influence and achievements, which came from his colleagues, officials at the National Science Foundation, NASA, and others). In addition, Zare supplied in writing his impressions of events related to the Mars rock. Some descriptions came from a magazine feature on Zare (James Shreeve, "The Light on Life," *Discover* [May 1997]: pp. 50–55), as well as from material supplied by Stanford, and from author interviews with associates of Zare's, including David Salisbury and Simon Clemett.

94 **A world-renowned laser** • The "towering figure" quote is from Nobel laureate Dudley Herschbach, of Harvard, under whom Zare studied for his

graduate degree. (See Janet Basu, "Fourth Rock from the Sun," *Stanford Today*, Nov.–Dec. 1996.)

94 **Zare had never** • Zare's unpublished written account.

94 **He had enlightened** • *Stanford Today*, Nov.–Dec. 1996.

95 **A few years earlier** • Zare's written account; see also James Shreeve, "The Light on Life," pp. 50–55.

95 **One of his grad** • Author interviews with Zare and Clemett; Zare's written account.

95 **Among those who** • Author interviews with McKay, Thomas, and Romanek.

96 **Thomas enjoyed her** • Interviews with Thomas and Romanek.

96 **Romanek and Thomas had** • Letter, with hand-drawn sketches of the samples, sent by Romanek and Thomas to Simon Clemett at Zarelab, November 18, 1994. The letter accompanied the sample shipment.

96 **Clemett found the whole** • Author interviews with Zare and Clemett.

96 **The Zarelab technique** • Zare worked on the technique initially with colleagues Yan Kuhn Hahn and Renato Zenobi.

96 **Clemett called it "chemistry** • Author interview with Clemett; quote also cited in Donald Goldsmith, *The Hunt for Life on Mars* (New York: Dutton/Penguin, 1997), p. 84.

96 **In simple terms** • Author interviews with Zare and Clemett, and written descriptions from them and from David Salisbury (then at Stanford News Service). The device adapted for the meteorite samples was the Microprobe Two-Step Laser Mass Spectrometer, sometimes called by the even knottier monicker Two-Step Laser Desorption Ionization Time-of-Flight Mass Spectrometer. The effect of the instrument's heat flash was to remove the whole target molecule intact, before the chemical bonds could break, and convert the molecule directly into gaseous form. (Nobody understood why this worked so well, Zare said.)

Then the second beam fired. The beam knocked electrons off a designated class of molecules, giving them an electrical charge. By creating an electric field within the vacuum chamber, the researchers could accelerate the selected molecules to velocities approaching a hundred miles per second—into a wall. A detector measured the travel times, enabling the researchers to then "weigh" the mass of each molecule. (The shorter the flight time, the lighter the molecular mass.) They would end up with a kind of census of the molecule types by weight. The ultraviolet laser could be tuned to pick out selected families of organic molecules: amino acids, DNA bases, or PAHs. Combining this selectivity with the mass measurements, the researchers could very accurately determine the relative abundance of various organic compounds.

The technique's unprecedented sensitivity posed a problem. At first, it tended to "sample itself," as Clemett put it. It measured minute amounts of pump oil, fingerprints, and other organic contaminants in the instrument itself. After much effort, the team developed ways of eliminating potential sources of contamination.

97 **When Clemett pushed** • Clemett demonstrated the device for the author. Clemett worked on the Mars meteorite with postdoctoral fellows Xavier Chillier and Claude Maechling.

97 **Kathie Thomas phoned** • Author interviews with Thomas, Romanek, Zare, and Clemett.

98 **Zare froze momentarily** • Zare's written account; author interview with Zare.

98 **As a boy, Zare** • Author interviews with Zare, Zare's written account; see also Shreeve, "The Light on Life," pp. 50–55.

98 **By the age of twenty-four** • *Stanford Today*, Nov.–Dec. 1996; Zare earned his graduate degree in the lab of Dudley Herschbach, who with two others won the Nobel Prize for chemistry.

99 **It was this spatial** • Author interviews with McKay and other team members.

100 **The investigators also** • Author interview with Clemett; see also Goldsmith, *The Hunt for Life on Mars*, pp. 88–90. Chemically, PAHs (polycyclic aromatic hydrocarbons) are made up of hydrogen and carbon atoms, arranged in structures called benzene rings. They have the distinctive odor associated with the colorless and highly flammable liquid called benzene (which is derived from coal and used as a solvent). Benzene rings are a popular favorite among all known forms of life. In fact, the field of organic chemistry is essentially the study of such ring-shaped carbon compounds. Living things exploit carbon's friendly chemical handshake—its unique ability to form large, complex molecules in which other elements are bonded to its atoms.

100 **They caught up with** • Michael Meyer interview with Steven J. Dick, NASA archives. Regarding exobiology and NASA, see Steven J. Dick, *The Biological Universe* (Cambridge: Cambridge University Press, 1996), pp. 356, 477–78. In the 1960s, NASA had started funneling modest funds to exobiology studies because of the agency's decision to search for life on Mars.

The work, concentrated at its Ames Research Center in California, was heavily oriented toward research on the origins of life.

100 **In their presentation** • Author interviews and e-mail exchanges with Kathie Thomas. She said, "I regret that Simon [Clemett] is not the first author" on the paper reporting the first significant Martian organics: K. L. Thomas, C. S. Romanek, S. J. Clemett, E. K. Gibson, D. S. McKay, C. R. Maechling and R. N. Zare, "Preliminary Analysis of Polycyclic Aromatic Hydrocarbons in the Martian Meteorite (SNC) ALH84001," (abstract), *Lunar and Planetary Science Conference*, vol. 26 (1995), pp. 1409–1410.

100 **The reaction was** • Vincent Kiernan, "The Mars Meteorite: A Case Study in Controls on Dissemination of Science News," *Public Understanding of Science* 9 (2000): pp. 21–22; Kiernan cites R. Cowen, "Mars Meteorite Poses Puzzling Questions," *Science News* (March 25, 1995): p. 180; "A Chip Off the Old Mars," *Sky and Telescope* (July 1995): p. 12; C. Byars, "Mars Meteorite Contains Carbon Compounds," *Houston Chronicle*, March 18, 1995, p. A34; and K. Davidson, "Meteorite May Hold Secret That There Was Life on Mars," *San Francisco Examiner*, March 16, 1995, p. A4. Everett Gibson, in an interview with the author, credited the *Chronicle*'s Byars with being the first reporter to hint that the group might be thinking in terms of possible Martian biology, based on his coverage of this meeting.

101 **They seemed to be** • Author interviews with McKay and Zare. See also Goldsmith, *Hunt for Life on Mars*, p. 114. The composition of the PAHs in the meteorite was consistent with what the scientists would expect from the fossilization of very primitive microorganisms. On Earth, PAHs were known to occur in thousands of forms, but in the Martian meteorite they were dominated by only about a half dozen different compounds. The simplicity of this mix, plus the lack of lightweight PAHs (such as naphthalene), made them substantially different from the PAHs measured in non-Martian meteorites.

101 **Zarelab had found** • In 1989, the British group led by Ian P. Wright and including Gibson's friend Colin Pillinger had reported the detection of "nonspecific carbonaceous material" in another Martian meteorite, EETA79001. (This was the SNC in which trapped gases had first been found to match the composition of Mars's atmosphere.) The findings represented the first tentative detection of Martian organic material, and they generated a column by Isaac Asimov (Los Angeles Times Syndicate, Aug. 28, 1989). However, many researchers dismissed the results on grounds that the material was not Martian but terrestrial contamination. Later,

when the McKay team published the Zarelab results, they would be criticized for not noting the British group's earlier work. Subsequently some of the same scientists who had dismissed the earlier findings would attack the new ones on similar grounds. See Steven J. Dick and James E. Strick, *The Living Universe* (New Brunswick, N.J.: Rutgers University Press, 2004), pp. 195–96.

101 **In early 1995** • Author interviews with Romanek, Gibson, and McKay.

101 **To the concern** • McKay in press conference, August 7, 1996, and in author interviews.

104 **The instrument McKay** • Author interviews with McKay and McCoy.

104 **When he looked at** • Author interviews with McKay, Gibson, and Romanek. The new microscope made it possible to look at freshly fractured surfaces of samples from the rock, which had never been exposed to harsh chemicals. It also made it possible to apply a thinner, more fine-grained coating from a wider menu of coating materials, thus reducing the possibility that the coating itself would create shapes that might be interpreted as indigenous.

105 **The little band knew** • Author interviews with McKay and other team members.

107 **McKay and the others** • Author interviews with McKay and other team members; see also McKay interview with Steven J. Dick, NASA archives.

107 **In Palo Alto, the** • Author interviews with Zare; Zare's written account.

107 **A year earlier** • Author interviews with McKay and other team members, and with Huntoon. Among others informed at JSC was division chief Douglas Blanchard.

108 **But the group had** • Author interviews with Gibson.

108 **Outside the cone of** • Author interviews with Tim McCoy and Ralph Harvey. McCoy, curator of meteorites at the Smithsonian's National Museum of Natural History, was a postdoc working at Johnson Space Center, in Building 31, between 1994 and 1996. Harvey, of Case Western Reserve University in Cleveland and head of the U.S. Antarctic Search for Meteorites, was a visiting faculty fellow at Johnson Space Center in 1996, as the McKay group was preparing to publish. He had the office next to Gibson's in Building 31. His NASA sponsor was Gordon McKay, David's brother. He was researching the same Martian meteorite as the McKay group and would soon become a leading critic of the McKay hypothesis.

108 **An odd breach** • Steve McVicker, "The Xerox Files," *Houston Press*, April 17–23, 1997.

109 **In the run-up** • Author interviews with Gibson and other team members. See also *Newsweek* (Feb. 17, 1997), p. 56, for Gibson quote. The event was the annual meeting of the Lunar and Planetary Institute. The retraction of the paper drew both criticism and praise. Abhijit Basu, one of the people McKay had mentored in the 1970s, was now a researcher and professor who viewed the withdrawal of this paper as the proper restraint of "a true scientist." It was his impression that McKay had withdrawn the paper because he needed more data before going public. Basu would tell the story to his students as an example of the way the process should work: through verifying and reverifying. There was no such thing as "truth" in science, he would say. It's all conjecture—until it's refuted.

109 **At a gathering** • Author interview with the Smithsonian's Tim McCoy (who worked as a postdoc in Building 31 in 1994–96). The Berlin event was a meeting of the Meteoritical Society, one of two major annual meetings of meteorite scientists, the other being the NASA-funded Lunar and Planetary Institute's gatherings in Houston.

109 **Later that summer** • Author interviews with Ralph Harvey and Chris Romanek.

110 **What no outsiders** • Author interviews with McKay and other team members.

111 **In the abstract** • Draft document, with edits, June 17, 1996, provided to the author by David Salisbury of Stanford.

111 **On April 4, 1996** • Author interviews with Gibson and members of the staff of *Science.* Regarding Sagan's participation, see also William Sheehan and Stephen James O'Meara, *Mars* (Amherst, N.Y.: Prometheus Books, 2001), pp. 295–98; and William Poundstone, *Carl Sagan* (New York: Henry Holt, 1999), p. 379.

111 **Two years earlier** • Carl Sagan's *Pale Blue Dot* (New York: Random House, 1994), p. 242; also cited in Poundstone, *Carl Sagan*, p. 379.

111 **All through the summer** • Author interviews with McKay and members of his team, and *Science* staff members.

111 **Most of the assessments** • Author interviews with Gibson and *Science* staff members.

112 **They knew from the comments** • See David Salisbury, "Life on Mars: Reviewing the Evidence 9 Months Later," *Stanford News Service*, May 28, 1997, at: http://news-service.stanford.edu/news/1997/may28/mars.html.

113 **James Hartsfield, Johnson** • Author interview with James Hartsfield, of Johnson Space Center public affairs, and other NASA public affairs offi-

cers, and related memos and drafts of press releases showing edits, including draft release and memo from July 19, 1996: "Meteorite Yields Strong Evidence of Life on Early Mars."

113 **Salisbury, for his part** • Author interviews with David Salisbury of Stanford public affairs, as well as related memos, drafts of press releases showing edits, and other documents.

113 **In an early draft** • The phrase "circumstantial evidence" favored by Salisbury made it into a NASA Video Advisory issued on August 6, 1996, but the final wording in the lead paragraph of the joint NASA-Stanford press release of August 7 was: "A NASA research team of scientists at the Johnson Space Center (JSC), Houston, TX, and at Stanford University, Palo Alto, CA, has found evidence that strongly suggests primitive life may have existed on Mars more than 3.6 billion years ago."

113 **Visiting scientist Ralph** • Author interview with Harvey.

CHAPTER SEVEN: THE GRAND INQUISITOR

115 **Dan Goldin, the** • Accounts of the "grilling" in this chapter are based primarily on author interviews with Goldin, McKay, Gibson, Wesley Huntress, and Laurie Boeder.

116 **Daniel Saul Goldin grew** • Author interviews with Goldin; transcripts of Goldin speeches and testimony before Congress. See also Kathy Sawyer, "The Man on the Moon: NASA Chief Dan Goldin and a Little Chaos Just Might Save the Space Program," *Washington Post,* July 20, 1994. Goldin was born on July 23, 1940.

117 **In 1962, when** • In 1962, Goldin earned a bachelor's degree in mechanical engineering from City College. Although he would work as a "rocket scientist," and refer to himself that way, and would continue his learning, he never formally acquired a higher degree. He would receive many honorary doctorates from universities.

117 **(He wasn't all work** • Goldin Speaking at a November 22, 1996, symposium at George Washington University.

117 **For example, his team** • Author interviews with Goldin. See also Theresa Foley, "Mr. Goldin Goes to Washington," *Air and Space* (April–May 1995): pp. 39–40.

118 **His prescriptions for** • Author interviews, as a *Washington Post* staff writer, with members of the space council under the first President Bush. See also Foley, "Mr. Goldin," pp. 36–43. Before he came to NASA, Goldin

was vice president and general manager of the TRW Space and Technology Group, in Redondo Beach, California.

118 **Many people thought** • See W. Henry Lambright, *Transforming Government: Dan Goldin and the Remaking of NASA*, Report for the PricewaterhouseCoopers Endowment for the Business of Government, March 2001. See also David Morrison, "Low-Rent Space," *National Journal*, April 29, 1995, pp. 1027–32. NASA's fiscal 1993 budget request was for $122 billion over five years; by the time of the 1995 request, that figure had fallen to $87 billion. For fiscal 1996, after prodigious efforts to meet that mark, the agency was stunned in January 1995 to receive a White House directive to cut another $5 billion. The budget process makes figures difficult to boil down, but cumulatively, the *National Journal* said, the agency had sustained a 36 percent "projected budget reduction" since 1993. That was a huge cut by any standard, and more than the agency had been led to expect and plan for.

118 **NASA was unusual** • National Aeronautics and Space Act of 1958; Walter A. McDougall, . . . *the Heavens and the Earth* (New York: Basic Books, 1985), pp. 195–96, 200, 206–7, 228.

118 **The new man arrived** • Author interviews with members of the White House space council under the first President Bush.

119 **Announcing himself as** • Sawyer, "The Man on the Moon."

119 **At the beginning, as** • Morrison, "Low-Rent Space," p. 1028.

119 **"Leadership is not** • Author interview with Goldin.

120 **Goldin made the new** • Author interviews with Goldin and White House officials; Morrison, "Low-Rent Space."

120 **Goldin believed that a** • Author interviews with Goldin and other NASA officials. See also Sawyer, "The Man on the Moon."

120 **Some who got** • Author interviews with various NASA officials. They cited examples of Goldin's inconsistencies, such as his promises to "empower" lower-level employees by delegating decisions and to encourage fearless risk taking, while instead he made decisions at the top and instilled fear. Wes Huntress, then NASA's chief space scientist, made the Jedi and Sith comparisons in an interview with the author.

120 **Goldin had waded** • Foley, "Mr. Goldin," p. 38. Goldin also complained that the peer-review system at NASA was governed by those already part of the space community. "It almost guarantees that you will lop off the new ideas. It guarantees some level . . . of technical excellence in the details, but it may be mediocrity in the concepts. . . . We have to change and open

up the whole process to allow new places, new faces, and revolutionary new ideas."

120 **He dismantled the** • Foley, "Mr. Goldin," pp. 36–43.

120 **He railed against** • Author interviews with Goldin; see also Goldin's speech, "A New Trajectory for NASA" (Washington Roundtable on Science and Public Policy, Oct. 15, 1996; reprinted by the George C. Marshall Institute). See also a January 6, 1999, deposition filed in U.S. Court of Federal Claims, *Northrop Grumman Corp. v. the U.S.* (provided by NASA Watch; see Web site: http://www.spaceref.com/nasa/sseic/01.06.99.goldin.html). Goldin said he had felt "shock and utter disgust" in the summer of 1992, as the new administrator, when he discovered contractor managers who "were only committed to the protection of their empires and their jobs and not committed to the end product for the United States government. . . . The inmates were running the asylum, . . . and everybody was vying for their own piece of the pie." Goldin concluded that "you cannot straighten out this kind of a mess with nice meetings, with teacups and doilies. . . . It was ugly, it was miserable, it was problematic, but it was without malice aforethought" on Goldin's part. At another point, Goldin said, "This is my life, this program. And I gave up my position in industry because I was so worried about this. . . . I was not just frustrated with industry, but I was frustrated with industry and myself. As hard as I was on everybody else, I was harder on myself because I felt I could not move as fast as I should have."

121 **Goldin and his top lieutenants** • January 6, 1999, deposition filed in U.S. Court of Federal Claims, *Northrop Grumman Corp. v. the U.S.* Goldin said that when he first took office, he discovered the shuttle was taking off in "conditions where that shuttle could have blown up on the launch pad. The contractors were making a tremendous fee . . . and every time there was a problem, they would add more inspectors." He brought in a legendary NASA hand, George Abbey, whom he credited with making the shuttle three or four times more reliable and yet reducing spending by a billion a year.

121 **As for the embattled** • The controversial plan involved a historic, White House–approved merger of the American and Russian spaceflight programs—former Cold War rivals joined in new purpose. Goldin's approach was partially vindicated in 1998, when the first components of the orbital research laboratory finally at least made it into low Earth orbit. On the day Goldin interrogated McKay and Gibson, U.S. astronaut Shannon Lucid was aboard the Russian facility *Mir* in orbit with two Russian cosmonauts.

121 **In August 1993** • Author interviews with Goldin; Morrison, "Low-Rent Space," p. 1029; Foley, "Mr. Goldin," pp. 38, 40; Lambright, *Transforming Government*, p. 20.

122 **Goldin acknowledged that** • Author interview with Goldin.

122 **Goldin's personal vision** • Goldin articulated his hopes for the long term on numerous public occasions, including a speech to the American Geophysical Union, May 26, 1994.

123 **In these advances** • Author interviews with Goldin and Wes Huntress, and internal NASA memos and planning documents provided by Huntress.

123 **Having already gone** • NASA's science office had been cut from about 120 to 65 people.

124 **Some at headquarters** • John Kerridge interview with Steven Dick, NASA archives.

124 **On July 12, 1996** • Author interview with Huntress; Don Savage, NASA public affairs officer, "memo for the record." The records of McKay and Gibson, like those of *Science*, indicate that the magazine informed them of the paper's acceptance on July 16. Don Savage's memo indicates, however, that NASA science official Edward Weiler called him at home earlier, on July 12, to tell him of "possible publication in *Science*." Weiler told the author he does not recall who at *Science* called him.

Although this may seem a minor point, the flow of information is always interesting to some. Asked about the notification, *Science* magazine deputy editor Brooks Hansen told the author in an e-mail, "The printed accept date on the paper is 16 July, and that is when we (the editorial office) would have informed the authors of formal acceptance. . . . The editorial office of *Science* does not routinely inform author's institutions of decisions, and not several days in advance of informing the authors." However, he said, the magazine or the public affairs office might in certain cases ("where the authors have requested or facilitated" such a move) discuss press relations with the author's institutions while a paper's acceptance is still pending.

125 **Goldin activated a** • Author interviews with Goldin and Huntress. It was the normal routine that, every morning, Goldin and each of his top lieutenants would be handed a three-by-five index card with that day's schedule written on it. In Huntress's view, a good day was one for which the card showed a lot of white space, which meant few formal meetings. But those days were rare. In order to contain the secret of the rock within the smallest possible circle, the nature of any meetings about the rock was not indicated on the little index cards.

125 **Just as Orson Welles** • *War of the Worlds* broadcast, Halloween night 1938.

125 **A key meeting took** • Author interviews with Huntress, Boeder, and Savage. There were a number of meetings, with a changing cast of players, and some mark the dates differently. Savage noted in his "memo for the record" a large meeting and teleconference on July 22, which included headquarters officials Huntress, Boeder, Ed Weiler, Jurgen Rahe, and Joe Boyce, with McKay and Gibson teleconferenced in from Houston, along with Johnson Space Center public affairs officers James Hartsfield, Steve Nesbitt, and Jeff Carr. Huntress's records indicate the meeting described here took place on July 26. It is possible that both are correct.

125 **They decided to** • Author interview with Huntress; also Michael Meyer interview with Dick, NASA archives.

125 **Boeder explained that** • Author interviews with Boeder and Savage at NASA, and *Science* staff members; also Savage memo. NASA public affairs staff had been in touch with counterparts at *Science* to discuss the timing of the publication and a related press conference. Possible dates of publication had ranged from August 2 (NASA's preference) to August 9 to August 16, but the final decision at the magazine was that publication was not possible earlier than August 16.

126 **On the morning of Tuesday** • Author interviews with Goldin and Huntress. The two space officials arrived at Panetta's office twenty minutes late, Huntress said, because a clerk had given the guard at the gate a list that misspelled his boss's name as "Golden." The guard was unyielding, but Goldin thundered his way into an anteroom, where he was able to call Panetta's office and arrange a rescue.

126 **In addition to his** • Author interviews with Goldin and Huntress, and "talking points" documents provided by Huntress.

127 **Among White House concerns** • Also, though few people besides Clinton were aware of it yet, the president was spending some of his time in phone calls designed to extricate him from a dangerous liaison with a young former White House intern.

128 **For Huntress, it was** • Author interview with Huntress.

128 **The NASA men** • Author interviews with Goldin and Huntress.

129 **On one wall** • Author interview with Gore.

129 **Gore greeted his** • Author interviews with Goldin and Huntress. "Wait a minute" quote first cited in Mimi Swartz, "It Came from Outer Space," *Texas Monthly* (Nov. 1996): p. 122.

131 **The NASA officials themselves** • Author interview with Huntress.

131 **When they threw** • NASA exobiology official Michael Meyer, in interview

with Dick, NASA archives. Meyer was skeptical of the images. They were visually convincing to many, but he said, with "my background in phyto-plankton—I'm used to looking at things under a microscope. I wasn't as excited as everybody else. . . . But it was a very nice set of evidence, so you know, go with it."

131 **Gibson, for one** • Author interview with Gibson.

131 **As the session** • Author interviews with McKay and Goldin.

131 **Chris Romanek, on his** • Author interview with Romanek.

132 **The workforce seemed** • "Interview: NASA Administrator Daniel Goldin," *Final Frontier* (July–Aug. 1994): p. 88.

132 **Now the Houston guys** • Author interview with Goldin. After grilling McKay and Gibson, Goldin revealed the secret of the rock to one "unauthorized" person—his terminally ill father, Louis. The elder Goldin had once earned a degree in biology but because of the Depression had been unable to get a job in the field. Now he was in a Florida hospital, dying of cancer of the spine. His son called that night and said, "Dad, I'm going to tell you some information that's gonna knock your socks off." He added, "Hang in there. . . . You're going to see your son on national TV, making this announcement." Goldin said later that the news kept the old man alive a little longer. Louis Goldin would die a few days after the press conference about the rock.

132 **Morris saw rich** • Author interview with Rick Borchelt, White House Office of Science and Technology Policy.

133 **Morris's plan was** • Author interviews with Goldin and Borchelt. See also transcript of Goldin talk at George Washington University Space Policy Institute symposium, "Life in the Universe: What Can the Martian Fossil Tell Us?" (Washington, D.C., George Washington University, Nov. 22, 1996), p. 88, in which he listed some of the hurdles: A mission to Mars would take from six hundred to one thousand days, and "we need to send somewhere between 500 and 1,000 metric tons" from Earth to Mars, depending on the approach. "Do you generate fuels and breathable air on Mars? Do you need retro rockets firing as you arrive or use a technique called aerobraking? Will you stay on the surface a month or a year and a half? What type of propulsion system will you use?" The current cost of lifting one pound to low Earth orbit was $10,000—one pound of anything. That meant the least a human expedition would cost was about $12 billion. NASA was working on technologies that (Goldin hoped) would get the project into the realm of feasibility, ultimately reducing the price tag to $200 per pound in perhaps

ten years. Goldin and his team would fail in that crucial effort. In 1999, a $1 billion–plus public-and-private project to develop the X-33, a successor to the space shuttle, ran up against insurmountable technical barriers. (See Lambright, *Transforming Government*, pp. 22–24.)

Goldin had been spelling out the hurdles on the road to Mars for years. See also, for example, his Wernher von Braun Memorial Lecture, April 29, 1993: "Colonizing Space: What Is Our Goal?" by Goldin and Dr. Alex Roland. (Occasional Paper Series, National Air and Space Museum, Smithsonian Institution); and Goldin's speech delivered on February 7, 1995, at the Space Transportation Association breakfast (Arlington, Va., *Space Trans*, March–April 1996), pp. 2–9, in which he expressed outrage that the nation had gone "for *twenty-five years* without a new rocket engine" because of space agency failures.

133 **On August 2, Rowlands** • Joel Achenbach, *Captured by Aliens* (New York: Simon & Schuster, 1999), p. 136. See also Howard Kurtz, "The Hooker, Line and Sinker," *Washington Post*, September 4, 1996, p. B1, at www.washingtonpost.com/wp-srv/local/longterm/tours/scandal/morris.htm.

CHAPTER EIGHT: "KLAATU BARADA NIKTO"

134 **The McKays had come** • The account of the McKay family foray into the hill country is based on author interviews with David McKay and Mary Fae McKay.

136 **Everett Gibson, back** • Author interview with Gibson. Pillinger was phoning from the Open University, Milton Keynes, UK.

136 **For one thing, the** • Author interview with Rick Borchelt, White House Office of Science and Technology Policy; also e-mail from Borchelt, Oct. 25, 2001.

136 **But the first person** • Author interviews with Leonard David.

137 **An unabashed space** • David worked both as a journalist and also, at various times, for the government. His work has been published in the *Financial Times*, *Sky and Telescope*, *Astronomy*, *Space News*, *Aerospace America*, and other publications. Among his numerous other affiliations, he was editor of the National Space Society's *Ad Astra* and *Space World* magazines. In the mid-1980s, he was director of research for the National Commission on Space. For NASA, he wrote for the publication *Spinoff*, and he conducted interviews and wrote scripts for NASA radio programming. He also worked on public NASA exhibits.

137 **For months now** · For an analysis of how the story became public, see Vincent Kiernan, "The Mars Meteorite: A Case Study in Controls on Dissemination of Science News," *Public Understanding of Science,* vol. 9 (2000): pp. 15–41.

137 **A few months after** · Author interviews with Leonard David; see also John Kerridge interview with Steven Dick, NASA archives.

137 **Kerridge and others** · John Kerridge interview with Steven Dick, NASA archives, pp. 9–15.

138 **As July turned to** · Author interviews with David and Savage; also Savage memo.

139 **The NASA headquarters** · Author interviews with Goldin, Huntress, Savage, Boeder, and others.

139 **The NASA group had** · Hartsfield e-mail, July 17, 1996, to public affairs officials at Johnson Space Center noted that the McKay-Gibson paper had just been accepted by *Science* and that NASA headquarters had approved a "tentative plan" to hold "a press conference on August 5, 6 or 7." At the same time, it stated that the publication date was "currently believed to be August 16." NASA headquarters was insisting that the press conference be held in Washington, not at Johnson Space Center. Hartsfield noted that among other preparations, he was readying photos and "video B-roll" (background material for television) of the meteorite lab and the meteorite itself in Building 31, and of McKay and Gibson at work on the more advanced microscope in Building 13. At that point, technicians were working to develop animation showing the hypothetical history of the meteorite as proposed by McKay et al.

140 **Nan Broadbent, communications** · Author interview with Broadbent.

140 **The magazine receives** · The magazine's conditions of acceptance call for papers that reveal "a novel concept of broad importance to the scientific community," which has never previously been disclosed to the public, and which is likely to stimulate further investigation and debate, according to AAAS Web site instructions for first-time authors. The paper then has to be approved by a member of the AAAS Board of Reviewing Editors, made up of around one hundred leading, active scientists. Finally, the manuscript goes to anonymous peer reviewers selected for their expertise in the field. (With the advent of the Internet, *Science* and other journals were developing ways of stepping up the pace on selected papers by such means as rapid electronic publication.)

140 **Journalists consider these** · Deborah Blum and Mary Knudson, eds., *A*

Field Guide for Science Writers (New York: Oxford University Press, 1998), p. 10.

141 **Ordinarily, the orchestrated** • Some scientists simply elected not to publish their papers in journals whose constraints they found onerous. This often meant turning to outlets with less prestige and influence and in some cases a less rigorous process of peer review. (Geoffrey Marcy, one of the leading discoverers of extrasolar planets, had the moxie and credibility to make this form of rebellion work for him.) In the end, whatever the outlet, the information became fodder for the machinery by which the science world did its self-policing. For a brief discussion of the embargo system as it worked in this and one other celebrated case, see Eliot Marshall, "Embargoes: Too Hot to Hold: Life on Mars and Cloned Sheep Couldn't Be Kept Under Wraps," *Science*, vol. 282 (Oct. 30, 1998): p. 862. For *Science* magazine's own description of its philosophy on the matter, see editorial in the same issue: Floyd E. Bloom, "Embracing the Embargo," p. 877.

141 **In any case, journalists** • Most of the research that made headlines and seemed to trumpet certainty was, to varying degrees, tentative. So when *was* a development ripe for public consumption? Was it when the researcher first completed the experiment? When the paper describing the results made it through the review process and got accepted by a journal? When the journal published it? When it was tested and supported in subsequent experiments by other groups? After specialists in the field accepted it by broad consensus?

The answer was any and all of the above, according to Boyce Rensberger, director of the Knight Science Journalism Fellowships at MIT and former science editor of *The Washington Post* (in an interview with the author). The scientist doing the work, or a journalist who found out about it, could elect to tell the world at any of these stages. In the normal course of events, the uncertainty connected with the research results would diminish, and the credibility of the assertions would increase, as the work passed successfully through each wicket. Ideally, any journalist reporting on the matter would forthrightly disclose—even emphasize—the degree of uncertainty. This sliding uncertainty scale, however, was often a source of confusion for citizens, bombarded as they were with new "discoveries" that were hailed one day and discarded the next. People were seeing snapshots of a mostly hidden process captured at varying stages.

141 **Dick Zare, for example** • Author interviews with Zare.

141 **Some at the Foundation** • Author interviews with Scott Borg, head of polar

programs at the National Science Foundation, and Curt Suplee, director of the foundation's public affairs.

142 **Like the scientists in** • Author interview with Broadbent; see also Kiernan, "Mars Meteorite," p. 22.

142 **One morning a few** • Author interview with Broadbent.

142 **In early summer** • Author interviews with Broadbent and Borchelt; see also Kiernan, "Mars Meteorite," pp. 22–24.

142 **Broadbent and her** • Author interview with Broadbent; Kiernan, "Mars Meteorite," p. 24. Regarding Sagan's participation as reviewer, see also William Sheehan and Stephen James O'Meara, *Mars* (Amherst, N.Y.: Prometheus, 2001), pp. 295–98, and William Poundstone, *Carl Sagan* (New York: Henry Holt, 1999) p. 379.

143 **From as early as** • Hartsfield e-mail, July 17, 1996.

143 **Because of the sensational** • Author interview with Don Savage; Savage memo; see also Kiernan, "Mars Meteorite," p. 23. Savage first contacted *Science* on July 17 to discuss the paper's publication date. On July 22, according to his memo, he participated in a teleconference with NASA headquarters officials Wes Huntress, Laurie Boeder, Edward Weiler, Jergen Rahe, and Joe Boyce; with Johnson Space Center's David McKay, Everett Gibson, public affairs officers James Hartsfield, Steve Nesbitt, and Jeff Carr; and with *Science* editor Brooks Hansen and public affairs official Diane Dondershine, to continue the planning. The next day, *Science* decided that the earliest possible publication date would be August 16.

143 **On the subject of leaks** • Author interviews with Savage and Boeder; Savage memo.

143 **When things later went** • Author interviews with Broadbent, Boeder, Savage, Goldin, and others involved.

143 **Early on Terrible Tuesday** • Author interview with Broadbent. Among the others involved in the action at *Science* that day were Broadbent's assistant, Diane Dondershine, and managing editor Monica Bradford.

144 **The magazine was now** • Author interviews with Broadbent and Boeder.

145 **She understood that the** • Author interview with Boeder.

145 **The agency's DNA** • National Space Act of 1958 (since amended).

145 **The next thing Nan** • Author interviews with Broadbent, Goldin, Huntress, Savage, and others.

146 **The AP's Paul Recer** • Kiernan, "Mars Meteorite," p. 26; author interviews with Broadbent and Savage.

146 **The *Science* staff** • Author interview with Broadbent; Kiernan, "Mars Me-

teorite," pp. 26–27. After Borchelt supported Broadbent's decision, Boeder subsequently refused to take Borchelt's calls.

146 **A horde of relentless** • The author was working on unrelated projects at her desk in the *Washington Post* newsroom when a flurry of phone calls and the AP dispatch alerted her. She and her editor, Curt Suplee, were part of the "horde" pressing for release of the meteorite paper.

146 **They ended up having** • Kiernan, "Mars Meteorite," p. 20.

147 **Keeping the secret** • Author interview with Zare; Zare's written account.

147 **A film crew from** • Author interviews with Zare and Salisbury.

147 **Salisbury fended them** • In advance of the planned press conference (which had now been moved up), Zare had gone to Condoleezza Rice, then provost of Stanford, to secure permission for Salisbury to accompany him to Washington, Zare told the author. When he told Rice the press conference would be about possible life on Mars and swore her to secrecy, he said, the future U.S. secretary of state seemed remarkably "unflapped."

147 **He was also appalled** • Author interview with Zare; Kiernan, "Mars Meteorite," pp. 28–29.

147 **At around three-thirty** • Author interviews with Zare, Salisbury, and Broadbent; Zare's written account; Kiernan, "Mars Meteorite," p. 29.

148 **During the flight** • Author interview with Zare; Zare's written account.

148 **A few weeks earlier** • J. William Schopf, *Cradle of Life* (Princeton, N.J.: Princeton University Press, 1999), p. 306.

149 **Goldin and his top** • Author interviews with Goldin, Huntress, and others.

149 **Schopf soon realized** • Schopf, *Cradle of Life*, pp. 305–6.

149 **Arriving at NASA** • Author interviews with Goldin, McKay, Boeder, Savage, and others; see also Schopf, *Cradle of Life*, p. 307–9. There was another unexpected attendee, according to several who were present that day. NASA officials expressed surprise at the arrival of Hojatollah Vali and were not sure what role to give him. Vali ended up sitting at the dais, but he did not say a word during the press conference. By all accounts, including that of Vali himself in an interview with the author, it was a bit awkward.

149 **First thing in that** • Author interview with Weiler. No one in the room that day took the trouble to introduce Schopf to Weiler or others present, Schopf told the author in an e-mail, May 6, 2005. Except for the McKay group and chief exobiologist Michael Meyer, Schopf said, "I had no idea who any of these folks were."

150 **Events had led Weiler** • Astrophysicist Weiler, chief of NASA's Hubble Space Telescope science team, had led the demoralized Hubble team

through months of ridicule, after the revelation that the observatory had been launched with a serious flaw, and on to the successful astronaut repair job in space. Weiler would subsequently become director of NASA's new Origins initiative, NASA's chief space scientist, and in 2004, director of NASA's Goddard Space Flight Center.

150 **Schopf was not the** • Author interviews with Savage and other participants.

151 **Some of those on** • Author interviews with Weiler, Boeder, and others.

151 **Schopf gave his** • Schopf, *Cradle of Life*, pp. 307–9. In an e-mail to the author, May 6, 2005, Schopf conceded that he might have misremembered the precise wording but insisted that the import was the same: Boeder wanted them to be "more positive." At least two scientists in the room that day remember Goldin and/or Boeder telling members of the McKay group not to "wimp out."

151 **She had been arguing** • Author interview with Boeder.

152 **As Weiler would recall** • Author interview with Weiler. McKay, when asked about the incident, would say he never felt anyone was pressuring him to change the content of his presentation, to say his findings were more certain than they actually were. He did feel pressure to improve his delivery.

CHAPTER NINE: IN THE BEAM

153 **Chris Romanek was** • Author interviews with Romanek provided the basis for this account of his wild dash to Washington for the press conference.

155 **A pumped Dan Goldin** • David Salisbury, draft of feature article, August 8, 1996.

155 **The countdown was** • The author attended the press conference and also interviewed all the main participants. See also videotape and transcript of the event (NASA archives).

156 **But first, the** • Transcript of comments by President Clinton, August 7, 1996, 1:15 P.M., South Lawn of the White House (http://www2.jpl.nasa.gov/snc/clinton.html). White House science office staff member Rick Borchelt told the author that adviser Dick Morris had urged the president to issue a statement the day before, when the story first began to break, but Borchelt and others had persuaded Clinton to wait until the story jelled. Hollywood would later incorporate a portion of Clinton's real-life Rose Garden commentary into the movie *Contact* (based on a 1985 novel by Carl Sagan). The director would make it seem as if the president was talking about the first radio signal received from an intelligent alien civilization.

Here is the complete transcript of the President's words:

President Clinton Statement
Regarding Mars Meteorite Discovery

THE WHITE HOUSE
Office of the Press Secretary
For Immediate Release
August 7, 1996

REMARKS BY THE PRESIDENT
UPON DEPARTURE

The South Lawn
1:15 P.M. EDT

THE PRESIDENT: Good afternoon. I'm glad to be joined by my science and technology adviser, Dr. Jack Gibbons, to make a few comments about today's announcement by NASA.

This is the product of years of exploration and months of intensive study by some of the world's most distinguished scientists. Like all discoveries, this one will and should continue to be reviewed, examined, and scrutinized. It must be confirmed by other scientists. But, clearly, the fact that something of this magnitude is being explored is another vindication of America's space program and our continuing support for it, even in these tough financial times. I am determined that the American space program will put its full intellectual power and technological prowess behind the search for further evidence of life on Mars.

First, I have asked Administrator Goldin to ensure that this finding is subject to a methodical process of further peer review and validation. Second, I have asked the Vice President to convene at the White House before the end of the year a bipartisan space summit on the future of America's space program. A significant purpose of this summit will be to discuss how America should pursue answers to the scientific questions raised by this finding. Third, we are committed to the aggressive plan we have put in place for robotic exploration of Mars. America's next unmanned mission to Mars is scheduled to lift off from the Kennedy Space Center in November. It will be followed by a second mission in December. I should tell you that the first mission is scheduled to land on Mars on July the 4th, 1997—Independence Day.

It is well worth contemplating how we reached this moment of discovery. More than 4 billion years ago this piece of rock was formed as a part of the original crust of Mars. After billions of years it broke from the surface and began a 16-million-year journey through space that would end here on Earth. It arrived in a meteor shower 13,000 years ago. And in 1984 an American scientist on an annual U.S. government mission to search for meteors on Antarctica picked it up and took it to be studied. Appropriately, it was the first rock to be picked up that year—rock number 84001. [Actually, it was the first to be *analyzed* that year.]

Today, rock 84001 speaks to us across all those billions of years and millions of miles. It speaks of the possibility of life. If this discovery is confirmed, it will surely be one of the most stunning insights into our universe that science has ever uncovered. Its implications are as far-reaching and awe inspiring as can be imagined. Even as it promises answers to some of our oldest questions, it poses still others even more fundamental.

We will continue to listen closely to what it has to say as we continue the search for answers and for knowledge that is as old as humanity itself but essential to our people's future.

Thank you.

156 **When the cameras** • As the press conference began, the space agency's sound system let out a high-pitched feedback keen, like an irate stepped-on cat. Goldin opened the proceedings by vowing not to fire anyone for the sound-system glitch.

157 **Wes Huntress, Goldin's** • For the published paper, see D. S. McKay, E. K. Gibson Jr., K. L. Thomas-Keprta, H. Vali, C. S. Romanek, S. J. Clemett, X. D. F. Chillier, C. R. Maechling, and R. N. Zare, "Search for Past Life on Mars: Possible Relic Biogenic Activity in Martian Meteorite ALH84001," *Science* (Aug. 16, 1996): pp. 924–30.

161 **McKay concluded with** • McKay et al., "Search for Past Life." See also T. Stevens and J. McKinley, "Lithoautotrophic Microbial Ecosystems in Deep Basalt Aquifers," *Science* (Oct. 20, 1995); and Michael Ray Taylor, *Dark Life* (New York: Scribner, 1999), pp. 96–97. A selection of comparison images—Earth and Mars—could be found at the following site as of early 2005: http://ares.jsc.nasa.gov/astrobiology/biomarkers/images.html.

As recounted in *Dark Life*, Todd Stevens, a microbial ecologist, had left some of his Columbia River samples with Thomas-Keprta. Anne Taunton, an intern working for McKay and Thomas-Keprta, studied one of them in the summer of 1996 (although she had been kept in the dark about the Mars rock hypothesis). When she boosted the magnification, she found the rock "packed with discrete colonies of fossilized rods and filaments. Some of the apparent bugs were only 30 or 40 nanometers in diameter and 150 nanometers long," well below the standard size range for bacteria.

When she delivered the photos of these shapes, Kathie Thomas-Keprta stared at them and said, "This is just what we needed. Oh, my gosh, I can't believe it." She even asked Taunton to cancel a planned trip. Taunton went to McKay and insisted that he tell her what was going on. He pulled out one of his own microscope images of wormlike rods resembling the ones she had found in the Columbia River basalt. "Umm, that one's from Mars," McKay told her. She was thrilled when McKay displayed her Columbia River sample images at the press conference. Stevens, even though he had provided the samples and written the earlier paper, remained skeptical of the nanobacteria theory.

162 **While the other scientists** • Zare's written account and interviews with author; also e-mail from Schopf to author, May 6, 2005. A few months after the press conference, Schopf said, he and Zare had a friendly chat at Wesleyan University in Connecticut. As Schopf recalled, "I explained to him about the problems with the potential origins of the organics [the PAHs] he had reported. He told me that had he known [this information earlier], he never would have been a coauthor [on the McKay group's *Science* paper]. Following that, to his credit, [Zare] declined the opportunity to coauthor works with the group."

Zare remembered the incident differently. Zare said he did not regret coauthoring the paper or renounce Zarelab's findings on the organics. But he did say he was disappointed that it was only during this encounter with Schopf that he learned for the first time about Schopf's earlier visit to Johnson Space Center and the reservations he expressed to the McKay group about its fledgling biological hypothesis. McKay and company had neglected to tell him. "That bothered me," Zare said. "But I've never disowned the paper. In retrospect, of course, you'd express a number of things in different ways."

Zare said that he had also been put off by what he saw as too little skepticism, "too much strong belief" on the part of some members of the McKay team, notably Everett Gibson, that "what they were doing had to be right."

While there had been various scenarios proposed for the origins of the PAHs in the rock, in Zare's view that remained a mystery. "To this day, I don't know what they're coming from."

Zare said he had declined to sign the papers produced by the Houston group after 1996 not because of anything Bill Schopf had told him but "because the work on the magnetites wasn't mine. It was their work. I gave them advice. I cheered them on." In their work published later on the magnetites, the McKay group thanked Zare for his help.

163 **Evidence of the rock's** • Ralph P. Harvey and Harry McSween Jr., "A Possible High-Temperature Origin for the Carbonates in the Martian Meteorite ALH84001," *Nature* (July 4, 1996): pp. 49–51.

164 **In the months and** • Author interviews with McKay; see also William Sheehan and Stephen James O'Meara, *Mars* (Amherst, N.Y.: Prometheus, 2001), p. 295, regarding Sagan's role.

165 **By early on August 7** • Vincent Kiernan, "Mars Meteorite," *Public Understanding of Science*, vol. 9 (2000): p. 33, citing AAAS 1996 Annual Report; author interviews with *Science* staff.

165 **NASA's Web site scored** • Ibid., 32–33, citing weekly report of the NASA public affairs office.

165 **Despite the unmannerly** • Ibid., pp. 29–35.

166 **Regardless of the content** • J. William Schopf, *Cradle of Life* (Princeton, N.J.: Princeton University Press, 1999), p. 324.

166 **An auction house in** • *New York Times*, Oct. 31, 1996, p. D6. And *USA Today*, Nov. 21, 1996, reported that, at the November 20 auction, the top bid of $1.1 million for a set of three Martian specimens failed to meet the minimum set by the unidentified seller.

166 **The iconic seventy-five-year-old** • David Culton, "Author of Mars Novel Skeptical of NASA's Claims," Gannett News Service, Aug. 8, 1996.

166 **Microsoft's chief strategic** • Nathan Myhrvold, "Mars to Humanity: Get Over Yourself," *Slate*, Aug. 14, 1996; summarized in *Time* (August 26, 1996): p. 64.

166 **Back at Building 31** • Author interview with the Smithsonian's Tim McCoy, who was working in Building 31 at the time.

167 **Everybody connected with** • Author interview with Gibson, regarding the statistics. Also in interviews with the author, David McKay and Mary Fae McKay described their participation in a special program on the Mars rock for the Discovery Channel. The cameras followed David to work, and came to the McKay home to capture the family fixing dinner, including David chopping mushrooms for his signature "roast on a grill" from a Presbyterian cookbook.

167 **Gibson would do more** • Author interview with Gibson. One occasion especially impressed Gibson: he and his wife were walking down a London street after he had given a TV interview when a woman came up to him, reached out, and grabbed his collar, saying, "You're the Mars man!"

167 **The rock also helped** • Morris's resignation caused an embarrassing distraction for the president just hours before the session of the Democratic convention where he accepted his party's unanimous nomination for a second term. Morris would acknowledge his bad behavior in a memoir, *Behind the Oval Office: Winning the Presidency in the Nineties* (New York: Random House, 1997).

CHAPTER TEN: SCHOPF SHOCK

169 **In the summer of** • Schopf described the China trip in a lecture at Goddard Space Flight Center, May 4, 2001; see also J. William Schopf, *Cradle of Life* (Princeton, N.J.: Princeton University Press, 1999), pp. 201–7.

169 **One day, Schopf** • Schopf, *Cradle of Life,* p. 204.

170 **His hosts could not** • In an e-mail to the author, Schopf dismissed the appellation "god of the Precambrian," used by some of his colleagues, calling it overblown.

170 **The encounter would** • Regarding the practical separation of the fields of planetary sciences and origin of life studies, see Steven J. Dick, *The Biological Universe* (Cambridge: Cambridge University Press, 1996), pp. 473–502.

170 **In 1960, as an** • Schopf, *Cradle of Life,* pp. 52–53.

170 **Scientists had found** • Ibid., p. 29. In the early 1900s, Charles Doolittle Walcott discovered the best-preserved Cambrian fauna in a formation in the Canadian Rockies; he called it the Burgess shale. It would be celebrated in Stephen Jay Gould's best-seller, *Wonderful Life* (New York: Norton, 1989).

171 **The skeptics had** • Schopf, *Cradle of Life,* pp. 72–75.

171 **In 1961, just as** • Ibid., pp. 183, 193–95. Though stromatolites had peppered the surface of the young Earth, they eventually became rare, because other life *did* evolve. Snails, for example, found the mat builders quite tasty. Shark Bay and a few other spots were too salty for the predators, thereby providing protected habitats where stromatolite populations could survive.

172 **Schopf was troubled by** • Ibid., chapter 1, especially p. 34. Another reason it took so long for doubts to be resolved about the true nature of stromatolites was the chasm between geologists and microbiologists. The latter had shown years earlier that bacterial communities indeed built formations layered in squishy or leathery rather than stony material. The microbiologists called these things bacterial mats, but they were quite similar to stro-

matolites. As Schopf would write, the two tribes—geologists and microbiologists—barely spoke the same language and "on most college campuses they even occupy separate 'homelands' . . . and as each prods its students to learn more and more about less and less, science becomes increasingly fragmented."

172 **At least one mentor** • Ibid., pp. 52–55.

172 **A colleague had discovered** • Ibid., pp. 35–44 and 55–64. The colleague who found the fossils—"long, thin, threadlike filaments and tiny, hollow, balloonlike balls"—was Stanley A. Tyler.

173 **For years, Barghoorn** • Ibid., pp. 56–61.

173 **Schopf acknowledged that** • Ibid., pp. 59–61. Barghoorn listed himself as the first author of the rushed-up paper, even though someone else (a colleague who had died recently) had done most of the initial groundbreaking work on the project all those years ago. (When Barghoorn offered to add Schopf's name to the authors' list, Schopf declined, feeling that his contribution had been too marginal. Instead, his mentor inserted a line acknowledging Schopf's "assistance.") Barghoorn's paper brought Schopf, a second-year graduate student, his first taste of public attention. He went on the lecture circuit. In an e-mail to the author, May 6, 2005, Schopf said of the incident that Barghoorn had "told me what to do. I did it (I did not approve, then or now, but as far as I could see, I had no choice in the matter). He was boss. I was not. Heck, I was a fledgling underling. I reported this in *Cradle* because I thought then, and now, that it was pivotal to the development of the field, and I wanted Professor Cloud [Preston E. Cloud, the author of the other paper] to get due credit for spurring Barghoorn into action."

174 **The strategy was this** • Ibid., 184–201. Stromatolites formed from sediment trapped and bound in shallow marine waters by layered colonies of mucilage-secreting one-celled microbes. When the colonies were alive, their layered mats would be leathery or gelatinous. But these microcommunities differed from one layer to the next. The animals that dominated the uppermost layers would (scientists thought) typically be cyanobacteria—formerly called blue-green algae—that lived by photosynthesis (converting the energy of sunlight to chemical energy), with oxygen as a by-product. Living in the layers below them would be colonies of a different color and lifestyle—for example, "switch-hitting" bacteria whose metabolisms operated either with or without oxygen, and below them bacteria that would actually be *poisoned* by the slightest trace of free oxygen. Below them would be yet other totally different breeds of bug.

The ideal thing was to find sites where water had filled the animal cells with deposits of the tiny mineral (quartz) grains that made up the chert. As the grains solidified over thousands of years, they would grow through the cell walls, preserving them as fossils instead of crushing them.

174 **In 1982 (the same** • Ibid., pp. 75–100.

174 **Schopf himself** • Schopf had the specimens archived at the facility in London where he had been working at the time, now called the Natural History Museum.

174 **He probed for** • The Apex fossils did not occur in stromatolites, Schopf noted in an e-mail to the author, although stromatolites from units of about the same age had been found to the west, at North Pole Dome.

174 **After several months** • J. William Schopf, "Microfossils of the Early Archean Apex Chert: New Evidence of the Antiquity of Life," *Science*, vol. 260 (1993): pp. 640–46.

175 **The opening of** • Schopf lecture at Goddard. See also Schopf, *Cradle of Life*, p. 3.

175 **Schopf's comments generated** • Schopf would also catch some grief over the appearance. In an e-mail, he told the author that following his appearance on the NASA stage, he had gotten "creamed" by various colleagues, including a Nobel laureate he would not name, "for not being more harsh" in his condemnation of the McKay interpretation. Schopf emphasized that he had not enjoyed being cast as the naysayer and had tried to avoid the task by recommending others to do it instead.

176 **Schopf was among** • The nanoscale animals had escaped the notice of the modern world, Robert Folk and others argued, because their size range was just below the resolution of an optical microscope and the mesh size of the filters microbiologists commonly use to strain bacteria from liquids.

Schopf maintained that the smallest *confirmed* microbes on Earth (of the genus *Mycoplasma*) were a good bit larger than some of the putative fossils in the Mars rock and yet were so small they contained only a fraction of the genes in most bacteria. These Earth microbes were so short of the usual complement of biological machinery that they could live only as parasites in the cells of other organisms.

176 **The basic cell** • Boyce Rensberger, *Life Itself* (New York: Oxford University Press, 1996) describes the cell, "the fundamental particle of life," and its workings in detail.

176 **They were smaller** • Transcript of NASA press conference. See also Schopf, *Cradle of Life*, p. 316.

177 **"This is half-baked** • Edward Anders, "Evaluating the Evidence for Past Life on Mars" (letter), *Science* (Dec. 20, 1996): pp. 2119–21. Donald Goldsmith, *Hunt for Life on Mars* (New York: Dutton/Penguin, 1997), pp. 101–107.

177 **In 1961, as students** • Christopher Wills and Jeffrey Bada, *The Spark of Life* (Cambridge, Mass.: Perseus, 2000), pp. 85–86; Goldsmith, *Hunt for Life on Mars*, p. 7; Charles Petit, "Pieces of the Rock," *Air and Space* (Apr.–May 1997), pp. 38, 40–41. The 1961 claim was made by chemist Bartholomew Nagy of Fordham University in the journal *Nature* (Nov. 18, 1961). Now meteorite specialists—who lumped themselves under the tongue-tangling label *meteoriticists* (pronounced meteor-WRIT-assists)—generally labored with a bit of an inferiority complex, in a provincial borderland stuck among the larger, older fields of geology, chemistry, and astronomy. Far removed from the fundable cachet of the surging genetics and biotech enterprises, they rested "at the bottom of the pecking order" for grant money. Hungry as they were to improve this state of affairs, many confessed to fears about what an Orgueil-class public relations debacle would do to their future prospects. Regarding the meteoriticists' fears, see also Allan Treiman's comment to Sharon Begley and Adam Rogers, "War of the Worlds," *Newsweek* (Feb. 10, 1997), p. 58.

177 **This and other discredited** • Cosmochemist and meteorite specialist John Kerridge, who for a time became one of the McKay group's most aggressive and visible critics, looked back through the literature and determined that a scientific "spasm" over evidence of life-forms in meteorites had erupted roughly once in every generation, or about every twenty to thirty years since the beginning of meteorite studies in the early nineteenth century. (See John Kerridge interview with Steven Dick, NASA archives.) Typically, a claim and the accompanying public flurry was followed by a counterassault that effectively drove the notion underground again. The question of extraterrestrial life ("Are we alone?") was of such sensational importance that somebody would take up the baton at the "slightest hint" of evidence. And by Kerridge's acerbic reckoning, the McKay claims were a tad overdue.

On Terrible Tuesday, the day the story leaked, the staff of ABC's *Nightline* had faxed Kerridge a copy of the McKay paper and asked for comment. Kerridge soon found himself besieged with media phone calls. "I have never been run off my feet that way; I mean it was just extraordinary," he would tell Dick later. "I mean I'd break off and go to the bathroom and come back and there'd be three call-waiting messages on my desk; I mean it was ludicrous. . . . God knows what it must have been like for McKay and company."

Reporters sought Kerridge out because, although he had never studied the Allan Hills rock, he had led a NASA study of ways to look for signs of ancient life on Mars. In numerous interviews (see, for example, Petit, "Pieces of the Rock," p. 40), he pronounced the McKay group's claims "just pitifully short of convincing," and "utterly unconvincing."

177 **In an e-mail to McKay** • Petit, "Pieces of the Rock," pp. 40–41.

178 ***Science* magazine published** • Edward Anders, "Evaluating the Evidence"; Petit, "Pieces of the Rock," pp. 36–41. Gibson, commenting on the Anders communication, preferred to see the glass as half full: "The first paragraph . . . is amazing. It congratulates us!" he told a reporter. "Knowing him, that he said anything positive is wonderful."

178 **Geologist Ralph Harvey** • Harvey's comments were published, respectively, in Petit, "Pieces of the Rock," p. 37; and J. K. Beatty, "The Messenger from Mars," *Sky and Telescope* (July 1997): p. 39.

178 **In the months** • Sharon Begley and Adam Rogers, "War of the Worlds."

179 **"I am appalled** • Petit, "Pieces of the Rock," p. 40.

179 **Gibson, ever the most** • Begley and Rogers, "War of the Worlds," p. 56.

179 **NASA's reputation as a** • Author interview with numerous scientists. It was a National Science Foundation official who used the phrase "entitlement agency."

179 **Two chemists claimed** • Goldsmith, *Hunt for Life on Mars*, pp. 6, 91; Deborah Blum and Mary Knudson, eds., *A Field Guide for Science Writers* (New York: Oxford University Press, 1998), pp. 12, 100.

179 **But the *way* it worked** • Blum and Knudson, eds., *A Field Guide for Science Writers*, pp. 12, 100.

180 **One meteorite expert** • Allan Treiman commenting to Begley and Rogers, "War of the Worlds," p. 58.

180 **At the time of the** • Records provided to the author by the office of NASA meteorite curator Carlton Allen.

180 **One researcher was** • Petit, "Pieces of the Rock," pp. 36–37.

180 **After the August** • Author interviews with Scott Borg, of the National Science Foundation, NASA officials, and others involved in the process; interagency press releases.

181 **As government committees** • Author interviews with committee officials and scientists, including McKay and Andrew Steele, one of the sample recipients.

181 **Not long after** • Author interviews with Treiman, McKay, and others.

181 **Treiman had been** • Author interviews with Treiman.

183 **"About thirty years ago** • Darwin to Henry Fawcett, Sept. 18, 1861. Cambridge University Darwin Correspondence Online Database: http://libpro13.lib.cam.ac.uk/perl/nav?pclass=calent&pkey=3257. For Darwin's comments on the manuscript of Fawcett's address, see "On the method of Mr. Darwin in His Treatise on the Origin of Species," *Rep. British Association for the Advancement of Science* (1861) part 2: 141–43.

183 **Carol Cleland of the** • Author interview with Cleland. See also Cleland, "Historical Science, Experimental Science, and the Scientific Method," *Geology* vol. 29 (2001): pp. 987–90; Cleland, "Project Report: Philosophical Issues in Astrobiology," *NAI, Year 4 Annual Report* (University of Colorado, Boulder); Carol Cleland and Christopher Chyba, "Does 'Life' Have a Definition?" in *Planets and Life: The Emerging Science of Astrobiology*, W. Sullivan and J. Baross, eds., chapter 7 (Cambridge: Cambridge University Press, forthcoming).

183 **Experimental researchers are** • Experimental researchers often use a control, a separate experiment (in which the experimental variables are not changed) that provides a neutral reference point for comparison. For historical researchers—as in the case of the dinosaur extinction, for example—running controlled experiments is clearly untenable.

184 **"Some [people] might draw** • Petit, "Pieces of the Rock," p. 39.

184 **Others thought the** • Steven J. Dick and James E. Strick, *The Living Universe* (New Brunswick, N.J.: Rutgers University Press, 2004), p. 191.

184 **NASA funded not only** • Author interviews with scientists and officials. See also Dick and Strick, *Living Universe*, p. 199.

185 **At the height of** • Author interview with Tim McCoy, manager of the Smithsonian meteorite collection. Privately held Martian meteorites, he said, commonly sold for $1,000 to $2,000 per gram.

CHAPTER ELEVEN: EXPLORATIONS

186 **The reserved scientist** • Author interviews with McKay.

186 **Nevertheless, the Clinton administration** • Author interviews with Dan Goldin, Wes Huntress, David McKay, and other participants and planners, as well as personal and official government documents form the main basis for the account of the vice president's December 11, 1996, meeting and related events. The author also attended a press briefing held by participants immediately after the December 11 meeting.

Among the NASA and White House draft planning documents provided to the author and/or available in the NASA history archives were:

- A Sept. 5 NASA memo, "Talking Points on the Summit," that suggests an outside group of experts similar to one mobilized to review the space station to take a " 'fresh look' at what opportunities and challenges the Mars discovery and the Europa discovery pose."
- A Sept. 5 NASA draft called "Administrator's Guidance Regarding Mars Exploration Strategy" that states as second of six points: "NASA will not use inconclusive scientific data from the meteorite as a basis to seek a budget increase for Mars-related missions." The other points emphasize the need to partner with other agencies and nations, engage the public, and rely on "appropriate" outside experts and scientific procedures to decide on "how and where we should go" to conduct space science research.
- NASA memo, Sept. 18, 1996, from NASA's Alan Ladwig, associate administrator for policy and plans, stating that the purpose of the summit was "to gain an understanding of the near-term funding requirements to sustain NASA's current programs and to discuss the long range future of the space program. A significant purpose . . . will also be to discuss how America should pursue answers to the scientific questions raised by the meteorite, as well as other recent findings on Europa, a moon of Jupiter."
- NASA document, Oct. 2, 1996, "Vice President's Symposium on Space Science: Strawman Plan," that states under the heading "Policy Context," "In early and mid-1996, the projected 5-year budget runout for the agency indicated a decline from approximately $14 billion to well less than $12 billion." Senator Barbara Mikulski, ranking minority member of the Senate Subcommittee on VA-HUD-Independent Agencies, suggested to the administration there be "a White House summit of congressional and administration leaders" to discuss the future of the space program, the document states. "After the announcement of possible evidence for ancient life in a Martian meteorite in August, the Administration announced that a summit would be held in December to discuss priorities for the space program in an era of expanding opportunities for scientific discovery but contracting discretionary budgets." Under "Technical Context," the document states, "The agency is eager to take advantage of this opportunity to explain its proposed research plans and to attract support for them in the current environment of enhanced interest."

• NASA memo, Nov. 8, 1996, from Alan Ladwig to more than three dozen NASA officials, noting that the presidential election was over and, in response to a Nov. 5 memo from the White House science office, outlining a series of weekly teleconferences to plan for the summit.

• Letter from George Brown Jr., ranking Democrat on the House Science Committee, to Vice President Gore, Nov. 18, 1996, urging the administration to address the issue of NASA's budget earlier in the process than usual—that is, at the summit planned for early January. Brown says, "I recognize that discussion of specific agency budgetary matters prior to the formal release of the Federal Budget is not normal practice, but I believe that this is a unique situation which requires the utmost candor if it is to be successful."

• Memorandum for the vice president (the date noted only as "November xx, 1996"), from science adviser John Gibbons and vice presidential aide Greg Simon, stating the purpose of the December 11 White House gathering: "to hear from leaders in the U.S. science community on what priorities the U.S. space program should embrace in light of the Mars meteorite and other recent discoveries in space science."

187 **The group steered clear** • NASA legislative affairs office memo (e-mail), Nov. 22, 1996, stated, "The VP has asked for this [symposium] specifically so he may 'chew on' some of the issues surrounding this [Mars rock] discovery. . . . The symposium will not touch on budget issues in a substantive manner."

187 **The other men and** • A document prepared for the vice president's symposium listing "NASA Suggested Participants," dated Nov. 20, 1996, shows only one woman among the dozen or so people at the table, with the rest of the attendees seated nearby. A scrawled notation suggests "more women, higher caliber" people at the table.

187 **Gould was delighted** • "Salon Interview with Gould," Salon.com, no. 33 (Sept. 23–27, 1996).

188 **Two years earlier** • The term *astrobiology* supplanted the term *exobiology*. Though people differed on the exact definitions, astrobiology was generally taken to be a more sweeping term, incorporating several existing disciplines and pioneering new ones. NASA defined *astrobiology* as "the study of the living universe" in its 1996 Strategic Plan. That study focuses on the relationship between the origins and evolution of life, on one hand, and

the origins and evolution of planets, stars, and the cosmos as a whole, on the other. For an extended discussion of the evolution of the term's meaning, see Steven J. Dick and James E. Strick, *The Living Universe* (New Brunswick, N.J.: Rutgers University Press, 2004), pp. 205–13.

188 **Aside from the profound** • Steven J. Dick, *The Biological Universe* (Cambridge: Cambridge University Press, 1996), pp. 1–2, 322.

189 **Richard Zare, for one** • Author interviews with Zare, and Zare's written commentary.

189 **The White House** • The meeting planners had prepared lists of questions to be addressed about life's origins, about the most promising places to search for extraterrestrial life, about the Mars rock, and about the theological, ethical, and philosophical implications. Among the latter, a proposed query in one NASA document showed the tortuous care of a staffer acutely attuned to political and religious sensibilities: "Does this exploration evict God from His/Her heavens to leave Him/Her only in the human heart?"

189 **Just before the vice** • NASA memo from Alan Ladwig, Dec. 10, 1996.

189 **Sagan was arguably** • William Poundstone, *Carl Sagan* (New York: Henry Holt, 1999), pp. 355, 379. An astronomer-educator had suggested naming the meteorite the Sagan rock, but because of the uncertainties about the claims and other factors, it never happened.

189 **He had sent** • Sagan, three-page letter, Oct. 26, 1996, addressed to Gibbons and Huntress (copied to Mark S. Allen, director of the National Research Council Space Studies Board), provided to the author.

190 **He eventually garnered** • Poundstone, *Carl Sagan*, pp. 355–58.

190 **In one of his** • William Sheehan and Stephen James O'Meara, *Mars* (Amherst, N.Y.: Prometheus Books, 2001), p. 298.

191 **A couple of weeks after** • Craig Venter, of the Institute for Genomic Research in Rockville, Maryland, collaborated with Carl R. Woese, of the University of Illinois, Urbana, who had first suggested in 1977 that certain microbes deserved their own separate branch. Other participants were from Johns Hopkins University. (See C. J. Bult et al., "Complete Genome Sequence of the Methanogenic Archaeon, *Methanococcus jannaschii*," *Science* [Aug. 23, 1996]: pp. 1058–73.)

192 **Now, as the Treaty** • Author interviews with Goldin, Huntress, and others.

192 **(If anyone but the** • Matt Ridley, *Genome* (New York: HarperCollins, 2000), p. 12, for example, says of living things, such as rabbits: "They do not defy the second law of thermodynamics, which says that in a closed system everything tends from order towards disorder, because rabbits are not closed systems.

Rabbits build packets of order and complexity called bodies but at the cost of expending large amounts of energy. In Erwin Schrödinger's phrase, living creatures 'drink orderliness' from the environment."

192 **McKay was no** • Author interview with McKay.

193 **The policymakers were eager** • Astrophysicists had shown that the cosmic inventory of carbon, oxygen, calcium, and iron was manufactured in the thermonuclear fires that burned in the hearts of earlier generations of stars. These microscopic building blocks of life were then distributed across space and time through another act of stellar generosity—star death by cataclysmic explosion. (Sagan was famously fond of saying, "We are all star stuff.") Bones, blood, and the very matter with which inquiring minds contemplated the vastness and their place in it were all a legacy from the stars.

"Walking down the corridors of an observatory, you see collections of carbon atoms hunched over silicon boxes, controlling distant telescopes of iron and aluminum in an attempt to trace the origin of the very substances of which they are made," astrophysicist Robert Kirshner wrote in "The Earth's Elements," *Life in the Universe*, a special issue of *Scientific American* (New York: Freeman, 1995), p. 19.

193 **The excitement surrounding** • Huntress, July 1996, advisory memo to Goldin. As NASA prepared to inform the White House about the rock, Huntress wrote: "The Federal Government should consider supporting the idea of 'Origins' as a principal [Dan—i.e. Kennedy-like?] goal for the agency. There are no large immediate resource requirements since the program is current technology-limited and investments in technology development will be required before any missions can be launched; sometime early in the next decade."

Under the heading "Political," Huntress wrote: "Dan, your call on this one—" and then, "The president could play on the popularity of the space program, and the idea of life in outer space [as demonstrated by the current immense popularity of science fiction in print and movies], by recognizing this discovery as one of the most amazing results of his Administration. He could announce that his new Administration after the election will establish a program in NASA to search for life's origins in this and other solar systems. America will send small inexpensive robotic missions to Mars at every opportunity to follow up on this discovery." The seeds of the Origins program were contained in a NASA precursor called TOPS (Toward Other Planetary Systems).

193 **As Huntress and Goldin** • One example of NASA's concern about the budget outlook was contained in a May 20, 1996, written summary of a May 16 hearing on the fiscal year 1997 NASA budget request (NASA archives). The Subcommittee on VA-HUD-Independent Agencies of the Committee on Appropriations held the hearing. For four years NASA funding had been essentially frozen at the 1992 level; the agency had taken a 36 percent reduction in the planned budget coincident with a 40 percent increase in productivity and a reduction in its payroll for employees and contractors by 30,000 people. See also Workshop Plan for Space Science Symposium, Oct. 23, 1996 (NASA archives): in early and mid-1996, the projected five-year NASA budget runout showed a decline from about $14 billion to less than $12 billion. The agency budget had peaked in 1966, early in the Apollo program, at a figure that, after adjustment for inflation, was the equivalent of at least double the current total.

193 **President Clinton, in** • White House Transcript, Remarks by the President, South Lawn, Aug. 7, 1996. The summit had been proposed initially at a May committee hearing by Senator Barbara Mikulski (Democrat of Maryland), whose state was home to thousands of NASA employees; see June 18, 1996, letter from Mikulski to Al Gore, expressing concern about the NASA budget outlook, referring to the "need to stand sentry to ensure that there is sustainable funding for core NASA missions in the outyear budgets" and calling for the summit "to discuss the future of space funding." See also "Commentary: Modest Summit Expectations," by space policy historian John Logsdon, of George Washington University, *Space News* (Dec. 9–15, 1996): p. 34.

The previous such summit on space had been held in 1993, resulting in the historic though controversial agreement to embrace Russia, America's former Cold War adversary and chief rival in space, as a partner in constructing the international space station.

193 **NASA and the other agencies** • In planning for the session, Gore had told advisers that he wanted to "chew on" issues surrounding the Mars rock claims, hear all the newest and best ideas for where the U.S. space program was headed, and be able to bargain in good faith with Congress during the budget battles to come. In their first meeting on the topic, on August 21, the vice president had indicated that he wanted to get fifty people (later reduced to about thirty) together. Once that had happened, the president and vice president planned to meet with the ultimate summit group (to include the president, vice president, White House science adviser, NASA's Goldin, key administration officials, and the congressional leadership) to

forge a consensus on the budget. (The attendees also agreed that there should be a premeeting with congressional staffers to "lay out best and worst case for 1-year and 5-year budget.")

See NASA legislative affairs document (copy of e-mail), 9:30 A.M., Nov. 22, 1996, distributed by Jeff Lawrence; document provided by Huntress, undated, titled "Status on Space Summit"; also author interview with Rick Borchelt of White House science staff.

193 **In late October** • "The Search for Origins: Findings of a Space Science Workshop," October 28–30, 1996. Claude Canizares of MIT, chairman of the National Research Council's Space Studies Board, led the workshop. Afterward, he called it "a remarkable conclave, because this very diverse group, many of whom had never met before, came to the realization that their several scientific disciplines were all converging on a core theme. . . . The study of Origins seeks answers, in scientific terms, to the fundamental questions about how we came to be, questions as old as human thought itself."

A related event took place on November 22, when George Washington University's Space Policy Institute and other space groups sponsored a public symposium on "the cultural, intellectual and scientific significance of the recent announcement of evidence of long-ago life on Mars." Glenn MacPherson of the Smithsonian's National Museum of Natural History, who decades earlier had first classified the Mars rock as coming from an asteroid, brought the museum's chunk of the meteorite for display. He and a security guard stayed throughout the day. An overflow crowd of 550 showed up; some had to watch on closed-circuit TV.

194 **As part of its** • "Planning Guidance for Space Summit," undated, prepared by NASA science office for meeting with vice president's staff. Also, author interviews with Goldin, Huntress, and others, and various other summit documents. See also John Logsdon, "Commentary: Modest Summit Expectations." In this case, NASA's goal was not the limitless budget of Apollo, or even a major increase. The prize would be mere predictability: an agreement by the White House and congress to end the pattern of frustrating budgetary fits and starts, and give NASA managers sufficient stability that they could dare to plan a program and follow that plan.

194 **The administration was grappling with** • John Logsdon, "Commentary: Modest Summit Expectations."

195 **The summit process served** • Michael Meyer interview with Steven Dick, NASA archives.

195 **When the meeting** • December 11 press briefing by Gibbons and other meeting participants.

195 **Within two months** • Author interviews with Dan Goldin; letter, Feb. 6, 1997, from Senator Mikulski, original instigator of the summit, to the vice president saying a final meeting was not necessary and expressing satisfaction at Clinton's budget for fiscal year 1998, which proposed a five-year funding level that would be "stable and sustainable." See also another perspective on the summit in NASA Watch online: http://www.nasawatch.com/nss/10.29.97.nss.summit.html.

Years later, in an interview with the author, Goldin would revise his glowing assessment of the summit process, expressing regret in hindsight that its participants had not taken a more practical, hard-nosed approach to the options. See also John Logsdon, written introduction to report on George Washington University Symposium, Nov. 22, 1996, "Life in the Universe: What Can the Martian Fossil Tell Us?" Logsdon said the summit was canceled not only because the anticipated budget cuts were abandoned but also "largely as a result of the public excitement over the Mars meteorite findings and NASA's putting forward a powerful strategy for the space science centered around the theme of exploring the origins of the solar system and the universe, and of life beyond earth. Key to this policy and budgetary decision was [the] early December 1996 meeting between Vice President Gore and a group of NASA officials, space scientists and other representatives of the public."

196 **And in the realm of** • Author interview with NASA science official Edward Weiler. He said the impetus for the budget turnaround was clearly boosted by the McKay team's 1996 claims. But given the long lead time in the budget-preparation process, the momentum had already been building, because of the flow of discoveries from the repaired Hubble Space Telescope, data from the injured but functioning Galileo spacecraft at Jupiter, and the payoff from other science investments of the 1980s.

196 **But far from being** • Author interviews with Goldin, Huntress, Weiler, and others at NASA. In November 1996, NASA announced the new Ancient Martian Meteorite Research Program, with funding of at least $1.5 million over two years for studies of the twelve known Martian meteorites.

196 **Buoyed by public excitement** • Dick and Strick, *Living Universe*, p. 179; author interviews with Goldin and Huntress. In 1998, the president approved and Congress funded the NASA Astrobiology Institute, a consortium of almost a dozen NASA-selected academic and research institutions.

197 **Pathfinder's assignment was** • Astrogeophysicist Chris McKay, March 31, 1999, lecture to Toronto Mars Society chapter.

198 **Data flowing from** • Images from NASA's Mars Global Surveyor, released in 2000, for example, revealed signs of relatively recent (less than a million years old) fluid flows in dozens of cliff walls and escarpments of craters, mesas, and troughs—sinuous channels in patterns that resembled spring-fed water-drainage areas on Earth, something like gully washes. This mystified scientists. For one thing, conditions on chill, arid modern Mars were such that any water reaching the surface supposedly could not remain liquid for very long.

198 **In 1997, in the** • Author interviews with Rummel.

198 **After the McKay team's** • Author interviews with Goldin; letter from MIT president Charles Vest to Goldin, Aug. 14, 1996. Vest suggested as one response to the Mars rock claims that NASA enlist U.S. research universities in studies of the potential for extraterrestrial life, possibly through the establishment of "centers of excellence" in needed research specialties. See also Goldin speech, Dec. 10, 1996, to the American Society for Cell Biology convention, in San Francisco; Goldin remarks to press following December 11, 1996, vice president's meeting, from author's notes; Goldin speech to the American Astronomical Society, Washington, D.C., Jan. 7, 1998.

198 **Goldin would also soon** • In November 1996, NASA announced its new meteorite research program. Also during this period, NASA and the National Science Foundation (which provided $1.3 million for the Mars Rock Special Research Opportunity) formed an edgy partnership on such projects as the Antarctic meteorite search and a campaign to investigate Earth's extremophile microbe populations. Ralph Harvey, the Case Western geologist and McKay critic, was surprised, he told the author, to find his threatened Antarctic search program saved, in effect, by new funding from NASA.

199 **In early 1997** • The Space Studies Board's report for the National Research Council, "Mars Sample Return: Issues and Recommendations" (Washington, D.C.: National Academy Press, 1997). See also National Research Council reports, "The Quarantine and Certification of Martian Samples," May 29, 2001; "Signs of Life," based on the April 2000 Workshop on life-detection techniques.

If scientists hoped to ever bring back extraterrestrial rocks, they realized, they would have to prepare what one called "the mother of all Environmental Impact Statements." See also Ron Cowen, "Scooping Up a Chunk of Mars," *Science News* (Apr. 25, 1998): pp. 265–67. Bill Schopf is quoted as saying, "I would be loath to have the public think that NASA was going to handle the Martian samples in the way that the lunar samples were handled. . . . NASA's feet have to be held to the fire." For other ongoing criticism of NASA as taking the issue too lightly, see Barry E. DiGregorio, "Rethinking Mars Sample Return," *Space News* (Feb. 22, 1999); Gilbert Levin, "Shed Light, Not Heat, on Mars," *Space News* (May 10, 1999).

199 **The anti-contamination people** • Nobody knew enough. NASA's anti-contamination specialists lacked experience in high-level biological containment. Other organizations, such as the army's infectious disease experts, had expertise in biological containment but not in "the biology of nonpathogenic microbes" and such. And nobody had ever built a facility that could simultaneously keep microbes from getting in from outside and out from inside.

200 **A central question for** • "The ALH84001 meteorite illustrated the need to distinguish between biotic and abiotic mechanisms at the nanometer scale," according to the National Research Council Space Studies Board's Committee on Planetary and Lunar Exploration (COMPLEX), in its report "On NASA Mars Sample-Return Mission Options," based on a meeting held October 16–18, 1996.

The group said they were "pleased that NASA has taken the opportunity provided by the increased attention to Mars exploration resulting from the McKay et al. paper . . . to accelerate planning of a program of Mars sample-return missions." However, they cautioned that the findings were "only suggestive" and that "COMPLEX believes it is inappropriate to predicate an important aspect of future martian studies on the unconfirmed results described in a single scientific paper," and that along with the meteorite findings NASA should include in its thinking "the new findings on microbial life in extreme terrestrial environments, . . . current understanding of the evolution of the martian and other planetary environments and recent findings about the existence of planets around other stars."

200 **And beyond the** • Michael Meyer interview with Steven Dick, NASA archives.

200 **The losses, eventually** • Report of the Mars Program Independent Assessment Team, headed by Tom Young, and Report of the Loss of the Mars Polar Lander and Deep Space 2 Missions, by the JPL Special Review Board, March 14, 2000.

Investigators concluded that the Mars Climate Orbiter probably burned up in the Martian atmosphere on September 23, 1999, because engineers famously mixed up imperial and metric units when calculating the craft's trajectory. Three months later, the Polar Lander, also carrying two surface microprobes, disappeared. Investigators found that a software error probably caused it to crash into the surface. The losses totaled about $360 million.

201 **Huntress, for one, was** • E-mail from Huntress to author, April 8, 2005.

201 **NASA officials pushed** • Just a few months earlier, in May 1999, David McKay had spoken optimistically of the possibility that a Mars sample return mission might be launched as early as 2005, saying, "If it all works out, we'll have roughly half a kilogram from one location, collected in 2003, and another half kilogram from another location, collected in 2005." (See McKay interview in *OE* [*Optical Engineering*], Report No. 185, May 1999.) Separate robots were to accumulate a trove of the samples for pickup by a return-to-Earth spacecraft.

CHAPTER TWELVE: AT DAGGERS DRAWN

202 **On August 7, 1996,** • Author interviews with Andrew Steele.

203 **In mid-September 1996** • Andrew Lawler, "Planetary Science: Mars Meteorite Quest Goes Global," *Science* (Sept. 20, 1996): pp. 1653–54.

203 **After several more attempts** • Author interviews with Steele and McKay.

203 **One chilly day in** • Author interviews with Steele.

204 **By measuring the beam's** • The group had to overcome a limitation in the instrument in order to study the meteorite sample, Steele told the author. The device was designed for the study of relatively flat surfaces, so the "needle could only move up and down a limited amount without special maneuvering." When it ran into a "mountain" instead of a gentle hill, the result could be either a false reading or a stuck arm. Steele's Portsmouth group was the first to work out the methods required to use the atomic force microscope on cells and other rough surfaces where the needle had to accommodate mountain peaks. For the Mars rock study, they used a combination of several instruments and techniques.

205 **Now he had a** • See A. Steele, D. T. Goddard, I. B. Beech, R. C. Tapper, D. Stapleton, and J. R. Smith, "Atomic Force Microscopy Imaging of Fragments from the Martian Meteorite ALH94001," *Journal of Microscopy*, vol. 189 (1998): pp. 2-7.

205 **The meeting, held in** • Author interviews with Steele, Gibson, and other participants; Allan Treiman, "Life on Mars Vigorously Debated at Conference," *Eos* (June 3, 1997). The meeting was the twenty-eighth Lunar and Planetary Science Conference, March 17–21, 1997.

205 **In the frenetic and stomach-churning** • Author interviews with David McKay and Mary Fae McKay.

206 **Gibson was energized** • Author interviews with Gibson. The account of debate highlights at the conference is based on author interviews with

Gibson, Steele, Treiman, and other participants; also various press accounts, including Richard Kerr, "Life on Mars: Martian 'Microbes' Cover Their Tracks," *Science* (Apr. 4, 1997): pp. 30–31, and J. Kelly Beatty, "Messenger from Mars," *Sky and Telescope* (July 1997): pp. 36–39.

206 **Over at the Gilruth Center** • Gibson and other members of the McKay group would defend their claims later that year: see Everett K. Gibson, Jr., David S. McKay, Kathie Thomas-Keprta, and Christopher S. Romanek, "The Case for Relic Life on Mars," *Scientific American*, vol. 277 (December 1997): pp. 58–65.

207 **But models developed by** • Ralph Harvey, of Case Western Reserve University, and Harry Y. McSween Jr., of the University of Tennessee, provided one of the strongest assaults on the low-temp argument. See Ralph Harvey and Harry Y. McSween Jr., "A Possible High-Temperature Origin for the Carbonates in the Martian Meteorite ALH84001," *Nature*, vol. 382 (July 4, 1996): pp. 49–51. They suggested the carbonates were formed by a rapid reaction between rock and boiling water during an impact on Mars by an asteroid or comet.

207 **Joseph L. Kirschvink** • Joseph L. Kirschvink et al., "Paleomagnetic Evidence of a Low-Temperature Origin of Carbonate in the Martian Meteorite ALH84001," *Science* (Mar. 14, 1997): pp. 1629–33.

208 **Within four years** • J. E. P. Connerny et al., "The Global Magnetic Field of Mars and Implications for Crustal Evolution," *Geophysical Research Letters* (Nov. 1, 2001): pp. 4015–18.

208 **"One thing is clear,"** • White paper by the McKay group, presented at the meeting.

208 ***Science* magazine would** • Richard Kerr, "Life on Mars: Martian Rocks Tell Divergent Stories," *Science* (Nov. 8, 1992): p. 918.

208 **As for the** • Steven J. Dick and James E. Strick, *Living Universe* (New Brunswick, N.J.: Rutgers University Press, 2004): p. 196.

208 **Simon Clemett, of Zarelab** • Simon J. Clemett et al., "Evidence for the Extraterrestrial Origin of Polycyclic Aromatic Hydrocarbons in the Martian Meteorite ALH84001," *Faraday Discussions (of the Chemical Society)* 109 (1998): pp. 417–36. Harvey told the author, "We found that, yes, indeed, there were these little wormy-looking shapes there, but they were a very small subset of a larger continuum of things that went from wormy little things to plates to mineral faces."

See also A. J. T. Jull et al., "Isotopic Composition of Carbonates in the SNC Meteorites Allan Hills 84001 and Nakhla," *Meteoritics*, vol. 30 (1995):

pp. 311–18. Tim Jull, of the University of Arizona, and his colleagues had reported evidence that the carbonate globules—with which the PAHs were closely associated—contained terrestrial contamination. Their research showed much more carbon 14 than should have been present, given that the isotope is radioactive and decays quickly. Something must have replaced it recently. The team's explanation was that terrestrial carbon penetrated and replenished the depleted supply of carbon 14, and this meant that terrestrial oxygen might have followed the same path.

209 **Others argued that** • J. P. Bradley, R. P. Harvey, and H. Y. McSween Jr., "No 'Nanofossils' in Martian Meteorite," *Nature* (Dec. 4, 1997): p. 454. Richard Zare, Clemett's boss at Zarelab, would also issue a statement (*Stanford News*, Jan. 14, 1998) noting that the Bada research had focused on amino acids in the meteorite, not PAHs. Because amino acids are water soluble, water saturation "provides a potential mechanism for carrying contaminated material into the meteorite's interior. Therefore, I do not find it all that surprising that these compounds may be terrestrial in origin." But, he went on, "Because the PAHs are highly insoluble, however, I do not feel that the Bada study can be extrapolated to them. Therefore, I conclude that this work does not shed any important new light on the origin of the PAHs in ALH84001."

209 **"You can't tell** • Elizabeth K. Wilson, "Bitten by the Space Bug," *Chemical and Engineering News* (Nov. 18, 1996): p. 29.

209 **At least one avid** • Cosmochemist John Kerridge, as quoted in Charles Petit, "Pieces of the Rock," *Air and Space* (Apr.-May 1997): p. 40.

210 **The peripatetic Ralph** • Kerr, "Life on Mars: Martian 'Microbes' Cover Their Tracks." Harvey and colleague John Bradley also found that most of the magnetite crystals were perfectly aligned in three dimensions with the surrounding carbonate in a way that argued against biological origins.

210 **Of all the challenges** • Author interview with McKay. See also Beatty, "Messenger from Mars," p. 39.

210 **None of these disagreements** • Author interviews with Steele.

211 **Mary Fae McKay was** • Author interviews with Mary Fae McKay.

211 **Some considered the** • For Blanchard comments, see Kathy Sawyer, "Digging into Data on Mars Life Claim; Research Community's Verdict Is Years Away," *Washington Post*, Mar. 20, 1997, p. A3.

212 **Some weeks after this** • David Salisbury, "Debate Over Evidence for Martian Life in Meteorite Rages On," *Stanford News*, May 29, 1997.

The opposition was expressing complaints similar to Zare's. Ralph Harvey, for one, told the author that he thought poorly chosen peer reviewers were allowing a number of unworthy papers to get published. (As the years and papers rolled by, he would come to feel strongly that too many people who knew next to nothing about meteorites, Bill Schopf included, were involved as referees on proposed papers and in other key roles in the debate. "I don't see much of a sign . . . that they chose really competent people to review them," he said in November 2001, following another round of papers on the rock.) Like Zare and others, he also felt that his involvement in the "media circus just derailed my research for a while." Other people were publishing ideas that Harvey and company had come up with earlier but failed to publish, he said, because "I had been spending a little bit too much time giving talks."

213 **The Brits would soon determine** • A. Steele et al., "Investigations into an Unknown Organism on the Martian Meteorite Allan Hills 84001," *Meteoritics and Planetary Science,* vol. 35 (2000): pp. 273–81; see also A. Steele, J. Toporski, D. Goddard, D. Stapleton, and D. S. McKay, "The Imaging of Terrestrial Microbial Contamination of Meteorites," *Microscopy and Analysis,* vol. 83 (2001), pp. 5–7; A. Steele et al., "F Atomic Force Microscopy Imaging of Fragments from the Martian Meteorite ALH84001," *Journal of Microscopy,* vol. 189 (1998), pp. 2–7; A. Steele et al., "Terrestrial Contamination of an Antarctic Chondrite (Abstract)," *Meteoritics and Planetary Science,* vol. 33 (1998): p. A149; A. Steele et al., "The Contamination of Murchison Meteorite," *Lunar and Planetary Science* vol. 30 (1999): Abstract 1293, Lunar and Planetary Institute, Houston (CD-ROM); A. Steele et al., "The Microbiological Contamination of Meteorites: A Null Hypothesis," *Lunar and Planetary Science,* vol. 31 (2000), Abstract 1670, Lunar and Planetary Institute, Houston (CD-ROM); Jan Toporski et al., "Contamination of Nakhla by Terrestrial Microorganisms," *Lunar and Planetary Science,* vol. 30 (1999), Abstract 1526, Lunar and Planetary Institute, Houston (CD-ROM).

213 **Steele delivered the** • Author interviews with Steele and McKay; see also A. Steele et al., "Imaging of the Biological Contamination of Meteorites: A Practical Assessment," Abstracts of the Thirtieth Lunar and Planetary Science Conference, Mar. 15–19, 1999, Lunar and Planetary Institute, Houston. The paper noted that the detection of the terrestrial organisms and their products in the rock "do not necessarily negate the possibility that it contains evidence for early life on Mars. However, it becomes more challenging to separate such evidence from the terrestrial contamination." The evidence also raised the possibility that the entire framework of frac-

tures leading deeper into the meteorite might have been contaminated—though no trace of Earth bugs deep inside the rock had been found in careful searches.

213 **Steele and his coworkers** • They used scanning electron microscopes of two types (typical and environmental) and an atomic force microscope.

214 **Also, another group** • Jeffrey Bada et al., "A Search for Endogenous Amino Acids in Martian Meteorite ALH84001," *Science* (Jan. 16, 1998): pp. 362–65. See also L. H. Burckle and J. S. Delaney, "Terrestrial Microfossils in Antarctic Ordinary Chondrites," *Meteoritics and Planetary Science*, vol. 34 (1999): pp. 475–78. This group reported detecting in the meteorite diatoms in cracks, which they said must have been blown inland over the ice, along with dust. They concluded that contamination with micrometer-sized organisms might be widespread in Antarctica.

214 **In the past, the** • A. Steele et al., "Imaging of the Biological Contamination of Meteorites: A Practical Assessment."

215 **As Steele would write in** • A. Steele et al., "Imaging of the Biological Contamination of Meteorites."

216 **Steele and company** • Author interview with Steele. As Steele's team continued sorting out what was what in the rock, they found a strange chemical signature that seemed to be a sign of biological activity—a so-called biomarker—because they saw it only when there was cellular debris or other biology present. After what Steele deemed a ridiculous amount of work, they finally figured it out. The mystery matter turned out to be—cue Dustin Hoffman—*plastic!* It seemed that before the curators sent out meteorite samples to researchers, they routinely heat-sealed the container bags. There had been a small leakage of the plastic into the samples. What's more, contaminating bacteria considered the plastic to be food. So the Allan Hills meteorite had not only been contaminated by bacteria, it had been contaminated by plastic and the plastic was nourishing the bacteria.

Plastic was not the only such oddity. Steele also discovered fungi growing on a sample inside a plastic bag. The team traced it to Building 31. It turned out the air-conditioning system had a heat exchanger on the roof, and the contamination—in the form of the fungus penicillium, a green bread mold—was leaking into the lab through the unfiltered ducts. NASA put in a HEPA filter, which was supposed to be changed every three months. In Steele's view, managers did not have a rigorous enough plan to see that this was done. Space center officials considered the problem corrected and reported that it had not affected critical laboratory results. In any case, Steele still considered the NASA curation facility the best in the world.

216 **These techniques, called** • Immunoassay tests were used to measure biological compounds in nearly all areas of clinical laboratory science and basic biomedical research. The technique depended on the specificity of antigen-antibody interactions and the sensitivity of chemical reactions such as enzymatic activity, radioactive decay, fluorescence, and luminescence.

217 **In any case, the** • See www.masse.co.uk/. The MASSE project was intended to seek evidence of terrestrial contamination, prebiotic chemicals, and organic biomarkers for extinct or extant life on Mars. It was to employ both in vivo and in vitro techniques to produce, isolate, and use antibodies, in combination with immunofluorescence and developing protein-array technology.

217 **One of the schmoozers** • Author interview with Kerridge at the cocktail party.

217 **Not far away from** • Michael Ray Taylor, *Dark Life* (New York: Scribner, 1999), pp. 240–41.

218 **In the last years of** • Richard Kerr, "Requiem for Life on Mars? Support for Microbes Fades," *Science* (Nov. 20, 1998): p. 1398; see also Dick and Strick, *Living Universe,* p. 196.

218 **Some (though not all** • Author interviews with the McKay group. See also Steven J. Dick and James E. Strick, *The Living Universe* (New Brunswick, N.J.: Rutgers University Press, 2004), p. 190.

218 **The rock was indeed** • Tim Jull, et al., of the University of Arizona, did a study of the isotopic composition of the carbon in the meteorite which Dick Zare said "does indicate a degree of terrestrial contamination that is much greater than I suspected was present." The study showed that just 20 percent of the organic material was extraterrestrial, but it did not indicate where this portion was located in the meteorite. Zare said that although he did not believe the research "refutes the basic contention in our original paper," it "does cast new doubt on our hypothesis." See David F. Salisbury, "Latest Research Casts New Doubt on Evidence for Fossil Life in Martian Meteorite," *Stanford News,* Jan. 14, 1998.

218 **The McKay regulars couldn't** • Author interviews with the McKay group.

Author interview with Allan Treiman, concerning the factions within factions. See also Allan Treiman, "Microbe in a Martian Meteorite? An Update on the Controversy," *Sky and Telescope,* vol. 97 (April 1999): pp. 52–58. (One example of the branching internal disagreements: regarding the temperature at which the carbonates—and the magnetic crystals in them—had formed, people who agreed that conditions were too hot for life differed about whether the rock was in those high temperatures for a long time—under high pressure, deep in the crust of Mars—or whether the rock had experienced heat and high pressure for only a few microseconds as the result of an impact shock, when an asteroid or comet slammed into Mars. Then there were others who agreed that the carbonate globules were deposited from water at relatively low temperatures, but that there was no biology involved at that time. The magnetite crystals and organics—the PAHs—formed later, they contended, as the carbonates decomposed, possibly in connection with volcanic events.)

219 **McKay group antagonist Ralph** • Author interview with Harvey.

219 **"You have to understand** • Author interview with Gibson.

219 **As the controversy** • In an author interview, Ralph Harvey argued the point from the opposite direction. If the McKay group's original premise was that all the evidence taken together, not any single line of evidence, made the case for biology in the rock, then as most of the lines of evidence weakened under attack, in his view the group's "holistic" case was "just gone."

219 **Steele was among those** • Author interview with Steele.

219 **Simon Clemett, for another** • Author interview with Clemett.

220 **Painful as it could get** • Derek Sears and William Hartman, "Conference on Early Mars," Houston, April 24–27, 1997, *Meteoritics and Planetary Science*, vol. 32 (1997): pp. 445–46. See also Dick and Strick, *Living Universe*, p. 195, which quotes the editorial.

220 **Steele would work in** • Steele was hired at Carnegie by Wes Huntress, the former top lieutenant to Dan Goldin who had watched the story of the rock unfold in official Washington and was now director of the institution's Geophysical Laboratory. Steele also secured a research fellowship at Oxford University, in addition to his affiliation with Portsmouth University. In the United States, he had a tie to Montana State University, which gave him access to instruments he needed. He also had consultant status with NASA contractor Lockheed Martin.

CHAPTER THIRTEEN: BINGO

222 **In fact, Thomas-Keprta** • Author interviews with Thomas-Keprta and other members of the McKay group were the basis for the account of this incident.

222 **Those seductive magnetic** • "Some would call this the smoking gun," McKay, Gibson, and Thomas-Keprta wrote, referring to the magnetic crystals, in a summation of the state of play on the rock presented at the July 1999 Fifth International Conference on Mars in Pasadena, California. See McKay et al., "Possible Evidence for Life in ALH84001," p. 3, at: http://www.lpi.usra.edu/meetings/5thMars99/pdf/6211.pdf.

223 **The work of the McKay** • A crystal is a solid form bounded by several flat, smooth planes, or faces. This form is adopted by a chemical compound (mineral) when passing, under certain conditions, from the state of a liquid or gas to that of a solid. There is no geometry to study when the compound is liquid or gas, because the atomic forces that bind the mass together in the solid state are not present.

224 **Sensitive instruments aboard** • J. E. P. Connerney et al., "The Global Magnetic Field of Mars and Implications for Crustal Evolution," *Geophysical Research Letters*, vol. 28 (Nov. 1, 2001): pp. 4015–18. The onboard magnetometer showed that the ancient upland crust of Mars's southern hemisphere was magnetized, an indication of a past global magnetic field. Planetary scientists then used established dating techniques to estimate that the Martian magnetic field had disappeared about 3.7 or 3.8 billion years ago.

225 **It was a point of** • Author interview with Ralph Harvey; Richard Kerr, "Martian 'Microbes' Cover Their Tracks," *Science* (Apr. 4, 1997): pp. 30–31; Dick and Strick, *Living Universe*, p. 193.

225 **Wary that this** • Among the reasons for ruling out terrestrial contamination in this population of magnetites, Gibson said, were the following: the crystals were deeply embedded in the carbonate globules in the meteorite, and magnetite-producing bacteria preferred low-oxygen environments, and therefore would not likely have lived in Antarctic meltwater.

225 **One reason was that the** • Author interview with McKay.

226 **"We take our cues** • Author interview with Thomas-Keprta.

226 **With Thomas-Keprta as lead** • Kathie L. Thomas-Keprta, D. A. Bazylinski, J. L. Kirschvink, S. J. Clemett, D. S. McKay, S. J. Wentworth, H. Vali, E. K. Gibson Jr., and C. S. Romanek, "Elongated Prismatic Magnetite Crystals in ALH84001 Carbonate Globules: Potential Martian Magnetofossils," *Geochimica et Cosmochimica Acta*, vol. 64 (Dec. 1, 2000): pp. 4049–81. The international research journal for geochemistry and cosmochemistry is sponsored by the Geochemical Society and the Meteoritical Society. (The prime mover behind its instigation in the late 1940s was Paul Rosbaud, a

colorful editor who spied on the Nazis for the British government during the war.)

226 **It would generate no** • Author interview with Allan Treiman.

226 **With the first magnetite** • Author interviews with Thomas-Keprta, Simon Clemett, and others in the group.

226 **But she had never fully** • Buseck was Regents' Professor of Geology and Chemistry at Arizona State University.

228 **In the early 1980s** • Chemist Stephen Mann, of the University of Bristol, United Kingdom.

228 **Thomas-Keprta, with major** • Simon Clemett, having fought his way through bouts of psychic gloom over the bile storm surrounding the group's claims, had left Stanford's Zarelab and eventually joined the ranks in Building 31, as an employee of contractor Lockheed Martin. Here, he was hard at work on a new, improved laser device and other technologies, to be installed in the microscope complex down the hall. Clemett's presence on-site certainly made the collaboration easier. He and Thomas-Keprta no longer had to rely on phone and e-mail for their constant communications, which had continued apace even during his rock-climbing vacation.

229 **The answer her team** • All known magnetite (Fe_3O_4) crystals are octahedral; that is, they have the shape of two four-sided pyramids put together at their bases. Magnetites from the magnet-producing *bacteria* MV-1 are elongated. What the Thomas-Keprta team was proposing was that, in the case of MV-1, this had been accomplished by the addition of six faces.

230 **On February 27, 2001, the** • Kathie L. Thomas-Keprta, Simon J. Clemett, Dennis A. Bazylinski, Joseph L. Kirschvink, David S. McKay, Susan J. Wentworth, Hojatollah Vali, Everett K. Gibson Jr., Mary Fae McKay, and Christopher S. Romanek, "Truncated Hexa-octahedral Magnetite Crystals in ALH84001: Presumptive Biosignatures," *Proceedings of the National Academy of Sciences*, vol. 98 (2001): pp. 2164–69. (Mary Fae McKay helped edit the paper.)

231 **Joseph L. Kirschvink, a** • Caltech press release Feb. 27, 2001.

231 **MV-1 expert Dennis Bazylinski** • Bazylinski said, after the first of the two magnetite papers was published, "The significance to astrobiology and geobiology is that many scientists have been searching for 'biomarkers' for life, that is, chemical, isotopic, and/or mineral indications that life was present, either in extreme habitats or in ancient materials on Earth and, of course, now in extraterrestrial materials. The need for biomarkers is obvi-

ous and these magnetite crystals might prove to be an excellent biomarker." See Teddi Baron, "Research Shows Life May Have Existed on Mars," *Inside Iowa State*, Jan. 26, 2001; available at: http://www.iastate.edu/Inside/2001/0126/mars.shtml.

231 **Their claims got** • E. Imre Friedmann, Jacek Wierzchos, Carmen Ascaso, and Michael Winklhofer, "Chains of Magnetite Crystals in the Meteorite ALH84001: Evidence of Biological Origin," *Proceedings of the National Academy of Sciences*, vol. 98 (2001): pp. 2176–81.

231 **One of the first** • Imre Friedmann interview with Steven Dick, NASA archives.

232 **What they saw were** • "Teeny backbones" quote from Kathy Sawyer, "New Findings Energize Case for Life on Mars," *Washington Post*, Feb. 28, 2001, pp. A3, 24.

232 **This time (in contrast** • Ibid., p. A3, 24. See also Steven J. Dick and James E. Strick, *The Living Universe* (New Brunswick, N.J.: Rutgers University Press, 2004), p. 198.

233 **Astrogeophysicist Chris** • Author interview with Chris McKay.

233 **He told the Associated Press** • In an author interview with Harvey, he said: "Now, one thing hopefully every party in this will admit is our own ignorance. . . . We don't have a catalog of all the different minerals that interact with biological organisms at all, so how could we possibly have the same catalog for all the nonbiological interaction? It is just an insurmountable task. . . . People are paying a lot more attention to the specific problem of magnetites, etc., but . . . we're just going to continue to bracket the possibilities. It's going to take a long time."

Harvey, among others, was quite skeptical of Friedmann's findings, saying Friedmann had rushed to apply a new technique before it had been adequately tested. He also told the author that the publisher of this latest round of magnetite papers had failed to make the best choices of outside referees.

Stanford's Dick Zare would later tell the author he, too, was skeptical of the Friedmann team's "choo choo train" interpretation. Another scientist involved in research on magnet-making microbes told the author privately that he saw "serious problems with that paper" and that "many were surprised" when it got published.

233 **Planetary scientist and** • E-mail to author, Mar. 2, 2001.

233 **Several studies soon** • D. J. Barber and E. R. D. Scott, "Origin of Supposedly Biogenic Magnetite in the Martian Meteorite Allan Hills 84001," *Proceedings of the National Academy of Sciences*, vol. 99 (2002): pp. 6556–61.

This analysis suggested that the planes of the atoms in the Martian magnetites were aligned with atomic planes in the surrounding carbonates in which they were embedded, and therefore they must have formed in the rock as the result of impact heating and decomposition of carbonate, not inside microorganisms. Kathie Thomas-Keprta and coworkers responded that, like other groups, this one had studied a different component of the meteorite and was "comparing their apples with our oranges."

See also M. R. McCartney, U. Lins, M. Farina, P. R. Buseck, and R. B. Frankel, "Magnetic Microstructure of Bacterial Magnetite by Electron Holography," *European Journal of Mineralogy*, vol. 13 (2001): pp. 685–89, and M. R. McCartney, U. Lins, M. Farina, P. R. Buseck, and R. B. Frankel, R. E. Dunin-Borkowski, et al., "Off-Axis Electron Holography of Magnetotactic Bacteria: Magnetic Microstructure of Strains MV-1 and MS-1," *European Journal of Mineralogy*, vol. 13 (2001): pp. 671–84. An Arizona State University team, including Buseck, put forth evidence that the sizes of magnets made by various Earth bacteria vary greatly, and do not necessarily match those in the Mars rock. They said there was not yet any reliable way to distinguish biologically produced magnetic crystals from nonbiological ones and recommended further statistical study.

233 **David McKay had envisioned** • Author interviews with David McKay and others in Building 31.

233 **As the principal investigator** • Author interview with Gordon McKay.

234 **The Blue Team contingent** • D. C. Goldin, Douglas W. Ming, Craig S. Schwandt, Howard V. Lauer Jr., Richard A. Socki, Richard V. Morris, Gary E. Lofgren, and Gordon A. McKay, "A Simple Inorganic Process for Formation of Carbonates, Magnetite, and Sulfides in Martian Meteorite ALH84001," *American Mineralogist*, vol. 86 (2001): pp. 370–75; see also Richard Kerr, "Oddities Both Lunar and Martian," *Science* (Mar. 31, 2000): pp. 2402–03. To re-create the features seen in the Mars rock, the group cooked up their own carbonate globules by heating a soup of bicarbonates mixed with rock chips. The bicarbonate decomposed, and the resulting carbonates were deposited in cracks in the rock fragments. The researchers changed the composition of the solution four times, to mimic changes in Martian groundwater that might have produced the black-and-white Oreo rims around the carbonate blebs. They delivered a heat shock similar to the asteroid impact or other trauma that had apparently occurred later in the rock's time on Mars. This caused iron carbonates to decompose into magnetic crystals. In other words, they reported, they had

created minerals "quite similar chemically and mineralogically" to those in the Mars meteorite.

234 **Thomas-Keprta was delighted** • At meetings where they would present what they considered compelling evidence on the magnetite, Thomas-Keprta and Clemett told the author, they sometimes encountered complaints from fellow scientists that their explanations of the intricacies of magnet-producing bacteria and the Euclidean nuances of crystallography were impossible to follow. Taking the pleas to heart—at the risk of drawing fire for "dumbing down" or committing "show business"—the two developed a two-dimensional model in red, green, and blue that could be printed out, cut out, and folded into the three-dimensional shape of the magnetite crystals—much like a design from a children's game book. They put together what they hoped was a more user-friendly presentation for fellow scientists, including a video with little crystals rotating one way and another, shapes emerging and fading.

235 **As the debate wore** • Author interview with Treiman.

235 **For some time, McKay** • Author interviews with McKay.

236 **At home, McKay's** • Interview with Mary Fae McKay.

CHAPTER FOURTEEN: SCHADENFREUDE

237 **All this week, Ames** • Regarding the impact of the McKay group announcement, see Steven J. Dick and James E. Strick, *The Living Universe* (New Brunswick, N.J.: Rutgers University Press, 2004), p. 195.

237 **In competition with** • Author interviews with attendees at the astrobiology conference.

238 **The jousting had erupted** • Martin Brasier et al., "Questioning the Evidence for Earth's Oldest Fossils," *Nature* (Mar. 7, 2002): pp. 76–81.

238 ***Nature* had published both** • J. William Schopf et al., "Laser-Raman Imagery of Earth's Earliest Fossils," *Nature* (Mar. 7, 2002): pp. 73–76. Schopf's UCLA group was working with a group at the University of Alabama, Birmingham, where the laser-Raman spectroscopy facility was located. In an e-mail to the author, May 6, 2005, Schopf expressed displeasure with the fact that *Nature* had pitted the papers against each other, back-to-back, even though his paper had been accepted first.

238 **Brasier had set out** • Brasier e-mail to author, Mar. 5, 2002.

238 **To some, this turn** • For mention of Schopf's nickname "Bull," see Rex Dalton, "Squaring Up Over Ancient Life," *Nature* (June 20, 2002): p. 784.

In an e-mail to the author, May 6, 2005, Schopf took umbrage at the idea that he deserved or had ever been described by such a nickname, adding that he had never heard anyone say it "in my presence."

239 **In a moment of** • Oliver Morton, "The Secret of a Rock," *Newsweek International* (Mar. 18, 2002): pp. 42–43.

239 **Bob Hazen, a Washington** • Author interview with Hazen over lunch. Robert M. Hazen was a research scientist at the Carnegie Institution of Washington's Geophysical Laboratory and a professor of earth science at George Mason University.

239 **For Steele, the** • Author interviews with Steele and McKay.

239 **In the course of his** • Schopf told the author in an e-mail, May 6, 2005, that he was being "inappropriately tarred" by Steele and others. In his 1992 *Science* paper, Schopf noted, he had "named these species as *incertae sedis* (in official taxonomic terminology, of uncertain, undetermined, taxonomic position)" so as not to prejudice others.

240 **Mary Fae admired** • Author interviews with Mary Fae McKay and David McKay.

241 **But some thought** • Author interviews with conference attendees on all sides of the question of biology in the rock. Robert Hazen, of the Carnegie Institution of Washington, had accused Schopf (see R. M. Hazen, "Review of *Cradle of Life: The Discovery of Earth's Earliest Fossils*, J. W. Schopf," *Physics Today on the Web*, October 1999, at: http://www.aip.org/pt/oct99/) of having been mean-spirited in his treatment of McKay et al.'s work on the meteorite ALH84001. In an e-mail to the author, May 6, 2005, Schopf noted Hazen's rebuke and related that, in April, Schopf had found himself in the wine-and-beer line with Hazen at an astrobiology conference in Boulder, Colorado. Schopf said he took the occasion to explain to Hazen that he had never intended to be mean-spirited but had structured the chapter based on a lecture he gave to nonscience students, "The point being to show that scientists are human." Schopf and Hazen both told the author afterward that they were not particularly pleased or satisfied by the exchange, which Hazen remembered slightly differently.

241 **On Tuesday, McKay joined the** • Thomas-Keprta had planned to attend the conference and present her latest data on the magnetic crystals, but she had twisted her leg while skiing. As the Brasier-Schopf show got rolling, Simon Clemett held a cell phone toward the stage. At the other end of the connection was the injured Thomas-Keprta, listening in from her bed of pain back in Houston.

242 **In his book, written** • J. William Schopf, *Cradle of Life* (Princeton, N.J.: Princeton University Press, 1999), pp. 76–78.

242 **As Brasier had studied** • Brasier e-mail to author; also Brasier talk at this astrobiology conference (Apr. 2002).

243 **Starting in July 1999** • E-mail from John Lindsay, of the Australian National University, to the author, May 2, 2005. Lindsay explained that the project had started as a cooperative one involving him and his university as well as Oxford and Brasier. (The two men had worked together on UNESCO projects in China and Mongolia in the early 1990s.) The work had begun with younger rocks (circa 800 to 500 million years old) and gradually moved to older and older formations. The 1999 field study plan was disrupted when Lindsay's wife collapsed with terminal cancer, but somehow the team managed to continue. At this point, the work on the Apex chert "was a side issue," Lindsay said, "not our main focus." At first, team members were unable to locate the Schopf sample site. "Schopf had not documented the site properly," Lindsay said. After his wife's death, Lindsay returned to the field and joined Martin Van Kranendonk. In the meantime, the Geological Survey of Western Australia had been able to relocate the Schopf sampling site from photos taken there by the Schopf team. Lindsay found the site using that information, did a reconnaissance of the geology, and resampled the chert. "It was quickly apparent that the sampling site was not on the Apex chert but was 90 meters down a hydrothermal black chert dike beneath the Apex chert. There was no way the 'microfossils' could have been deposited in a shallow water beach setting, nor could they be cyanobacteria. Whatever they were, they were part of a high-temperature hydrothermal vent setting and formed at depth." Lindsay noted that the authors of the resulting *Nature* paper "were never all together at one time—thank goodness for e-mail." Lindsay had since moved to Houston, where he was at the Lunar and Planetary Institute and working with the McKay group. For more information on upcoming work on the Australian microfossils and related matters, see Lindsay's Web page at: www.lpi.usra.edu.

243 **By late September 2000** • Brasier e-mail to author.

243 **Now, in the first** • Schopf said in an e-mail to the author, May 6, 2005, that, actually, there has been continued disagreement as to the type of environment that existed at that site, hydrothermal or otherwise, but that it has been mapped by at least four sets of scientists. He added, "I am agnostic. They will work it out."

244 **Schopf counterattacked. First** · Rex Dalton, "Microfossils." In addition, Steele, in an interview with the author, described one of Schopf's writings as a seventeen-page "diatribe" objecting to the Brasier group findings. He said Schopf reprimanded the museum curator for giving out the samples.

245 **Schopf sent Brasier** · In an e-mail to the author, May 6, 2005, Schopf said he had sent Brasier "a copy of our manuscript out of courtesy. . . . I did what I thought was honest and right. I gave him our data and showed that he had been mistaken. Got nailed for it. But I can live with that. I would do it again. Help the science, if you can." Also see Dick and Strick, *Living Universe,* p. 199, for a succinct summary of the Brasier-Schopf dispute.

245 **Brasier, by contrast, compared** · Brasier e-mail to author.

246 **On the question of branching** · Although Schopf criticized the Brasier group's use of the confocal microscopy technique, Everett Gibson told the author in an e-mail that Schopf embraced the approach at the April 2005 NASA Astrobiology Meeting at the University of Colorado, Boulder. "Schopf went out of his way to praise confocal microscopy as a 'new and powerful' technique to study the fossil record," Gibson said. "It seems that unless he is using the technique, it isn't worthy. . . . An amazing switch on his part."

246 **The shapes Brasier had** · Morton, "Secret of a Rock," p. 43, uses the metaphor of the AIDS ribbon.

249 **As Schopf had acknowledged** · Schopf, *Cradle of Life,* p. 98.

249 **Brasier had gone to** · Schopf told the author in an e-mail, May 6, 2005, that "despite Dr. Brasier's assertion, I know of no paper he has ever published on modern microbes, modern microbial assemblages, fossil microbes, fossil microbial assemblages. He has, to my knowledge, never discovered any new Precambrian microorganisms, never described such organisms, never named such organisms," nor studied fossils preserved in a manner relevant to the Apex fossil dispute. "To the best of my knowledge," Schopf said, Brasier's work, rather, has concerned more complex Phanerozoic fossil microorganisms.

251 **For his part, Brasier** · Author interview with Brasier and e-mail exchange with Brasier, early 2002.

251 **Almost three months later** · Rex Dalton, "Microfossils."

The dispute would continue as to whether or not the fossil shapes branched in a way that suggested they were not a product of biology. But Schopf, for one, would later declare the matter closed. In an e-mail to the author, May 6, 2005, he declared that "the branching business has been solved, and published (with photos also of the mineralic artifacts described in her notes as branching things by Bonnie Packer.)" See J. W. Schopf, "Earth's Earliest Biosphere: Status of the Hunt," in P. G. Eriksson, W. Altermann, D. R. Nelson, W. U. Mueller, and O. Cateneanu (eds.), *The Precambrian Earth: Tempos and Events*, Developments in Precambrian Geology 12 (Amsterdam: Elsevier, 2004), pp. 516–39. "In truth," Schopf went on, "Bonnie Packer never showed me pictures of any branching fossils, simply because none of the fossils (not one) branch."

Schopf also pointed out in the e-mail that he and Bonnie Packer had published another paper (J. W. Schopf and B. M. Packer, "Early Archean [3.3 billion to 3.5 billion year old] Microfossils from Warrawoona Group, Australia," *Science*, vol. 237 [1987]: pp. 70–73) in which they reported "the finding of microscopic fossils both in the Apex chert and in a second chert unit of about the same age some 40 km to the west. It is incorrect to give the impression that she (or anyone else who studied those fossils at that time) had any misgivings whatever as to their biogenicity."

254 **"Okay, let me explain** • Schopf told the author in an e-mail, May 6, 2005, that Steele and Brasier "have consistently misrepresented my views" as to the question of whether those early microfossils represented cyanobacteria or not. However, Steele and others told the author that Schopf has—if not in his published paper in 1993, then in numerous talks and presentations—given the impression that his findings suggested fossil cyanobacteria at an unexpectedly early period. Still others say they believe that Schopf, in those instances, was simply responding to pressure from people eager to know what those early fossils had looked like, and "pond scum" had seemed a good bet.

Schopf pointed to the following passages from his paper—"Microfossils of the Early Archean Apex Chert: New Evidence of the Antiquity of Life," *Science*, vol. 260 (1993): pp. 640–46—to support his argument that he had left the question open at least in that initial description (boldface added by Schopf):

- (p. 643): "Because the affinities of these fossils in the Procaryotae [nonnucleated prokaryotic bacteria] **cannot be demonstrated unequivocally, I formally describe them as prokaryotes *incertae sedis*** [viz., prokaryotic microorganisms of uncertain systematic position]; and because the phylogenetic relations between them and the much (1,300 to 2,800 million years) younger, predominantly cyanobacterial fossil taxa to which they bear specific resemblance **are therefore un-**

determined, they have not been referred to previously described Proterozoic species."

- (p. 644) reads as follows: [after discussion of lines of evidence consistent with oxygen-producing photosynthesis] "These additional lines of evidence, however, are not conclusive; all the latter, which necessarily incorporates model-dependent uncertainties, **would be equally consistent with the presence of solely anoxic bacterial photosynthsizers.** Moreover, it is conceivable that the external similarity of the Apex microorganisms to younger oxygen-producing oscillatoriaceans masks significant differences of internal biochemical machinery; **thus, their morphology may provide a weak basis on which to infer paleophysiology.** To address this issue, additional data are needed."

260 **A few weeks later the journal** • Rex Dalton, "Microfossils"; see also Richard Kerr, "Reversals Reveal Pitfalls in Spotting Ancient and E.T. Life," *Science* (May 24, 2002): pp. 1384–85. See also Dick and Strick, *Living Universe*, p. 199, which notes that Jill Pasteris, of Washington University, an expert in the laser-Raman spectroscopy used by Schopf in his defense, raised objections to Schopf's interpretation. See also Kenneth Chang, "Oldest Bacteria Fossils? Or Are They Merely Tiny Rock Flaws?" *New York Times*, Mar. 12, 2002, p. D4. That account quotes Harvard scientist Andrew Knoll as saying that many paleontologists had long doubted whether Schopf's shapes were fossils. After the Brasier paper, he said, "I think the preponderance of evidence is that they are not."

260 **Others agreed** • Dick and Strick, *Living Universe*, p. 200.

260 **Andrew Steele would come** • Author interview with Steele.

261 **Even as researchers** • Author interview with Schopf; e-mail from Schopf to author, May 6, 2005. See also Schopf's abstract, "The Archean Fossil Wars: Why Science Has Won," prepared for the January 2005 Gordon Research Conference in Ventura, California.

In any case, Schopf took issue with those who saw the Apex fossils as his foremost achievement. "The Apex fossils do not rate," he said in the e-mail. "Heck, in 1965, Professor Barghoorn and I reported the presence of fossils 3.1 billion years old; others followed suit; the Apex assemblage is of interest only because it is old and morphologically relatively varied." Through fieldwork and interaction with local scientists, Schopf had also been able to discover the first or oldest Precambrian stromatolitic fossils in China, India, Australia, South Africa, and South America, and could point to numerous other accomplishments.

CHAPTER FIFTEEN: DOWN THE RABBIT HOLE

263 **McKay happened** • Author interview and lab sessions with McKay.

263 **But much of the time** • Roughly every two years, Earth and Mars come as close as about 40 million miles (65 million kilometers) apart. When they are on opposite sides of the sun in their orbital tracks, they can be over 200 million miles apart. See William K. Hartmann, *A Traveler's Guide to Mars* (New York: Workman, 2003), p. 174.

263 **In detailed three-dimensional** • For one attempt to show the true colors of Mars, see *National Geographic,* Feb. 2001, p. 39. A graphics team corrected an early uncalibrated color image from the Viking 2 lander, tweaking brightness and contrast, to show "how Mars might look to a human visitor"—at least in that location, on that day in 1976.

264 **Decades of Mars** • Recent studies from orbit had turned Mars from a "flatland" world into a three-dimensional realm, with a "layer-cake" structure, dust-driven weather, and a clear water cycle. (Author interviews and e-mail exchange of July 5, 2000, with NASA's chief Mars scientist, James Garvin.) The branching channels and other evidence that water once flowed on Mars all dated from the same ancient period, called Noachian (after the biblical Noah of the great flood), which ended about 3.5 billion years ago. In the late 1990s, the Mars Global Surveyor detected thousands of seepage sites that might be linked to surface water runoff in the modern era.

264 **The rovers were** • Mars science team leader Steven Squyres, of Cornell, in press briefing, Jan. 3, 2004; see also Squyres et al., "Opportunity at Meridiani Planum," *Science* (Dec. 3, 2004): pp. 1633–1844, in which eleven reports by 122 authors describe results from the Opportunity mission. "Liquid water was once intermittently present at the Martian surface at Meridiani, and at times it saturated the subsurface," wrote Squyres, Ray Arvidson, and others. "We infer conditions at Meridiani may have been habitable for some period of time in Martian history." The rovers were also turning up welcome evidence of minerals known (on Earth) to be good at preserving evidence of ancient life for vast periods of time.

264 **In keeping with** • Mars mission scientist Maria Zuber of MIT, Carnegie Institution lecture, "Climate Change on Mars," Nov. 4, 2004, and author interview.

264 **(In early 2005, European** • See John B. Murray, et al., "Evidence from the Mars Express High Resolution Stereo Camera for a Frozen Sea Close to Mars' Equator," *Nature*, vol. 434 (March 17, 2005): pp. 352–56. The find-

ing would require further confirmation to show that the signal was not coming from hydrated minerals rather than water ice.

264 **At the same time** • Author interview with Meyer, who took the reins of the Mars exploration program in late 2004. For more on the Martian wobble hypothesis, see also Hartmann, *A Traveler's Guide to Mars*, pp. 236, 248–49, 297–98, and 352. Earth's seasons are determined by the tilt of its axis—the line through the north and south poles—as it circles the sun. Earth and Mars are currently tilted at about 23.5 and 25 degrees, respectively, relative to the sun. Scientists believe Earth may have been spared the disruptive wobble by the steadying influence of its moon. (The two Martian moons are too tiny.) Some scientists also suspect that the Martian dip, initially even wilder, may have been dampened by the effects of a huge concentration of lava in the bulging Tharsis region, which deformed the planet early in its history.

265 **But were the signs** • Studies of sudden, catastrophic floods on Earth concluded that similar water action—brief and massive—formed the great river channel of Ares Vallis—at whose mouth Pathfinder had landed in 1997—and many of the other Martian outflow channels. See Hartmann, *Traveler's Guide to Mars*, pp. 222–23. Hartmann cites studies by Arizona hydrologist Vic Baker and U.S. Geological Survey scientist Michael Carr.

There were also possible signs of persistent water action on early Mars: a meandering streambed—a feature that takes a long time to form—for example. And studies of the Martian watershed showed that if Mars had abundant water, it would have drained into the northern hemisphere, which is flatter than the American Midwest. The only comparably flat areas on Earth are the abyssal plains beneath its oceans. (A well-known image of a quasi shoreline was still considered equivocal, however, because the "bathtub ring" could have been left by lava flow instead of water.)

265 **Key pieces of** • Zuber lecture. If there was a Martian ocean, it should have left a chemical signature—such as calcium carbonate all over the surface.

265 **Such highly acidic waters** • Author interview with NASA's Michael Meyer. For a brief further discussion of the missing carbonates, see Hartmann, *A Traveler's Guide to Mars*, pp. 147–48.

266 **The bad news for** • Squyres et al., "Opportunity at Meridiani Planum."

266 **Alternative approaches to the** • Author interview with McKay; see also NASA Ames press release, September 22, 2003, on the Rio Tinto project. Called MARTE (Mars Analog Research and Technology Experiment), it was

led by Carol Stoker, of the Ames center, with support from the Spanish government.

266 **He had joined a** • The team leader was Marjorie Chan, head of the University of Utah geology department. See Chan, et al., "A Possible Terrestrial Analogue for Haematite Concretions on Mars," *Nature*, vol. 429 (June 17, 2004): pp. 731–34.

267 **The new plan called** • Author interview with NASA Mars chief Meyer.

267 **Besides, the Allan Hills** • Mars Exploration Program Analysis Group (MEPAG), "Scientific Goals, Objectives, Investigations and Priorities," July 16, 2004, at website http://64.233.161.104/search?q=cache:FKd3NK5E7wlJ:mepag.jpl.nasa.gov/goals/MEPAG-goals-approved-716-04.doc.

268 **Farther down the hall** • Gibson had gotten interested in the topic almost a quarter century earlier, as a member of an Antarctic meteorite search team.

269 **One explanation was** • For discussion of possible sources of methane on Mars, see David Tenenbaum, "Mystery Methane Maker: Wanted Dead or Alive," *Astrobiology Magazine*, July 27, 2005, at: <http://www.astrobio.net/news/article1660.html>

270 **The previous year, in** • Author interviews with Steele and Hans Amundsen of the University of Oslo, Norway; and lectures by them, Oct. 4, 2004, at the Carnegie Institution, Washington, D.C. Amundsen was the leader of the multidisciplinary Arctic Mars Analogue Svalbard Expedition (AMASE). The project included scientists from Carnegie; the University of Leeds; Universidad de Burgos, Spain; GEMOC; Macquarie University, Australia; the Jet Propulsion Laboratory; the Lunar and Planetary Institute in Houston; and Penn State University.

270 **Several types of** • Unable to investigate the real Mars in person, scientists were seeking accessible analogues. For example, researchers camped out with a robot named Zoe in northern Chile's Atacama Desert—one of the most arid regions on Earth—to study the rare life there with the goal of developing life-detection criteria for Mars and other worlds. (See NASA and Carnegie Mellon University, Pittsburgh, Sept. 24, 2004, press release. See also "Seeking Earthly Clues to Alien Life," Astronomy.com, accessed Oct. 24, 2002.) At the Eureka Weather Station on Ellesmere Island in Canada's Arctic Nunavut territory, 690 miles from the North Pole, engineers planned to test the Mars Drill, a futuristic drilling rig being developed at Johnson Space Center for use on Mars or Earth's moon. (NASA was con-

sidering a drilling mission to Mars possibly in 2018.) And the activist citizens group the Mars Society was setting up Mars research bases on Devon Island, in the Canadian Arctic, and other spots.

270 **As Steele's Svalbard** • Geophysicist Hans E. F. Amundsen, University of Oslo, at Royal Norwegian Embassy session on "The Search for Life in the Solar System," in cooperation with the Carnegie Institution, Oct. 4, 2004.

271 **"It's assumed we'll look** • The Svalbard team did achieve what Steele, reporting in September 2004, called "a major milestone." They showed that, assuming Mars life is at all similar to earth life, "we'll be able to find even a single cell." They had deployed a suite of life-detection instruments, such as Steele's off-the-shelf package that included standard genetic techniques and, among other things, protein microarrays that could test for the presence of hundreds or thousands of biological building-block molecules at once. In 2005, the immunology technology pioneered by Steele and his coworkers was accepted for flight aboard the European Space Agency's proposed ExoMars mission. Methods and devices developed by the Svalbard project would also compete for roles in NASA's 2009 and 2011 Mars missions.

272 **The research focused** • "Small World," *Astrobiology Magazine* (May 13, 2004). See also E. Olavi Kajander, I. Kuronen, and N. Ciftcioglu, "Fetal Bovine Serum: Discovery of Nanobacteria," *Molecular Biology of the Cell,* suppl. vol. 7 (1996): p. 517a; E. O. Kajander, I. Kuronen, K. Akerman, A. Pelttari, and N. Ciftcioglu, "Nanobacteria from Blood, the Smallest Culturable Autonomously Replicating Agent on Earth," *Proceedings of the Society for Optical Engineering* no. 3111 (1997): pp. 420–28.

272 **In fact, the lead** • Dick and Strick, *Living Universe,* p. 194, referring to microbiologist Olavi Kajander's inability to get published before the meteorite controversy focused public attention on the issue of nanobacteria.

273 **But McKay, not** • Author interview with McKay.

273 **Some experts saw** • Steven J. Dick and James E. Strick, *The Living Universe* (New Brunswick, N.J.: Rutgers University Press, 2004), pp. 194–95.

274 **People had published** • Analysis of the papers provided by Allan Treiman.

274 **Spurred by the rock** • "Astrobiology Isn't a Dirty Word Anymore," *Scientist* (Jan. 19, 2004), p. 44. Goldin became NASA's longest-serving administrator, resigning in October 2001 with many accomplishments and a few significant failures—the balance tilting in his favor overall, by most assessments. (See W. Henry Lambright, *Transforming Government: Dan Goldin and the Remaking of NASA,* Report for the Pricewaterhouse Coopers Endowment for the Business of Government, March 2001.) In late 2003, Goldin sur-

faced in the news again when Boston University, in an internal power struggle, paid him $1.8 million to give up his job as its president—before he even started to work there. (See Sara Rimer, "Boston U. Pays Leader to Quit Before Starting," *New York Times,* Nov. 1, 2003, p. A1; Sara Rimer, "After the Tumult, Boston U. Wonders Where It Goes from Here," *New York Times,* Nov. 19, 2003, p. B9; and Sara Rimer, "1.8 Million Check for a Job Not Done Jolts Boston University," *New York Times,* Nov. 6, 2003, p. A29.)

274 **The controversy over** • Author interviews with Goldin and Huntress, who moved from NASA to the Carnegie Institution to be head of the Geophysical Laboratory. See also Dick and Strick, *Living Universe,* p. 195, and "Astrobiology Isn't a Dirty Word Anymore." Philosopher of science Carol Cleland said the McKay group's 1996 claim, "more than anything else, was the birth of astrobiology as a serious subject." Noting that the meaning of the rock remained in dispute, she said, "it made people realize how easy it could be for life to be happening elsewhere."

274 **The hostilities unleashed** • This summary of the rock's ripple effects is drawn from author interviews with numerous scientists on all sides of the controversy. See also Dick and Strick, *Living Universe,* pp. 195, 199, 201. See also interview with Ed Scott, Hawaii Institute of Geophysics and Planetology, a skeptic on the McKay group claims, "Mars Meteorite's Link to Life Questioned," at www.space.com/scienceastronomy/solarsystem/mars_meteor_020514.html.

275 **In the persistence** • Science historian Iris Fry cited in Dick and Strick, *Living Universe,* p. 199. (They also cite the "sociology of science" quote, which comes from science historian and philosopher Iris Fry, *The Emergence of Life on Earth: A Historical and Scientific Overview* [New Brunswick, N.J.: Rutgers University Press, 2000], p. 221.)

275 **On some far-off day** • Author interviews and lab sessions with McKay. NASA was considering replacing the long-awaited Mars sample return, tentatively scheduled for launch in 2013, with an Astrobiology Field Laboratory to work in place on the surface of Mars, NASA astrobiologist Michael Meyer, newly appointed Mars exploration chief, said in a talk before the D.C. Science Writers Group, Rayburn House Office Building, Oct. 28, 2004. In June 2005, he told the author that a decision between the sample return and the field lab was to be made in early 2006.

SELECTED BIBLIOGRAPHY

Achenbach, Joel. *Captured by Aliens: The Search for Life and Truth in a Very Large Universe.* New York: Simon & Schuster, 1999.

Blum, Deborah, and Mary Knudson, eds. *A Field Guide for Science Writers.* New York: Oxford University Press, 1997.

Broad, William J. *The Universe Below: Discovering the Secrets of the Deep Sea.* New York: Simon & Schuster, 1997.

Cassidy, William A. *Meteorites, Ice and Antarctica: A Personal Account.* Cambridge: Cambridge University Press, 2003.

Chaikin, Andrew. *A Man on the Moon: The Voyages of the Apollo Astronauts.* New York: Viking, 1994.

Collins, Michael. *Carrying the Fire: An Astronaut's Journeys.* New York: Farrar Straus and Giroux, 1974.

Dick, Steven J. *The Biological Universe: The Twentieth-Century Extraterrestrial Life Debate and the Limits of Science.* Cambridge: Cambridge University Press, 1996.

Dick, Steven J., and James E. Strick. *The Living Universe: NASA and the Development of Astrobiology.* New Brunswick, N.J.: Rutgers University Press, 2004.

Fry, Iris. *The Emergence of Life on Earth: A Historical and Scientific Overview.* New Brunswick, N.J.: Rutgers University Press, 2000.

Goldsmith, Donald. *The Hunt for Life on Mars.* New York: Dutton / Penguin, 1997.

Goldsmith, Donald, and Tobias Owen. *The Search for Life in the Universe.* 3rd ed. Sausalito, Calif.: University Science Books, 2002.

Gould, Stephen Jay. *Wonderful Life: The Burgess Shale and the Nature of History.* New York: W. W. Norton, 1989.

Hartmann, William K. *A Traveler's Guide to Mars: The Mysterious Landscapes of the Red Planet.* New York: Workman, 2003.

Jakosky, Bruce M. *The Search for Life on Other Planets.* Cambridge: Cambridge University Press, 1998.

Kaharl, Victoria A. *Waterbaby: The Story of Alvin.* New York: Oxford University Press, 1990.

Kranz, Gene. *Failure Is Not an Option: Mission Control from Mercury to Apollo 13 and Beyond.* New York: Simon & Schuster, 2000.

Lewis, John S. *Rain of Iron and Ice: The Very Real Threat of Comet and Asteroid Bombardment.* Reading, Mass.: Addison-Wesley, 1997.

Mawson, Sir Douglas. *Home of the Blizzard: Being the Story of the Australasian Antarctic Expedition.* London: W. Heinemann, 1915.

McDougall, Walter A. . . . *the Heavens and the Earth: A Political History of the Space Age.* New York: Basic Books, 1985.

McSween, Harry Y., Jr. *Stardust to Planets: A Geological Tour of the Solar System.* New York: St. Martin's, 1993.

Morton, Oliver. *Mapping Mars: Science, Imagination, and the Birth of a World.* New York: Picador, 2002.

Murray, Bruce. *Journey into Space: The First Three Decades of Space Exploration.* New York: Norton, 1989.

Murray, Charles, and Catherine Bly Cox. *Apollo: The Race to the Moon.* New York: Simon & Schuster, 1989.

Poundstone, William. *Carl Sagan: A Life in the Cosmos.* New York: Henry Holt, 1999.

Rensberger, Boyce. *Life Itself: Exploring the Realm of the Living Cell.* New York: Oxford University Press, 1996.

Ridley, Matt. *Genome: The Autobiography of a Species in 23 Chapters.* New York: HarperCollins, 1999.

Sagan, Carl. *Cosmos.* New York: Ballantine, 1985.

——. *Pale Blue Dot.* New York: Random House, 1994.

Schopf, J. William. *Cradle of Life: The Discovery of Earth's Earliest Fossils.* Princeton, N.J.: Princeton University Press, 1999.

Sheehan, William, and Stephen James O'Meara. *Mars: The Lure of the Red Planet.* Amherst, N.Y.: Prometheus Books, 2001.

Taylor, Michael Ray. *Dark Life: Martian Nanobacteria, Rock-Eating Cave Bugs, and Other Extreme Organisms of Inner Earth and Outerspace.* New York: Scribner, 1999.

Wilhelms, Don E. *To a Rocky Moon: A Geologist's History of Lunar Exploration.* Tucson: University of Arizona Press, 1993.

Wolfe, Tom. *The Right Stuff.* New York: Bantam, 1980.

ADDITIONAL SOURCES

For the paper containing the original claims, see: David S. McKay, Everett K. Gibson, Kathie L. Thomas-Keprta, Hojatollah Vali, Chris S. Romanek, Simon J. Clemett, S. D. F. Chillier, C. R. Maechling, Richard N. Zare. "Search for Past Life on Mars: Possible Relic Biogenic Activity in Martian Meteorite ALH84001," *Science* 273 (August 16, 1996): 924–30.

More information on Martian meteorites is at NASA Johnson Space Center's Astromaterials Curation Web site; the "Meteorites from Mars" link provides information for the general public (http://curator.jsc.nasa.gov/curator/antmet/marsmets/contents.htm); Charles Meyer, *Mars Meteorite Compendium*, NASA/JSC (2001), provides detailed technical/scientific information on the Web at: (http://curator.jsc.nasa.gov/curator/antmet/mmc/mmc.htm). For detailed information specific to the Allan Hills meteorite ALH84001, see http://www-curator.jsc.nasa.gov/curator/antmet/mmc/84001.pdf.

More information about some of the events discussed in Chapter Two can be found in NASA's Apollo 11 Lunar Surface Journal, at: http://www.hq.nasa.gov/office/pao/History/alsj/a11/a11.launch.html.

Information about the National Science Foundation's Antarctic Search for Meteorites program (ANSMET) is at: http://geology.cwru.edu/~ansmet/.

UNPUBLISHED WORK CITED

Transcripts of a series of interviews housed at the NASA headquarters history archives, conducted in 1997 by Steven J. Dick, a space historian and author then at the U.S. Naval Observatory (subsequently NASA's chief historian). In the interviews, he asked key scientists about astrobiology and the controversy surrounding the Martian meteorite.

Various documents, including e-mails, memos, notes, and rough drafts of press releases provided to the author by participants in the events connected with the Martian meteorite controversy, in addition to a variety of documents contained in NASA archives.

INDEX

ABOUT THE AUTHOR

Award-winning journalist KATHY SAWYER covered space science and technology for *The Washington Post* for nearly two decades, beginning in 1986 with the *Challenger* accident, which killed seven astronauts. Her work has also been published in magazines such as *National Geographic* and *Astronomy*. A native of Nashville, she began her career as a feature writer at the *Tennessean* newspaper there. Twice nominated for the Pulitzer Prize, she is the recipient of honors including the National Headliner Award for Best Domestic News Reporting for coverage of the war in Vietnam and other subjects, and the David N. Schramm Award for journalism in the field of high-energy astrophysics, given by the American Astronomical Society.

ABOUT THE TYPE

The text of this book was set in Filosofia. It was designed in 1996 by Zuzana Licko, who created it for digital typesetting as an interpretation of the sixteenth-century typeface Bodoni. Filosofia, an example of Licko's unusual font designs, has classical proportions with a strong vertical feeling, softened by rounded droplike serifs. She has designed many typefaces and is the cofounder of *Emigre* magazine, where many of them first appeared. Born in Bratislava, Czechoslovakia, Licko came to the United States in 1968. She studied graphic communications at the University of California at Berkeley, graduating in 1984.